AF412499

Modeling and Mechanics of Granular and Porous Materials

Gianfranco Capriz
Vito N. Ghionna
Pasquale Giovine

Editors

Birkhäuser
Boston • Basel • Berlin

Gianfranco Capriz
Dipartimento di Matematica
Università di Pisa
I-56127 Pisa
Italy

Vito N. Ghionna
Dipartimento di Meccanica e Materiali
Università "Mediterranea" di Reggio Calabria
I-89060 Reggio Calabria
Italy

Pasquale Giovine
Dipartimento di Meccanica e Materiali
Università "Mediterranea" di Reggio Calabria
I-89060 Reggio Calabria
Italy

Library of Congress Cataloging-in-Publication Data

Modeling and mechanics of granular and porous materials / Gianfranco Capriz, Vito N. Ghionna, Pasquale Giovine, editors
      p. cm – (Modeling and simulation in science, engineering & technology)
  Includes bibliographical references.
  ISBN 0-8176-4241-2—ISBN 3-7643-4241-2
    1. Granular materials–Mathematical models. 2. Porous materials–Mathematical
  models. I. Capriz, G. (Gianfranco) II. Ghionna, Vito N., 1939- III. Giovine, Pasquale,
  1959- IV. Series.

  TA418.78 .M62 2002
  620.1'16'015118–dc21
                                          2002018593
                                                 CIP

AMS Subject Classifications: 70A05, 70G99, 70K70, 74-02, 74C15, 74E20, 74F10, 74F20, 74L05, 74L10, 74S99, 76S05, 76T25

Printed on acid-free paper.
©2002 Birkhäuser Boston

Birkhäuser 

ISBN 0-8176-4241-2   SPIN 10836005
ISBN 3-7643-4241-2

Typeset by the editors.
Printed and bound by Hamilton Printing Company, Rensselaer, NY.
Printed in the United States of America.

9 8 7 6 5 4 3 2 1

Birkhäuser Boston • Basel • Berlin
A member of BertelsmannSpringer Science+Business Media GmbH

Preface

Soils are complex materials: they have a particulate structure and fluids can seep through pores, mechanically interacting with the solid skeleton. Moreover, at a microscopic level, the behaviour of the solid skeleton is highly unstable. External loadings are in fact taken by grain chains which are continuously destroyed and rebuilt. Many issues of modeling, even of the physical details of the phenomena, remain open, even obscure; de Gennes listed them not long ago in a critical review.

However, despite physical complexities, soil mechanics has developed on the assumption that a soil can be seen as a continuum, or better yet as a medium obtained by the superposition of two and sometimes three continua, one solid and the other fluids, which occupy the same portion of space. Furthermore, relatively simple and robust constitutive laws were adopted to describe the stress-strain behaviour and the interaction between the solid and the fluid continua.

The contrast between the intrinsic nature of soil and the simplistic engineering approach is self-evident. When trying to describe more and more sophisticated phenomena (static liquefaction, strain localisation, cyclic mobility, effects of diagenesis and weathering,.....), the naïve description of soil must be abandoned or, at least, improved.

Higher order continua, incrementally non-linear laws, micromechanical considerations must be taken into account. A new world was opened, where basic mathematical questions (such as the choice of the best tools to model phenomena and the proof of the well-posedness of the consequent problems) could be addressed.

The aim of this treatise is to bring together contributions of scientists with different backgrounds, namely engineers and mathematicians, people who are interested in soil behaviour and experts in particulate material modeling, so that open and non-conventional discussions can lead to fruitful collaboration and exchange of knowledge.

July, 2002
Gianfranco Capriz, Pisa
Vito N. Ghionna and Pasquale Giovine, Reggio Calabria

Contents

3 Thermodynamic Modeling of Granular Continua Exhibiting Quasi-Static Frictional Behaviour with Abrasion

Nina P. Kirchner and Kolumban Hutter   **63**

4 Modeling of Soil Behaviour: from Micro-Mechanical Analysis to Macroscopic Description

Roberto Nova   **85**

5 Dynamic Thermo-Poro-Mechanical Stability Analysis of Simple Shear on Frictional Materials

Ioannis Vardoulakis   **129**

II   Flow and Transport Phenomena in Particulate Materials   157

6   Mathematical Models for Soil Consolidation Problems: a State of the Art Report

Davide Ambrosi, Renato Lancellotta and Luigi Preziosi   **159**

7   Flow of Water in Rigid and Non-Rigid, Saturated and Unsaturated Soils

Peter A.C. Raats   **181**

III   Numerical Simulations   243

9   Continuum and Numerical Simulation of Porous Materials in Science and Technology

Wolfgang Ehlers   **245**

10   A Mathematical and Numerical Model for Finite Elastoplastic Deformations in Fluid Saturated Porous Media

Lorenzo Sanavia, Bernhard A. Schrefler and Paul Steinmann   **293**

*Modeling and Mechanics
of Granular and Porous
Materials*

Part I

Mechanics of Porous Media

Chapter 1

Constitutive Equations and Instabilities of Granular Materials

Félix Darve and Farid Laouafa

ABSTRACT Constitutive equations for geomaterials constitute a very intricate field. In the first part of this chapter, a synthetic view of constitutive formalism is presented. An intrinsic classification of all existing constitutive relations is deduced. Then examples of incrementally non-linear relations are given and some applications follow. A numerical study of the so-called "yield surfaces" is presented, and is followed by a discussion on the validity of the principle of superposition for incremental loading. Finally the question of bifurcations and instabilities in geomaterials is investigated. Essentially because of the non-associative character of geomaterial plastic strains, a large domain of failure with various modes of ruptures is exhibited.

1.1   Introduction

For many years the study of the mechanical behaviour of geomaterials and its description by constitutive relations has been developed in the framework characterized by isotropic linear elasticity (Hooke's law) and by solid friction (Coulomb's law). However, since the end of the 1960s the development of more powerful numerical methods such as the finite element method and the use of high-performance computers has brought to the fore a question that is becoming a crucial one: what constitutive relation for geomaterials must be introduced into a computer code? After thirty years of development, the choice is large and the state of affairs confused; we will try to classify the various existing constitutive relations into some general classes with respect to their structure.

Any user of finite element codes must be able to characterize the capacities of any constitutive model implemented in order to interpret correctly the numerical results obtained and also in order to know if the physical phenomena that the user considers to have an important influence on the behaviour of the engineering work being analysed can be effectively taken into account by the constitutive relation used.

The first part of this chapter is devoted to a presentation of the incremental formulation of constitutive relationships. Two main reasons have made such an incremental presentation indispensable. The first one is physical and is the fact that, as soon as some plastic irreversibilities are mobilized inside the geomaterial, the global constitutive functional, which relates the stress state $\boldsymbol{\sigma}(t)$ at a given time t to the strain state $\boldsymbol{\epsilon}(t)$ history up to this time, is a priori very difficult to formulate explicitly since this functional is singular at all stress-strain states (or more precisely non-differentiable, as we shall show). An incremental formulation enables us to avoid this fundamental difficulty. The second reason is numerical and stems from the fact that geomaterial behaviour, and the modeling of engineering works generally, exhibit many non-linearity sources which imply that the associated boundary value problem must be solved by successive steps linked to increments of boundary loading. Such finite element codes need therefore, the constitutive relationships to be expressed incrementally.

1.2   Principle of determinism

In all that follows, we will assume generally small transformations, an approximation often justified in civil engineering by the fundamental uncertainty of the spatial variation of mechanical properties. Thermo-mechanical effects will also be neglected by noting that the influence of temperature variations on geomaterials behaviour is generally linked with particular situations, which are often not without interest if we think of the freezing of soils to increase their strength or on the contrary of increasing the temperature to accelerate the primary consolidation of clays. Finally we assume that geomaterials are simple media in the sense of Truesdell & Noll [35], a hypothesis very widely accepted for the description of the mechanical behaviour of media.

1.2.1   Principle of determinism in the large

The first expression of the principle of determinism is obtained by writing that the stress state $\boldsymbol{\sigma}(t)$ at a given time t is a functional of the history of the tangent linear transformation up to this time t. It implies that it is necessary to know all the loading path in order to deduce the associated response path.

From a mathematical point of view, this is stated by the existence of a stress functional $\mathcal{F}$:

$$\boldsymbol{\sigma}(t) = \mathcal{F}\left[\mathbf{F}(\tau)\right], \qquad -\infty < \tau \leq t, \tag{1.1}$$

where $\mathbf{F}(\tau)$ is the *tangent linear transformation* at time τ; also called *deformation gradient*. The deformation gradient $\mathbf{F}$ is the jacobian matrix of the position $\phi(X, \tau)$ of the material point X at time τ.

The existence of such a functional, and not a function, is related to an essential physical characteristic: for irreversible behaviours the knowledge of the strain $\epsilon(t)$ at time t does not enable us to determine the stress and *vice versa*. For example, we can think of viscous materials where a given level of stress can be related to an infinite number of different strain states.

This functional $\mathcal{F}$ will be called linear if:

$$\forall \, \lambda \in \mathbb{R} : \mathcal{F}\left[\lambda \mathbf{F}(\tau)\right] \equiv \lambda \, \mathcal{F}\left[\mathbf{F}(\tau)\right],$$
$$\forall \, \mathbf{F}_1, \mathbf{F}_2 : \mathcal{F}\left[\mathbf{F}_1(\tau) + \mathbf{F}_2(\tau)\right] \equiv \mathcal{F}\left[\mathbf{F}_1(\tau)\right] + \mathcal{F}\left[\mathbf{F}_2(\tau)\right].$$

In such a case, the material response to a sum of histories will be simply equal to the sum of the responses to each history. This constitutes Boltzmann's principle and it is the basis of linear viscoelasticity theory, but it is not at all valid in elastoplasticity theory where, when one doubles for example the strain, the stress is obviously not doubled, due to the non-linear behaviour.

This stress-strain relationship, which must be studied in the framework of non-linear functionals, is moreover non-differentiable as soon as there exist some plastic irreversibilities. Owen & Williams [33] showed in fact that the assumptions of non-viscosity and of differentiability of the stress functional $\mathcal{F}$, imply that there is no internal dissipation. In other words a non-viscous material whose constitutive functional is differentiable is necessarily elastic.

Therefore, if one wants to describe the behaviour of anelastic materials (geomaterials are essentially of this kind) by using a stress-strain relationship, this relationship must be formulated by a non-linear and non-differentiable functional.

There is clearly a need, therefore, to study constitutive relations using an incremental formulation and no longer a global one. Note that incremental form tends to rate form when the 'time' increment tends to zero. In the following, the terms increment or rate are employed in the same sense.

We are now going to introduce an incremental formulation using a second statement of the principle of determinism.

1.2.2 *Principle of determinism in the small*

Let us consider a deformed solid, and let us assume that only quasi-static loading will be applied to it at a continuously varying rate.

The second principle of determinism, which can be called 'in the small' to distinguish it from the first one 'in the large', is obtained by writing that a *small* load applied during time increment dt induces a *small* uniquely determined response.

We denote by $d\boldsymbol{\epsilon} = \mathbf{D}dt$, the incremental strain tensor of order two equal to the product of strain rate tensor of order two, $\mathbf{D}$ (symmetric part of transformation rate $\mathbf{L}$: $\mathbf{L} = \dot{\mathbf{F}}\,\mathbf{F}^{-1}$) and the time increment dt, and by $d\boldsymbol{\sigma} = \hat{\boldsymbol{\sigma}}dt$ the incremental stress tensor, equal to the product of an objective time derivative of Cauchy stress tensor $\boldsymbol{\sigma}$ and dt. Thus the second determinism principle implies, from a mathematical point of view, the existence of a tensorial function $\mathbf{H}$ relating the three quantities:

$$\mathbf{H}_\chi \left(d\boldsymbol{\epsilon}, d\boldsymbol{\sigma}, dt\right) = 0. \tag{1.2}$$

What are the properties of this tensorial function $\mathbf{H}$? The first remark concerns the fact that $\mathbf{H}$ depends on the previous stress-strain history. This history is generally characterized by some scalars and tensorial variables which will appear as parameters χ in relation (1.2). These parameters describe, as far as possible, the actual deformed state of the solid. Following the various constitutive theories, they are sometimes called 'memory variables', 'hardening parameters', 'internal variables', *etc.*

Secondly, $\mathbf{H}$ must satisfy the objectivity principle; this means that $\mathbf{H}$ must be independent of any movement of the observer relative to the solid. Thus $\mathbf{H}$ is an isotropic function of all its arguments: $d\boldsymbol{\epsilon}$, $d\boldsymbol{\sigma}$ and also the state tensorial variables χ, which characterize its present deformed state. But it is an anisotropic function of $d\boldsymbol{\epsilon}$ and $d\boldsymbol{\sigma}$.

Finally, $\mathbf{H}$ is essentially a non-linear function as soon as there are some plastic irreversibilities, since, if $\mathbf{H}$ was linear, the functional $\mathcal{F}$ would be differentiable, a property which excludes the existence of non-viscous irreversibilities.

On the contrary, for linear viscoelasticity this function $\mathbf{H}$ is linear and takes the form

$$d\boldsymbol{\epsilon} = \mathbf{M}\,d\boldsymbol{\sigma} + \mathbf{C}dt,$$

where $\mathbf{M}$ is the (constant) elastic tensor of order four and $\mathbf{C}$ the creep rate tensor of order two of the material.

We remark here also that this determinism principle, which has very wide application in physics by connecting notions of cause and effect, is not always satisfied, particularly in cases such as bifurcation situations. In such conditions a continuous variation of state variables can induce, at the bifurcation point, a sudden change in the evolution of the system with loss of uniqueness as a result of local imperfections not taken into account by the analysis (see examples given by Darve [7]). At such a bifurcation point an identical cause in terms of state variables can have different effects; such a situation is common in geomaterial behaviour when a shear band appears, by localisation of plastic strains, while the deformation, up to this state, has been following mainly a diffuse mode (Vardoulakis *et al.* [37]).

1.3   Application to non-viscous materials

In formulating constitutive relations it is often more convenient to replace the stress tensor $\boldsymbol{\sigma}$ and strain tensor $\boldsymbol{\epsilon}$, of second-order, by two vectors of

$\mathrm{I\!R}^6$ defined in a six-dimensional related space:

$$\underline{\sigma} = \begin{pmatrix} \sigma_{11} \\ \sigma_{22} \\ \sigma_{33} \\ \sqrt{2}\,\sigma_{23} \\ \sqrt{2}\,\sigma_{31} \\ \sqrt{2}\,\sigma_{12} \end{pmatrix} \quad \text{and} \quad \underline{\epsilon} = \begin{pmatrix} \epsilon_{11} \\ \epsilon_{22} \\ \epsilon_{33} \\ \sqrt{2}\,\epsilon_{23} \\ \sqrt{2}\,\epsilon_{31} \\ \sqrt{2}\,\epsilon_{12} \end{pmatrix}.$$

We will utilize Greek indices, ranging from 1 to 6, in such a case and reserve Latin indices, ranging from 1 to 3, to characterize tensorial components in the original three-dimensional space. The $\sqrt{2}$ coefficients, which appear in the definition of these vectors, allow us to preserve the original metric unchanged (isometric mapping). For example,

$$\|\boldsymbol{\sigma}\|^2 = \sigma_{ij}\,\sigma_{ij} = (\sigma_{11})^2 + (\sigma_{22})^2 + (\sigma_{33})^2 + 2(\sigma_{23})^2 + 2(\sigma_{31})^2 + 2(\sigma_{12})^2 = \sigma_\alpha \sigma_\alpha.$$

The constitutive relation (1.2) will be written in the new notation as

$$\mathrm{H}_\chi(\mathrm{d}\epsilon_\alpha, \mathrm{d}\sigma_\beta, \mathrm{d}t) = 0,$$

where now H is a vectorial function of an $\mathrm{I\!R}^6$ function of 13 variables.

Now let us particularize, in this section, the class of materials considered and restrict the study to the case of non-viscous materials or to cases of loading for which a given geomaterial exhibits negligible viscosity. This means that the loading rate (characterized by time gradation on loading path) has no influence on material constitutive behaviour: at whatever rate the given loading path is followed, the response path remains unchanged. In other words the class of behaviour considered is rate independent.

This restriction of the constitutive law implies that the constitutive function H, which relates dϵ and dσ, is independent of the time increment dt during which the incremental loading has been applied. Therefore H is independent of dt and, in an equivalent manner if **H** is regular, one can now study the vectorial function **G**:

$$\mathrm{d}\epsilon_\alpha = \mathbf{G}_\alpha(\mathrm{d}\sigma_\beta). \tag{1.3}$$

1.3.1   Homogeneity of **G**

We have just seen that the behaviour of a non-viscous material is independent of the loading rate. This means that, in order to verify the one-to-one relation on a given loading path and its related unique response path, there is a dependency condition between the two rate vectors D and $\dot{\sigma}$. For instance, if the loading path is followed twice as fast, the response path will also be followed at exactly twice the rate.

From a mathematical point of view this independence of non-viscous behaviours on loading rates implies the identity

$$\forall \lambda \in \mathbb{R}^{+} : \mathbf{G}_{\alpha}(\lambda \, d\sigma_{\beta}) = \lambda \, \mathbf{G}_{\alpha}(d\sigma_{\beta}), \qquad (1.4)$$

which states that if the stress rate is multiplied by any positive scalar λ, the strain rate response is also multiplied by the same scalar λ.

This is the first property of $\mathbf{G}$: $\mathbf{G}$ is a homogeneous function of degree 1 in $d\sigma$ with respect to the positive values of the multiplying parameter. This homogeneity property must not be confused with that of 'positively homogeneous' functions, which is given by

$$\forall \lambda \in \mathbb{R} : f(\lambda \, d\sigma) = |\lambda| \, \mathbf{G}(d\sigma).$$

1.3.2   Non-linearity of $\mathbf{G}$

Let us recall that we want to describe plastic irreversibilities, and consider the incremental strains, which are produced by reversed an incremental load $-d\sigma$ applied after $d\sigma$. If the behaviour is purely elastic, this stress cycle would induce a strain cycle equally closed (with no permanent strain). Therefore for elastic behaviour,

$$\forall \, d\sigma \in \mathbb{R}^{6} : \mathbf{G}(-d\sigma) = -\mathbf{G}(d\sigma).$$

In a more general manner, $\mathbf{G}$ is linear for elasticity since a regular correspondence relationship between σ and ϵ implies the existence of a linear relation between $d\sigma$ and $d\epsilon$. In other words we know, in fact, that the principle of superposition for incremental loading is valid in elasticity and implies

$$\forall \, d\sigma^{(1)}, d\sigma^{(2)} \in \mathbb{R}^{6} \times \mathbb{R}^{6} : \mathbf{G}\left(d\sigma^{(1)} + d\sigma^{(2)}\right) = \mathbf{G}\left(d\sigma^{(1)}\right) + \mathbf{G}\left(d\sigma^{(2)}\right).$$

This incremental linearity, which is a characteristic property of elasticity if we consider non-viscous materials, must obviously not be confused with an eventual global linearity. The existence of such a global linear relationship between σ and ϵ describes linear-elastic behaviour, while an incrementally linear relation is characteristic of all elastic constitutive relations.

Since we want to describe the behaviour of essentially inelastic media such as geomaterials, we need therefore to take into account non-linear function $\mathbf{G}$.

1.3.3   Anisotropy of $\mathbf{G}$

The last property of $\mathbf{G}$ that we will demonstrate is its anisotropy. If we consider all the arguments of $\mathbf{G}$, which means the six components of $d\sigma$ as well as the memory parameters χ characterizing the previous strain history of the medium, the objectivity principle implies that $\mathbf{G}$ is an isotropic function of the whole of its tensorial arguments. In fact a constitutive relation, describing the behaviour of a material, must be independent of any frame of reference used to define or formulate it, this frame possibly being in motion with respect to the material. This expresses a very obvious physical reality: whatever the movement of an observer looking at the deformation of a homogeneous sample under loading, the relation between $d\sigma$ and $d\epsilon$, which describes the incremental mechanical properties of this sample, must remain invariant.

However, let us now consider a deformed anisotropic sample of a geomaterial and carry out an experiment which consists of rotating the principal axes of incremental loading with respect to the sample. We do not obtain the same incremental response merely rotated in the same manner.

So nature requires us to consider functions $\mathbf{G}$ anisotropic with respect to $d\sigma$. This anisotropy is directly linked to the geometrical meso-structure of the material, which is gradually modified by the strain (particularly irreversible) history. We have seen that this history can be characterized by scalar and tensorial state parameters.

In simple cases this anisotropy is a priori directly imposed by the choice of these state parameters. If we consider, for example, only scalar memory parameters (such as void ratio) defined independently in any frame, owing to the objectivity principle the function $\mathbf{G}$ will be an isotropic function, which is not supported by experiments.

If we add to these scalar memory parameters one single tensor variable (such as the stress tensor), it is easy to see that $\mathbf{G}$ will be an orthotropic function of $d\sigma$, the axes of orthotropy being identified with the principal axes of stress. In this case, it means that $\mathbf{G}$ is invariant by symmetry with respect to any plane containing two principal stress directions.

In the more general case of state variables with at least two non-commutating tensorial variables of second-order, anisotropy is a priori anything whatever. Orthotropy then becomes a constitutive assumption, which must be considered as an approximation to the real behaviour of the material for a class of loading in which stress and strain principal axes rotate.

Having described the three main properties of $\mathbf{G}$, we will now focus on the first one (the homogeneity of degree 1) to see the mathematical consequences of such property. Let us recall for this purpose Euler's identity for homogeneous regular functions of degree 1 by writing it for a function of two variables, for example, as:

$$\forall\, x, y \in \mathbb{R} \times \mathbb{R} : f(x, y) = x\frac{\partial f}{\partial x} + y\frac{\partial f}{\partial y},$$

where partial derivatives $\frac{\partial f}{\partial x}$ and $\frac{\partial f}{\partial y}$ are homogeneous functions of degree 0. Therefore the six functions G_α, which are homogeneous functions of degree

1 of the six variables $d\sigma_\beta$, satisfy the identity

$$G_\alpha(d\sigma_\beta) = \frac{\partial G_\alpha}{\partial(d\sigma_\beta)}d\sigma_\beta, \quad \alpha,\beta \in \{1,2,\ldots,6\}^2,$$

with summation on the repeated index β. If we note,

$$M_{\alpha\beta}(d\sigma_\beta) = \frac{\partial G_\alpha}{\partial(d\sigma_\beta)}, \quad \alpha,\beta \in \{1,2,\ldots,6\}^2,$$

we obtain the constitutive relation

$$d\epsilon_\alpha = M_{\alpha\beta}(d\sigma_\gamma)d\sigma_\gamma, \quad \alpha,\beta,\gamma \in \{1,2,\ldots,6\}^3,$$

where the 6×6 functions $M_{\alpha\beta}$ are homogeneous of degree 0 of the six variables $d\sigma_\gamma$. Therefore they are in an equivalent manner functions only of the direction of $d\sigma$, characterized by unit vector $\underline{u}$:

$$u_\gamma = \frac{d\sigma_\gamma}{\|d\sigma\|}, \quad \gamma \in \{1,2,\ldots,6\},$$

with

$$\|d\sigma\| = \sqrt{d\sigma_{ij}d\sigma_{ij}} = \sqrt{d\sigma_\alpha d\sigma_\alpha}, \quad i,j \in \{1,2,3\}^2, \quad \alpha \in \{1,2,\ldots,6\};$$

finally we obtain $(\alpha,\beta,\gamma \in \{1,2,\ldots,6\}^3)$

$$d\epsilon_\alpha = M_{\alpha\beta}(u_\gamma)d\sigma_\gamma. \tag{1.5}$$

Equation (1.5) is the general expression for all rate independent constitutive relations.

This constitutive matrix $\mathbf{M}$, which then can be called 'tangent constitutive matrix', should be utilized to solve bifurcation problems by strain localization into shear bands [7] and also preferably used in finite element computations. This matrix in fact describes, at best, the material behaviour near a given incremental loading direction.

It is clear that a given constitutive relation will be characterized by only one tangent constitutive tensor after its definition has been given. But from this tangent tensor it is possible to build an infinite number of other constitutive tensors, which can be called 'secant constitutive tensors' and which describe exactly the same constitutive behaviour. The relation (1.5) will now allow us to demonstrate a classification of all the existing rate independent constitutive relations with respect to their intrinsic structure.

1.4   Main classes of rate independent constitutive relations

The incremental non-linearity, which is also that of plastic irreversibilities, results, as we have just seen, from the constitutive tensor $\mathbf{M}$ varying with the direction of the incremental loading. The mode of the chosen directional variation is very much influenced by the constitutive model used. It is precisely this mode of description of incremental non-linearity that we are going to take as a guide for this review of main classes of existing constitutive relationships.

First of all we need to define the notion of 'tensorial zone' [12]. We will call a tensorial zone any domain in the incremental loading space on which the restriction of $\mathbf{G}$ is a linear function. In other words the relation between $\mathbf{d\epsilon}$ and $\mathbf{d\sigma}$ in a given tensorial zone is incrementally linear.

If we denote by Z the tensorial zone being considered, the definition will be written as:

$$\forall \boldsymbol{u} \in Z : \mathbf{M}(\boldsymbol{u}) \equiv \mathbf{M}^z.$$

In the zone Z, the constitutive relation is characterized by a unique tensor $\mathbf{M}^z$.

If $\boldsymbol{u}$ belongs to Z, $\lambda \boldsymbol{u}$ belongs also to Z for all real positive values λ. Therefore a zone is defined by a set of half-infinite straight lines, whose apex is the same and is at the origin of the incremental loading space. Tensorial zones then comprise adjacent hypercones, whose common apex is this origin.

What does the constitutive relation become on the common boundary of two (or several) adjacent tensorial zones?

If $\mathbf{M}^{z_1}$ and $\mathbf{M}^{z_2}$ are constitutive tensors attached respectively to tensorial zones Z_1 and Z_2, we must obviously satisfy the continuity requirement of the response to the loading directions $\boldsymbol{u}$:

$$\forall \boldsymbol{u} \in Z_1 \cap Z_2 : \mathbf{M}^{z_1} \boldsymbol{u} \equiv \mathbf{M}^{z_2} \boldsymbol{u}$$

or, equivalently,

$$\forall \boldsymbol{u} \in Z_1 \cap Z_2 : (\mathbf{M}^{z_1} - \mathbf{M}^{z_2}) \boldsymbol{u} \equiv 0. \qquad (1.6)$$

Relation (1.6) can be called a 'continuity condition' for zone changes. This condition forbids, in particular, an arbitrary choice of constitutive tensors in two adjacent tensorial zones.

We will see further that conventional elastoplastic relations satisfy this condition by means of a class of neutral loading paths, while other relations do not satisfy it. Gudehus [20] studied in this way the continuity of several constitutive relations for axisymmetrical loading paths with fixed principal axes, as did Di Benedetto & Darve [16] with rotating principal axes.

We have chosen this criterion of the number of tensorial zones to classify the main classes of non-viscous constitutive relations, since this number characterizes the structure of the relation. In fact the directional dependency of the constitutive tensor $\mathbf{M}$ with respect to the incremental loading can be described either in a discontinuous manner, by a finite number of different tensors related to the same number of tensorial zones (we will call such relations 'incrementally piecewise linear'), or in a continuous manner by a continuous variation of the constitutive tensor with the direction of the incremental loading (relations called 'incrementally non-linear').

1.4.1 Constitutive relations with one tensorial zone

The first class of relations that we are going to look at is related to the simplest assumption that only one tensorial zone exists. Therefore

$$\forall \boldsymbol{u} : \mathbf{M}(\boldsymbol{u}) \equiv \mathbf{M}.$$

We have here the class of elastic laws, since there is a unique linear relation between $d\epsilon$ and $d\sigma$:

$$d\epsilon_\alpha = M_{\alpha\beta}d\sigma_\beta \quad \alpha, \beta \in \{1, 2, \cdots, 6\}^2.$$

These elastic laws may be isotropic or anisotropic, linear or non-linear (global non-linearity). The isotropic and linear case corresponds to the well-known Hooke's law with its two constitutive constants: Young's modulus and Poisson's ratio.

1.4.2 Constitutive relations with two tensorial zones

We now find laws with two tensorial zones, which can then be called 'loading zones' and 'unloading zones'. These two tensorial zones are separated by a hyperplane in $d\sigma$ space or in $d\epsilon$ space.

These laws may be either of hypoelastic type with a unique loading-unloading criterion (for example, models proposed by Guélin [21]) or of elastoplastic type with one regular plastic potential (for example, models proposed by Mroz [31], Prevost [34] or Dafalias & Herrmann [5]).

The model developed by Duncan & Chang [19] is a non-linear isotropic hypoelastic model with a specific loading-unloading criterion. It is easy to verify that such a model cannot be continuous at the frontier between the two zones; we only mention it here as a reminder.

The classical elastoplastic relations utilize an elastic matrix $\mathbf{M}^e$, related to the unloading zone, and an elastoplastic matrix $\mathbf{M}^{ep}$, related to the loading zone.

Let us study the continuity condition for this last case. The assumption of the additive decomposition of the incremental strain into its elastic part (reversible) and its plastic part (irreversible) allows us to write

$$d\epsilon = d\epsilon^e + d\epsilon^p.$$

The plastic incremental strain $d\epsilon^p$ is given by a flow rule which is often specified in terms of a plastic potential g:

$$d\epsilon^p = \frac{\partial g}{\partial \sigma} \, d\lambda,$$

where $d\lambda$ is called the plastic multiplier. Let us recall the continuity condition (1.6) at the interface of the elastic zone and the elastoplastic zone, which may be written here as

$$(\mathbf{M}^e - \mathbf{M}^{ep})\, \boldsymbol{u} = 0$$

or

$$\mathbf{M}^e \mathrm{d}\sigma - \mathbf{M}^{ep}\mathrm{d}\sigma = \mathrm{d}\epsilon^e - \mathrm{d}\epsilon = \mathrm{d}\epsilon^p = 0.$$

This identity must be satisfied for any $\mathrm{d}\sigma$ belonging to the hyperplane, which is the frontier between the loading and unloading zones.

The loading condition is obtained by writing that the incremental stress is directed outwards from the elastic limit $f(\sigma) = 0$, defined by a yield function $f(\sigma, \chi)$, where χ is the set of internal variables. This function governs the onset of plastic deformations. It follows that

$$\mathrm{d}\sigma \cdot \frac{\partial f}{\partial \sigma} > 0.$$

In the same manner, the unloading condition is given by

$$\mathrm{d}\sigma \cdot \frac{\partial f}{\partial \sigma} < 0.$$

The equation of the hyperplane, the frontier between the two zones, in $\mathrm{d}\sigma$ space, is then

$$\mathrm{d}\sigma \cdot \frac{\partial f}{\partial \sigma} = 0.$$

An incremental path that satisfies this condition is called 'neutral loading'. The continuity condition $\mathrm{d}\epsilon^p = 0$ must then be satisfied by all neutral loadings

$$\mathrm{d}\sigma \cdot \frac{\partial f}{\partial \sigma} = 0.$$

By developing the consistency condition $\mathrm{d}f = 0$, which means that the elastic limit is displaced by the stress point on loading (or in other words that the stress state must remain on the yield surface), we can demonstrate that $d\lambda$ is proportional to $\mathrm{d}\sigma \cdot \frac{\partial f}{\partial \sigma}$. So the continuity condition is always fulfilled owing to the existence of neutral loading.

This large class of elastoplastic constitutive relations with a unique plastic potential is itself divided into several subclasses following the assumptions which allow precise expression of a plastic potential.

First of all, this plastic potential $g(\sigma)$ may or may not coincide with the elastic limit surface $f(\sigma)$. Where there is such a coincidence, the elastoplastic material is called 'associated', otherwise it is 'non-associated'. From experience, geomaterials have to be considered as non-associated, except for saturated undrained clays studied under total stresses.

This difference between associated and non-associated flow rules has a major consequence in terms of stability and bifurcation analysis. This will be developed in Section 1.8.1.

The plastic potential may also vary with the strain history and be dependent on state variables (called also 'hardening parameters') which characterize the memory of geomaterial.

1.4.3 Constitutive relations with four tensorial zones

To this class belong the hypoelastic constitutive relations with two loading-unloading criteria proposed by Davis & Mullenger [15] and elastoplastic models with a double plastic potential (Lade [26], Loret [29], Nova & Wood [32] and Vermeer [38], for example). This means that, after a given strain history, the direction of the plastic incremental strain vector is no longer defined in a unique way by the normal to a unique plastic potential (eventually a function of the history) but may coincide with two normals to two different potentials following the direction of the present incremental stress vector.

With respect to each of these two plastic mechanisms, the incremental stress may be considered to be a loading or an unloading. Therefore we obtain the following four possibilities:

1. $\frac{\partial f_1}{\partial \sigma} \cdot d\sigma > 0$: loading for criterion 1; p_1,

2. $\frac{\partial f_1}{\partial \sigma} \cdot d\sigma < 0$: unloading for criterion 1; e_1,

3. $\frac{\partial f_2}{\partial \sigma} \cdot d\sigma > 0$: loading for criterion 2; p_2,

4. $\frac{\partial f_2}{\partial \sigma} \cdot d\sigma < 0$: unloading for criterion 2; e_2.

Finally, four different constitutive tensors can be obtained:

$$\mathbf{M}^{e_1 e_2}, \mathbf{M}^{p_1 e_2}, \mathbf{M}^{e_1 p_2}, \mathbf{M}^{p_1 p_2}.$$

Each of them is associated with a certain tensorial zone in $d\sigma$ space. The various continuity conditions can be satisfied by applying both of the consistency equations, which enable us to obtain two arbitrary plastic multipliers $d\lambda_1$ and $d\lambda_2$ proportional to $\frac{\partial f_1}{\partial \sigma} \cdot d\sigma$ and to $\frac{\partial f_2}{\partial \sigma} \cdot d\sigma$ respectively. The neutral loading paths then belong to two hyperplanes whose equations are given by

$$\frac{\partial f_1}{\partial \sigma} \cdot d\sigma = 0, \quad \frac{\partial f_2}{\partial \sigma} \cdot d\sigma = 0.$$

These two hyperplanes enable us to define four tensorial zones in $d\sigma$ space. The reasoning and presentation will be the same for hypoelastic models with two loading-unloading criteria.

1.4.4 Constitutive relations with eight tensorial zones

By further increasing the number of tensorial zones we now find incrementally octolinear relations with eight tensorial zones.

Two models belong to this class: the elastoplastic model with three plastic potentials proposed by Aubry *et al.* [1] and the incremental octolinear relation developed by Darve and Labanieh [12]. The basis of the law with three plastic potentials is essentially the same as for the models with a double potential, which we have presented above. Concerning the incremental octolinear relation, the basis of this model derives from the assumption of eight tensorial zones in dσ space. It satisfies all the different continuity conditions for fixed principal stress and fixed principal strain axes [12] and has been generalized [20] to arrive at an 'incrementally non-linear of second-order' constitutive relation, which will be presented in Section 1.5.

1.4.5 Constitutive relations with an infinite number of tensorial zones

We could say that they have an infinite number of tensorial zones, since each direction of dσ space is linked with a given tangent constitutive tensor which varies in a continuous manner with this direction. They may be of three different types. The endochronic models (Valanis [36], Bazant [2]) take into account non-linearity by introducing an intrinsic 'time' which is always increasing. The hypoplastic models by Kolymbas [25] or Chambon *et al.* [3] assume a priori a direct non-linear relation between dϵ and dσ. Thirdly we find models based on a non-linear interpolation between given constitutive responses, the non-linearity being linked to the kind of interpolation rule used (Darve [6]). Another model has been proposed by Dafalias [4] in the framework of his 'bounding surface' theory of plasticity.

The interest in incrementally non-linear constitutive relations is based mainly on the fact that it is not necessary to postulate the existence of either an elastic limit surface (since any purely elastic domain has disappeared) and the linked loading-unloading condition or a plastic potential. In fact the non-linearity of the incremental relation allows a direct description of the different behaviours in 'loading' and in 'unloading'. For this reason the incrementally non-linear relations may be closer to the physics that governs deformation of geomaterials (see Table 1.1).

1.5 Incrementally non-linear constitutive relations of second-order

Let us recall relation (1.5):

$$d\epsilon_\alpha = M_{\alpha\beta}(u_\gamma)d\sigma_\gamma \tag{1.7}$$

NTZ[(*)]	Constitutive tensors	Classes of constitutive relations
1	$\mathbf{M}$	Elasticity Hyperelasticity Hypoelasticity in its strict sense
2	$\begin{vmatrix} \mathbf{M}^e \\ \mathbf{M}^{ep} \end{vmatrix}$ or $\begin{vmatrix} \mathbf{M}^+ \\ \mathbf{M}^- \end{vmatrix}$	Elastoplasticity with 1 plastic potential or Hypoelasticity with 1 load-unload crit.
4	$\begin{vmatrix} \mathbf{M}^{e_1 e_2} \\ \mathbf{M}^{p_1 e_2} \\ \mathbf{M}^{e_1 p_2} \\ \mathbf{M}^{p_1 p_2} \end{vmatrix}$	Elastoplasticity with 2 plastic potentials
4	$\begin{vmatrix} \mathbf{M}^{++} \\ \mathbf{M}^{-+} \\ \mathbf{M}^{+-} \\ \mathbf{M}^{--} \end{vmatrix}$	Hypoelasticity with 2 load-unload crit.
8		Elastoplasticity with 3 plastic potentials Octolinear incremental relation
∞	$\mathbf{M}\dfrac{\mathrm{d}\sigma}{\|\mathrm{d}\sigma\|}$	Endochronic models Incrementally non-linear relations

TABLE 1.1. The main classes of non-viscous constitutive relation.
(*)NTZ: Number of Tensorial Zones.

with

$$u_\gamma = \frac{\mathrm{d}\sigma_\gamma}{\|\mathrm{d}\sigma\|}, \quad \gamma \in \{1, 2, \cdots, 6\}.$$

We consider polynomial series expansions for 36 functions $M_{\alpha\beta}$ of six variables u_γ:

$$M_{\alpha\beta}(u_\gamma) = M^1_{\alpha\beta} + M^2_{\alpha\beta\gamma}u_\gamma + M^2_{\alpha\beta\gamma\delta}u_\gamma u_\delta + \cdots. \tag{1.8}$$

From equation (1.7) and equation (1.8) it follows that:

$$\mathrm{d}\epsilon_\alpha = M^1_{\alpha\beta}\,\mathrm{d}\sigma_\beta + \frac{1}{\|\mathrm{d}\sigma\|}M^2_{\alpha\beta\gamma}\,\mathrm{d}\sigma_\beta\,\mathrm{d}\sigma_\gamma + \cdots. \tag{1.9}$$

The first term of equation (1.9) describes the elastic behaviour and both the first terms are incrementally non-linear constitutive relations of second-order.

In this paper we will consider only loading paths which are defined in fixed incremental stress-strain principal axes. In such a case the expression of the model is the following (for a more general presentation see Darve and Dendani [11] or Darve [8]):

$$\begin{pmatrix} d\epsilon_1 \\ d\epsilon_2 \\ d\epsilon_3 \end{pmatrix} = \frac{1}{2}[\mathbf{N}^+ + \mathbf{N}^-]\cdot\begin{pmatrix} d\sigma_1 \\ d\sigma_2 \\ d\sigma_3 \end{pmatrix} + \frac{1}{2\|d\sigma\|}[\mathbf{N}^+ - \mathbf{N}^-]\cdot\begin{pmatrix} d\sigma_1^2 \\ d\sigma_2^2 \\ d\sigma_3^2 \end{pmatrix} \quad (1.10)$$

with
$$\|d\sigma\| = \sqrt{d\sigma_i\, d\sigma_i}, \quad i \in \{1,2,3\}.$$

The two matrices $\mathbf{N}^+$ and $\mathbf{N}^-$ have the following form (for more details see Darve and Dendani [11] or Darve [8]):

$$\mathbf{N}^+ = \begin{bmatrix} \dfrac{1}{E_1^+} & -\dfrac{\nu_{12}^+}{E_2^+} & -\dfrac{\nu_{13}^+}{E_3^+} \\[2mm] -\dfrac{\nu_{21}^+}{E_1^+} & \dfrac{1}{E_2^+} & -\dfrac{\nu_{23}^+}{E_3^+} \\[2mm] -\dfrac{\nu_{31}^+}{E_1^+} & -\dfrac{\nu_{32}^+}{E_2^+} & \dfrac{1}{E_3^+} \end{bmatrix}, \quad \mathbf{N}^- = \begin{bmatrix} \dfrac{1}{E_1^-} & -\dfrac{\nu_{12}^-}{E_2^-} & -\dfrac{\nu_{13}^-}{E_3^-} \\[2mm] -\dfrac{\nu_{21}^-}{E_1^-} & \dfrac{1}{E_2^-} & -\dfrac{\nu_{23}^-}{E_3^-} \\[2mm] -\dfrac{\nu_{31}^-}{E_1^-} & -\dfrac{\nu_{32}^-}{E_2^-} & \dfrac{1}{E_3^-} \end{bmatrix}, \quad (1.11)$$

where E_i and ν_{ij} are, respectively, tangent moduli and tangent Poisson's ratio on 'generalized triaxial paths'. These paths correspond to the conventional triaxial paths (the two constant lateral stresses are equal), but on 'generalized triaxial paths' the lateral stresses are fixed but independently. The superscript (+) means 'compression' in the axial direction for these paths and the superscript (-) means 'extension'.

The behaviour of the studied material for these specific paths is assumed to be given by laboratory triaxial tests and described by analytical expressions. It is in these expressions that the constitutive constants appear as well as the state variables (stress tensor and void ratio) and the memory parameters (which are of two type: of discontinuous and continuous nature). Thus $\mathbf{N}^+$ and $\mathbf{N}^-$ depend on state variables and memory parameters.

It is clear from equation (1.10) that this relation is homogeneous of degree one with respect to $d\sigma$. Thus, it describes a rate independent behaviour. Equation (1.10) is also non-linear in $d\sigma$, which means that it can describe plastic irreversible strains: for an elementary stress cycle $(d\sigma, -d\sigma)$ the irreversible strain is equal to

$$\begin{pmatrix} d\epsilon_1 \\ d\epsilon_2 \\ d\epsilon_3 \end{pmatrix} = \frac{1}{\|d\sigma\|}[\mathbf{N}^+ - \mathbf{N}^-]\cdot\begin{pmatrix} d\sigma_1^2 \\ d\sigma_2^2 \\ d\sigma_3^2 \end{pmatrix}.$$

The recoverable strain does not have an elastic nature since it is equal to

$$\begin{pmatrix} d\epsilon_1 \\ d\epsilon_2 \\ d\epsilon_3 \end{pmatrix} = \frac{1}{2}[\mathbf{N}^+ + \mathbf{N}^-]\cdot\begin{pmatrix} d\sigma_1 \\ d\sigma_2 \\ d\sigma_3 \end{pmatrix} - \frac{1}{2\|d\sigma\|}[\mathbf{N}^+ - \mathbf{N}^-]\cdot\begin{pmatrix} d\sigma_1^2 \\ d\sigma_2^2 \\ d\sigma_3^2 \end{pmatrix}.$$

It means that it is not possible to decompose in an additive manner the strain into an elastic part and a plastic one. Elasticity and plasticity are intrinsically mixed into the constitutive relation.

Now we will see how equation (1.10) degenerates in the one dimensional case and in the cases of an elastic material and a perfectly plastic one.

For the one dimensional case (1.10) degenerates into the scalar expression

$$d\epsilon = \frac{1}{2}\left(\frac{1}{E^+} + \frac{1}{E^-}\right) d\sigma + \frac{1}{2}\left(\frac{1}{E^+} - \frac{1}{E^-}\right)|d\sigma|. \qquad (1.12)$$

It is easy to verify that relation (1.12) is able to describe any rate independent one-dimensional behaviour with one single expression (1.12). If one wants to interpret (1.10) with conventional elastoplastic concepts, it is necessary to introduce a loading-unloading criterion and we obtain

$$\begin{cases} d\sigma \geq 0 \quad : \quad d\epsilon = \dfrac{1}{E^+}d\sigma & \text{(loading)}, \\[3mm] d\sigma \leq 0 \quad : \quad d\epsilon = \dfrac{1}{E^-}d\sigma & \text{(unloading)}, \end{cases}$$

If the behaviour of the material is elastic, we obtain the same behaviour for 'compressions' and 'extensions' (as defined previously) which implies

$$\mathbf{N}^+ = \mathbf{N}^-.$$

Relation (1.10) degenerates into

$$\begin{pmatrix} d\epsilon_1 \\ d\epsilon_2 \\ d\epsilon_3 \end{pmatrix} = [\mathbf{N}] \cdot \begin{pmatrix} d\sigma_1 \\ d\sigma_2 \\ d\sigma_3 \end{pmatrix}. \qquad (1.13)$$

Equation (1.13) is exactly the general expression of anisotropic non-linear elasticity in fixed incremental stress-strain principal axes.

Finally, it is also possible to exhibit the equations of perfect plasticity as a degenerating case. Let us write relation (1.10) in the form

$$\begin{pmatrix} d\epsilon_1 \\ d\epsilon_2 \\ d\epsilon_3 \end{pmatrix} = [\mathbf{N}(u)] \cdot \begin{pmatrix} d\sigma_1 \\ d\sigma_2 \\ d\sigma_3 \end{pmatrix}. \qquad (1.14)$$

By inverting (1.14) we obtain

$$d\sigma = \mathbf{N}^{-1}(u)\,d\epsilon.$$

Conditions for perfect plasticity imply

$$\begin{cases} \mathrm{d}\sigma & = & 0, \\ \|\mathrm{d}\epsilon\| & : & \text{undetermined.} \end{cases}$$

It follows therefore that

$$\det \mathbf{N}^{-1}(u) = 0, \tag{1.15}$$

which represents the plastic condition.

$$\mathbf{N}^{-1}(u)\,\mathrm{d}\epsilon = 0 \tag{1.16}$$

is a generalized flow rule, since the solution of equation (1.16) while equation (1.15) is verified gives the direction of $\mathrm{d}\epsilon$ but not its intensity. Condition (1.16) is directionally dependent with respect to $\mathrm{d}\sigma$ which reflects the fact that the yield surface is locally deformed into a vertex.

The incrementally 'octolinear' model is given by the relation

$$\begin{pmatrix} \mathrm{d}\epsilon_1 \\ \mathrm{d}\epsilon_2 \\ \mathrm{d}\epsilon_3 \end{pmatrix} = \frac{1}{2}\left[\mathbf{N}^+ + \mathbf{N}^-\right] \cdot \begin{pmatrix} \mathrm{d}\sigma_1 \\ \mathrm{d}\sigma_2 \\ \mathrm{d}\sigma_3 \end{pmatrix} + \frac{1}{2}\left[\mathbf{N}^+ - \mathbf{N}^-\right] \cdot \begin{pmatrix} |\mathrm{d}\sigma_1| \\ |\mathrm{d}\sigma_2| \\ |\mathrm{d}\sigma_3| \end{pmatrix}. \tag{1.17}$$

The incremental structure of this model is the same as an elastoplastic relation with three plastic potentials. The constitutive constants for both the incremental non-linear model and octolinear models are the same.

1.6   Application to the analysis of yield surfaces

In the elastoplastic theory, yield surfaces are classically defined as an elastic limit surface in the six-dimensional physical stress space. All stress states inside the yield surface can be reached without any inelastic strain. Experimentally it has appeared that it was very difficult to detect the stress state, along a given stress path, where the first plastic strain is developing, because (particularly for geomaterials) the yield surface of materials before any loading is defined by a very small domain (Hicher [22]) and the process of inducing plastic strains along a given loading path is gradual and continuous, particularly for the first applied loading. For that reason many authors have chosen a stain intensity criterion in order to obtain an objective procedure to plot the graph of the yield surfaces and to compare them together upon rational bases.

The same difficulty is met within our model since plastic strains always coexist with elastic ones, even if for small strains, $\mathbf{N}^+$ and $\mathbf{N}^-$ are very close and the behaviour thus rather elastic. The principle of the numerical procedure which is chosen in order to simulate the experimental procedure

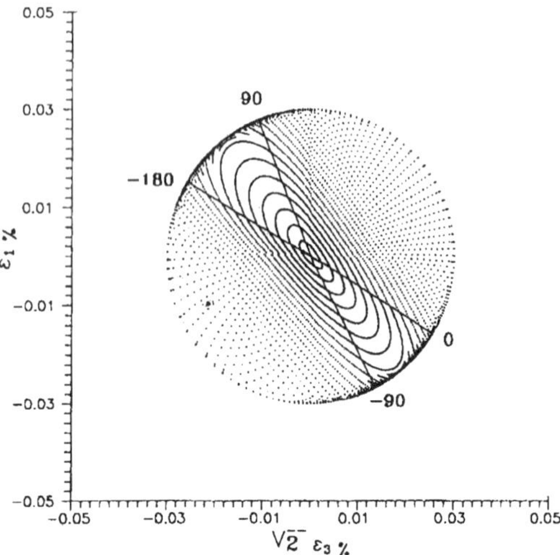

FIGURE 1.1. Principle of numerical simulation of the experimental procedure to obtain yield surfaces (in the strain space). The strain intensity criterion is equal to: $\|\Delta\epsilon\| = 3 \cdot 10^{-4}$ (Méghachou [30]).

is depicted in Figures 1.1 and 1.2. After a given loading path (which is in this example a stress isotropic loading) at a given stress state (here $\sigma_1 = \sigma_2 = \sigma_3 = 100\mathrm{kPa}$), we consider the family of axisymmetrical stress paths ($\sigma_2 = \sigma_3$) which are directed in all the axisymmetrical stress directions; then we stop the computation in the successive directions when the strain intensity criterion is reached: $\|\Delta\epsilon\| = \|\Delta\epsilon\|_{limit}$. Here,

$$\|\Delta\epsilon\|_{limit} = \sqrt{(\Delta\epsilon_1)^2 + 2(\Delta\epsilon_3)^2} = 3 \cdot 10^{-4}.$$

The stress paths are radial and rectilinear, while the strain responses are obviously not necessarily rectilinear even if the strain levels are rather small. The integration of the constitutive model on these rectilinear stress paths is stopped when the strain intensity criterion is fulfilled. 360 radial stress paths are considered, each one differing from the previous one by one degree. The figure obtained in the associated strain bisector plane is a circle, because of this same limit value, while in the stress bisector plane we obtain numerically the trace of the 'yield surface'. The yield surface is clearly more elongated in the isotropic direction since the stiffness is the largest in this direction. Angle signs are retained in the transformation but not angle values, as one can see in Figures 1.1 and 1.2. With this tool one can investigate the evolution of yield surfaces with a stress-strain history, particularly in order to exhibit the main hardening mechanism: kinematic, isotropic or rotational.

Figures 1.3 and 1.4 depict examples of such results for loose sand (Figure 1.3) and for dense sand (Figure 1.4) in axisymmetrical cases. In these cases

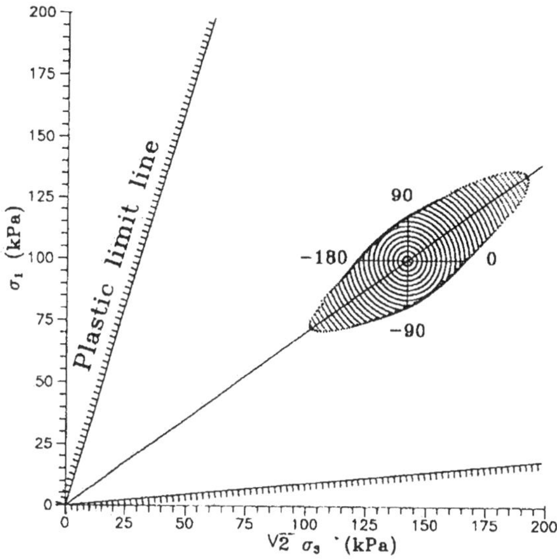

FIGURE 1.2. Principle of numerical simulation of the experimental procedure to obtain yield surfaces (in the stress space). The strain intensity criterion is equal to: $\|\Delta\epsilon\| = 3 \cdot 10^{-4}$ (Méghachou [30]).

the numerical modeling involves an integration of the model along some proportional stress paths, which are represented by continuous lines and correspond to constant lateral pressure paths and constant mean pressure paths, until reaching some limit stress states indicated on the figures by the black points at the end of lines. From these stress states reversals were simulated by the model until the stress states indicated by cross points were reached. It was from these last stress states that the numerical procedure to plot the yield surface was applied.

The 'plastic limit lines', which are plotted on Figure 1.3 and Figure 1.4, correspond to the classical and usual Mohr–Coulomb plastic criterion. A plastic limit surface is introduced inside the constitutive model as a stress surface which bounds the stress states that can be reached by any loading path. When the stress state is approaching this limit surface, at least one of the tangent moduli E_i tends to zero. At this time no constitutive softening is introduced inside the model. It means that the experimental characterizing of the plastic limit surface necessitates tests maintaining a homogeneous stress-strain field as long as possible in order to delay the strain localisation phenomenon. Three main conclusions are exhibited by both Figures 1.3 and 1.4:

1. the main mechanism which influences the evolution of the yield surfaces is a kinematic hardening,

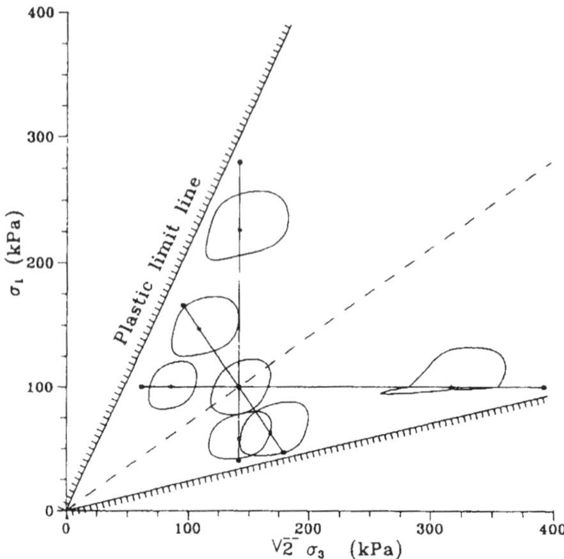

FIGURE 1.3. Numerical simulation of yield surfaces in a bisector stress plane for loose Hostun sand. The material is loaded until the black points at the end of continuous lines are reached, then unloaded to the cross points before plotting the yield surface (Méghachou [30]).

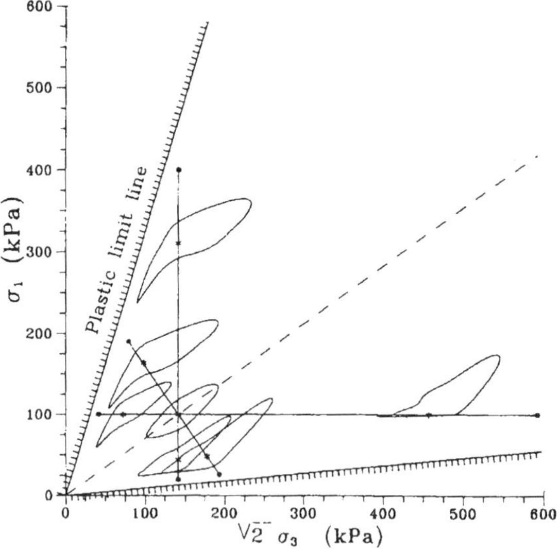

FIGURE 1.4. Numerical simulation of yield surfaces in a bisector stress plane for dense Hostun sand. The material is loaded until the black points at the end of continuous lines are reached, then unloaded to the cross points before plotting the yield surface (Méghachou [30]).

2. there is a small isotropic hardening, but the other largely more important factor is the influence of the shape of the plastic limit surface which curves the yield surfaces when they are approaching this limit surface,

3. a loose sand has rounder yield surfaces than the dense one.

Figure 1.5 depicts yield surfaces in a deviatoric stress plane (the mean pressure is equal to 300 kPa) for dense Hostun sand. The initial yield surface is precisely a circle because of the initial overall isotropy. Each yield surface has been plotted after a stress loading path has been followed until the black points seen on the figure at the end of continuous lines. The same method has been adopted as previously but without stress reversals. At the stress states (marked by the black points) 360 radial rectilinear stress paths are applied (each one differs from the adjacent ones by one degree in direction). The loading is stopped when the strain intensity criterion is reached. As previously the main hardening is kinematic. A quite astonishing point is the fact that the form of the plastic limit surface influences very fast, from the isotropic state, the yield surfaces whose form reflects and anticipates in a quite faithful way the pattern of the limit surface.

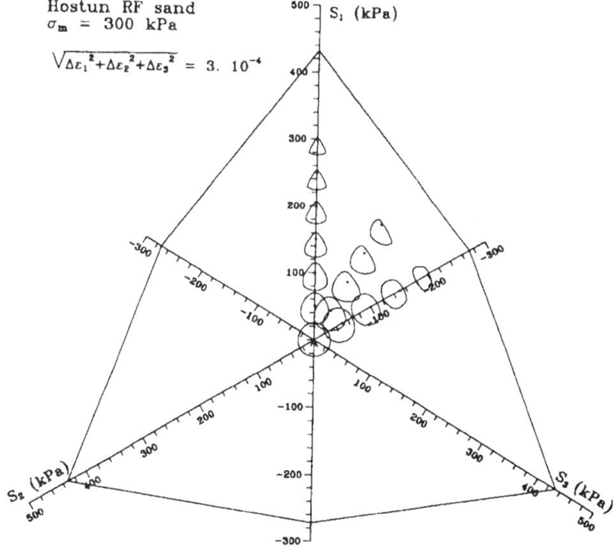

FIGURE 1.5. Numerical simulation of yield surfaces in a deviatoric stress plane for dense Hostun sand. The material is loaded until the black points are reached, before plotting the yield surface (Méghachou [30]).

1.7   Application to the principle of superposition for incremental loading

As recalled briefly in Section 1.2, this 'principle' implies that the material response to a sum of incremental loading is equal to the sum of the associated incremental responses. This 'principle' can be considered as being formulated in a comparable way to Boltzmann's principle. While the last one is applied to a finite loading (which means a certain loading history), the first one considers only incremental loading. It can be said that the first one is a principle of superposition 'in the small', while Boltzmann's principle is a principle of superposition 'in the large'.

Boltzmann's principle implies that the behaviour is linear elastic or visco–elastic, since the constitutive functional must be linear. The principle of superposition for incremental loading implies that the constitutive incremental function is linear, which implies that the described behaviour is non-linear elastic or non-linear visco-elastic.

If we consider only elasto-plastic strains and assume the general validity of the principle, the constitutive operator $\mathbf{G}$ (equation (1.3)) is linear and thus describes only elastic behaviour (linear or non–linear elasticity). More precisely the principle is strictly verified inside the same tensorial zone:

$$\mathbf{G}\left(\mathrm{d}\sigma^1 + \mathrm{d}\sigma^2\right) = \mathbf{G}\left(\mathrm{d}\sigma^1\right) + \mathbf{G}\left(\mathrm{d}\sigma^2\right),$$

if and only if $\mathrm{d}\sigma^1$ and $\mathrm{d}\sigma^2$ belong to the same tensorial zone.

Experimentally it is well known that with servo–controlled machines this principle is approximately verified. In fact such apparatuses are not able to follow exactly the required loading path but allow us to approach the path considered by a sequence of approximations. The experimental responses are generally considered as satisfying as soon as the approximations become 'small' enough.

For incrementally non–linear constitutive relations the 'principle' of superposition for incremental loading is never satisfied since the constitutive incremental function is non-linear. Therefore it is interesting to compare the responses provided by the model for rectilinear proportional paths and for piecewise linear approximated paths with bends.

The first example (Figure 1.6) is constituted by a 'drained triaxial' loading performed on dense Hostun sand, which means a path with a constant lateral pressure. This path is decomposed into several pieces divided into two parts: one is an incremental isotropic loading ($\mathrm{d}\sigma_2 = \mathrm{d}\sigma_3$) and the other is a constant mean pressure loading ($\mathrm{d}\sigma_1 + 2\mathrm{d}\sigma_3 = 0$). The length of the 'steps' varies and five cases have been considered: 1, 2, 4, 8 and 16 kPa.

Two main conclusions are exhibited by results depicted on Figure 1.6:

1. when the length of the 'steps' decreases, the response diagrams converge to certain asymptotical behaviour,

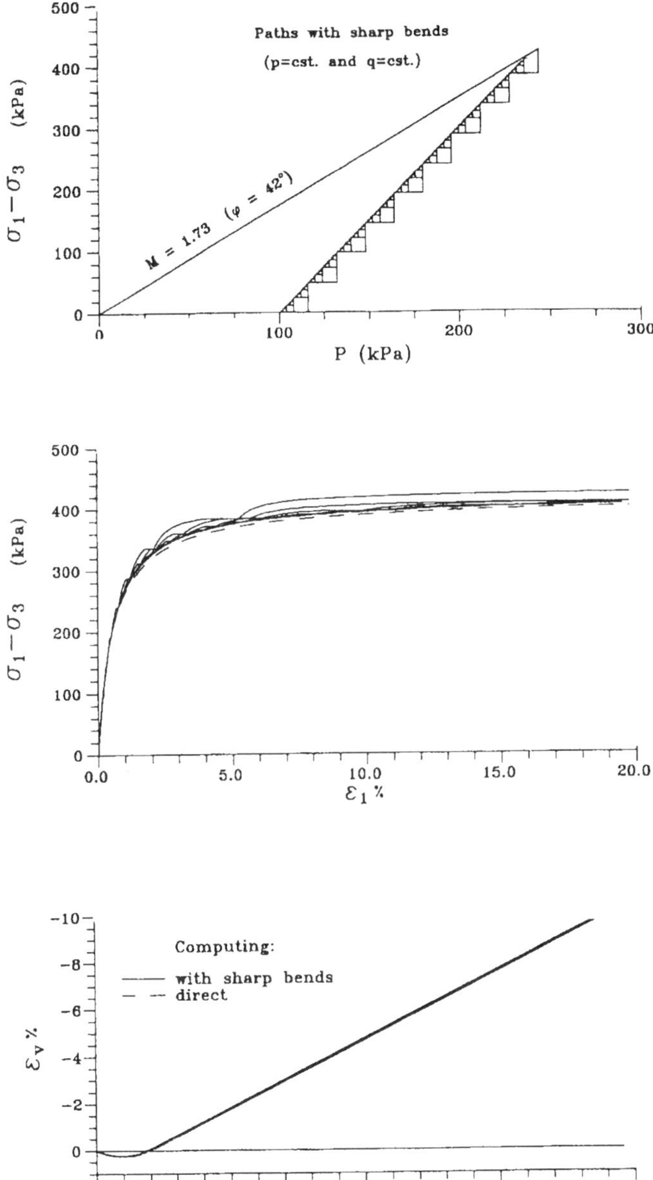

FIGURE 1.6. Comparison between numerical responses obtained from the incrementally non-linear model in the case of a rectilinear proportional loading (with a constant lateral pressure) – dash lines – and of 'staircase' paths-continuous lines. Dense Hostun sand (Méghachou [30]).

2. these converged response curves are not confounded with the responses to the proportional loading, but remain satisfactorily close.

It is also interesting to notice that the total length of the rectilinear proportional path and the decomposed paths (even for 'small' lengths of 'steps') are completely different (the ratio of the lengths is equal to $\sqrt{2}$). The same conclusions can be made by analyzing the results depicted in Figure 1.7 in the case of loose Hostun sand.

The same method is also considered in order to investigate the behaviour under an 'undrained' loading path (in fact isochoric loading) performed on loose Hostun sand. The following decomposition has been adopted in axisymmetric conditions:

$$\left\{ \begin{array}{llll} \dfrac{q}{p'} & = & \text{constant and} & d\epsilon_1 < 0 \\ p' & = & \text{constant and} & d\epsilon_1 > 0 \end{array} \right\}$$

with deviatoric stress $q = \sigma_1 - \sigma_3$ and the mean effective pressure $p' = \dfrac{\sigma_1' + 2\,\sigma_3'}{3}$.

As the first path $\left(\dfrac{q}{p'} = \text{const.} \right)$ is dilatant and the second one contractant, it is thus possible to fulfill the isochoric condition:

$$\Delta\epsilon_1 + 2\,\Delta\epsilon_3 = 0.$$

The decomposition is depicted in Figure 1.8. The first path is applied until a given path length is reached:

$$\sqrt{(\Delta p)^2 + (\Delta q)^2} = 2\,\text{kPa}.$$

The results are depicted in Figures 1.9 and 1.10 and show that there is no significant influence of the kind of decomposition on the quantitative responses. Figure 1.10 allows us to verify that when the length of the sharp bends is decreasing the response curves converge and that the asymptotic diagram is rather close (but not coinciding) to the response issued from the rectilinear proportional isochoric strain path:

$$\left\{ \begin{array}{l} d\epsilon_1 > 0 \\ d\epsilon_2 = d\epsilon_3 \\ d\epsilon_1 + 2\,d\epsilon_3 = 0. \end{array} \right.$$

The four chosen lengths are equal to

$$\sqrt{(\Delta p)^2 + (\Delta q)^2} = 1, 2, 4, 8\,\text{kPa}.$$

Finally the computation is stopped only because of an artefact: for small levels of stress both the decomposition paths become contractant.

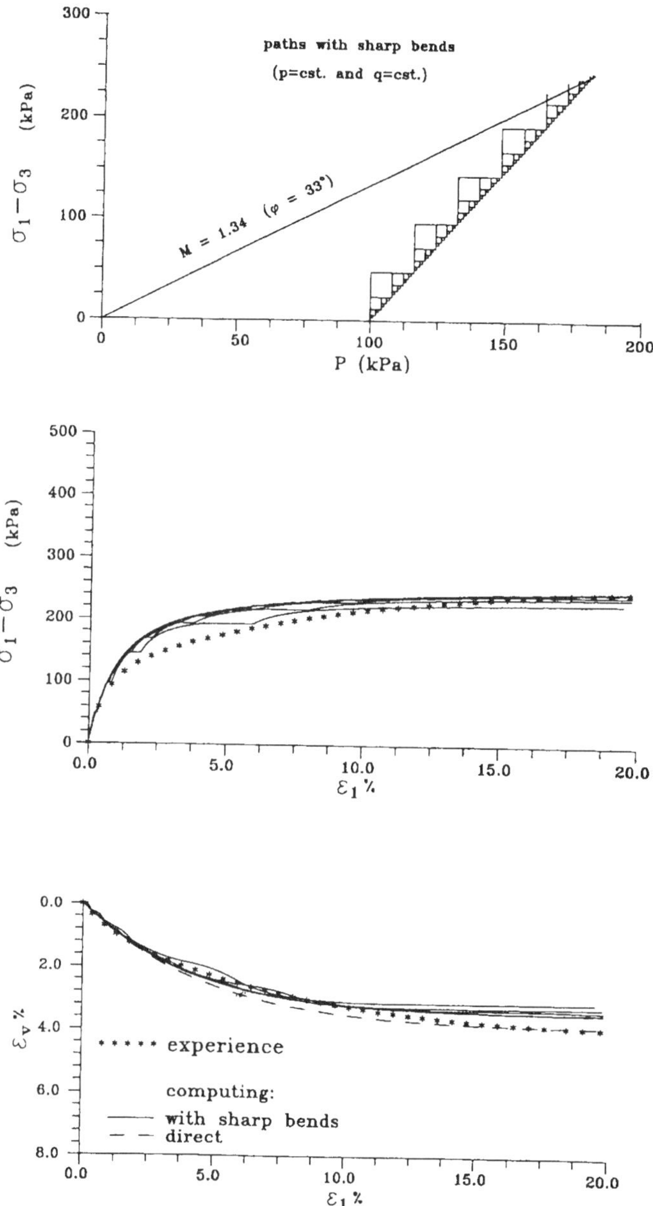

FIGURE 1.7. Comparison between numerical responses obtained from the incrementally non-linear model in the case of a rectilinear proportional loading (with a constant lateral pressure) – dash lines – and of 'staircase' paths-continuous lines. Loose Hostun sand (Méghachou [30]).

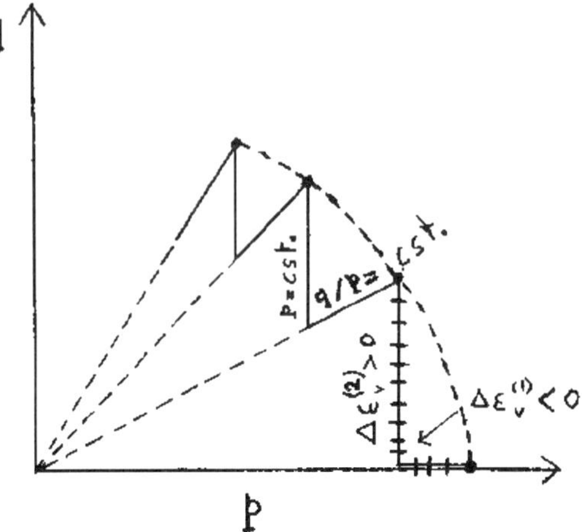

FIGURE 1.8. The considered decomposition of the loading path in the case of an isochoric loading.

1.8 Application to instabilities and bifurcations of granular materials

1.8.1 Material stability

Non-associated materials such as soils exhibit clearly constitutive instabilities strictly inside the plastic failure surface. Theoretical and numerical investigations have been performed for different soils (loose sand and dense sand) and for different stress-strain paths to show such features. Let us recall the definition of constitutive or material stability. A universal theory of structural or mechanical stability does not exist, as yet. Roughly speaking, stability means that small perturbations lead to small responses. The constitutive stability concept in the Lyapunov sense can be stated as follows:

For a given rate independent material, a stress-strain state (σ, ϵ) for a given strain history, is called stable, if any small change of the incremental loading dl leads to a small change of the incremental response dr.

More formally:

$$\forall \varepsilon > 0 \quad \exists \eta = \eta(\varepsilon) > 0 \quad \text{such that} \quad \|dl\|_\alpha < \eta \;\Rightarrow\; \|dr\|_\alpha < \varepsilon, \quad (1.18)$$

where $\|.\|_\alpha$ denotes a norm in the considered normed space.

While we have just seen above a definition of stability, this question can be investigated using the well-known Hill's second-order work criterion (Hill [23]). This sufficient condition, based on the scalar product of the rate of strain and the associated rate of stress, states that if this product re-

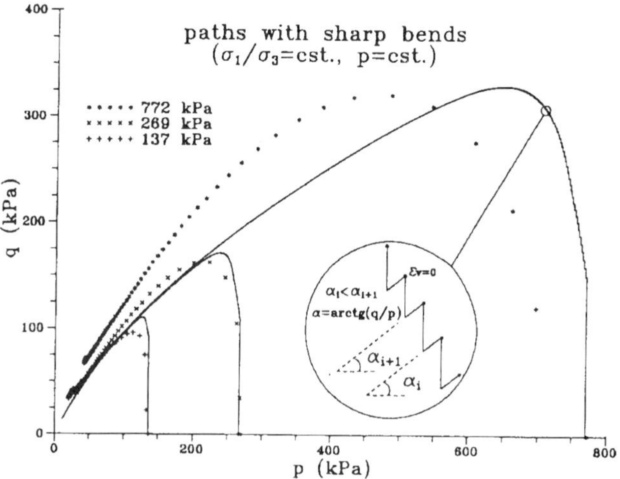

FIGURE 1.9. Simulation of an isochoric loading path by a path decomposed into radial unloading ($q/p' = $ const.) and constant mean pressure loading ($p' = $ const.). Three different initial isotropic stresses are considered and the material is loose Hostun sand (Méghachou [30]).

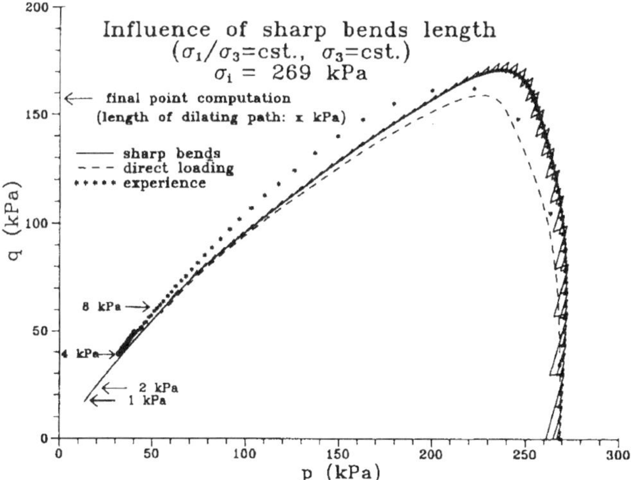

FIGURE 1.10. Study of the convergence of response curves when the decomposed path's length is decreasing from 8 to 1 kPa. The responses to the isochoric proportional strain path are plotted with dashed lines (Méghachou [30]).

mains strictly positive, then stability is guaranteed. For a rate independent material it follows, in incremental form:

$${}^{t}\mathrm{d}\epsilon \cdot \mathrm{d}\sigma > 0 \quad \forall\, \mathrm{d}\epsilon \in \mathrm{I\!R}^{n} - \{0\} \quad \text{with} \quad \mathrm{d}\sigma = \mathbf{F}(\mathrm{d}\epsilon). \tag{1.19}$$

The second-order work for non-associated materials is only a function of the symmetric part of the constitutive operator $\mathbf{P}$, since

$$\begin{aligned} \mathrm{d}^2 W &= {}^{t}\mathrm{d}\epsilon \cdot \mathrm{d}\sigma \\ &= {}^{t}\mathrm{d}\epsilon \cdot \mathbf{P} \cdot \mathrm{d}\epsilon \\ &= {}^{t}\mathrm{d}\epsilon \cdot \mathbf{P}^{s} \cdot \mathrm{d}\epsilon \end{aligned} \quad \text{with} \quad \left\{ \begin{aligned} \mathrm{d}\sigma &= \frac{\partial \mathbf{F}}{\partial(\mathrm{d}\epsilon)}\, \mathrm{d}\epsilon = \mathbf{P}\, \mathrm{d}\epsilon \\ \mathbf{P}^{s} &= \tfrac{1}{2}\left(\mathbf{P} + {}^{t}\mathbf{P}\right) \end{aligned} \right. . \tag{1.20}$$

Thus for incrementally linear constitutive relations condition (1.19) is equivalent to

$$\det \mathbf{P}^{s} > 0. \tag{1.21}$$

In axisymmetric conditions and in plane strain conditions, the sign of $\mathrm{d}^2 W$ has been clearly linked to the values of the determinant of the symmetric part of $\mathbf{P}$: $(\mathrm{d}^2 W \leq 0) \subset (\det \mathbf{P}^{s} < 0)$ in the general case of incrementally non-linear constitutive relations (Darve & Laouafa [13]).

Equations (1.19) or (1.21) are exclusive conditions, which means that if these sufficient conditions are fulfilled, then stability is guaranteed. In associated elasto-plasticity this condition implies that the hardening modulus remains positive when stability is fulfilled. The loss of stability coincides with the plastic flow and the stress state is located on the plastic limit surface. The loss of stability for non-associated materials corresponds to some material features which allow some specific failure modes inside the plastic limit condition, as for example the liquefaction phenomenon [9]. The incrementally non-linear constitutive model [8] can capture such peculiar features, in drained, undrained or more general conditions.

We have shown that there exists an instability surface strictly inside the limit surface. The results obtained in the case of undrained axisymmetric loading performed on loose sands [10] have been extended to the case of dense sand in axisymmetric conditions [14].

We have extended these previous results for the case of plane strain conditions (Figure 1.16) and for the radial stress path in the deviatoric plane (Figures 1.14 and 1.15). Two different densities have been considered which correspond to both cases of loose and dense sands, and two different constitutive relations have been utilized: the octo-linear model which is incrementally piecewise linear and the non-linear model of the second-order which is thoroughly incrementally non-linear.

The instability surface in the stress space defines a set of potential instability states, which means that at these stress-strain states there exists at least one direction of the incremental stress or the incremental strain leading to a nil or negative value of the second-order work. The stability is lost, and the instability will become effective if the loading direction coincides with an unstable direction and if the nature of the loading corresponds to a dead loading. In other words if the perturbating loading is stress (or force) controlled, then the deformation of the body is not kinematically constrained. The strain response will be arbitrarily large and thus it will not fulfill the Lyapunov definition.

The mechanism of catastrophic landslides, especially the initiation of the sliding process, shows a sudden and quick movement of the soil, which denotes clearly a change of the mechanical state from a stable state to an unstable one due to a small perturbation.

In a mass of soil the perturbation can have several origins. It could be an hydraulic cause due for instance to a strong rainfall, a dynamic perturbation caused by seismic waves, a shock or an explosion or a static additional loading (house constructions; trackways opening, etc.).

Classical plastic failure

Let us consider a drained triaxial compression path. For a constant radial stress σ_3 the curve $(\sigma_1 - \sigma_3)$ versus ϵ_1 shows a peak. This behaviour is observed experimentally for dense soils. In this special loading case, the plastic failure criterion is fulfilled at the maximum of the axial stress σ_1 (σ_1 peak). In this case the second-order work is as follows:

$$\mathrm{d}^2 W = {}^t\mathrm{d}\sigma \cdot \mathrm{d}\epsilon = \mathrm{d}\sigma_1 \, \mathrm{d}\epsilon_1 + 2 \, \mathrm{d}\sigma_3 \, \mathrm{d}\epsilon_3 = \mathrm{d}\sigma_1 \, \mathrm{d}\epsilon_1$$

and we can observe that the second-order work vanishes at this peak. It is negative after the peak in the softening regime. If we write the constitutive relation in this special case, we obtain

$$\begin{pmatrix} \mathrm{d}\sigma_1 \\ \mathrm{d}\sigma_3 \end{pmatrix} = \mathbf{A} \begin{pmatrix} \mathrm{d}\epsilon_1 \\ \mathrm{d}\epsilon_3 \end{pmatrix}.$$

At σ_1 peak and with the stress constraint condition $\sigma_3 = $ const., it follows the so-called plasticity criterion

$$\det \mathbf{P} = 0 \tag{1.22}$$

and the flow rule is given by

$$\mathbf{A} \begin{pmatrix} \mathrm{d}\epsilon_1 \\ \mathrm{d}\epsilon_3 \end{pmatrix} = \begin{pmatrix} 0 \\ 0 \end{pmatrix}.$$

Undrained loose sand

Now the loading path is an undrained triaxial compression. A typical behaviour observed experimentally is characterized by a maximum of $q = \sigma_1 - \sigma_3$ in the curve $q = \sigma_1 - \sigma_3$ versus axial strain ϵ_1 for loose sand.

The undrained condition is performed by imposing a volumetric constraint $\epsilon_v = \epsilon_1 + \epsilon_2 + \epsilon_3 = 0$. Subject to this kinematic condition the second-order work takes the expression

$$\mathrm{d}^2 W = (\mathrm{d}\sigma_1 - \mathrm{d}\sigma_3) \cdot \mathrm{d}\epsilon_1$$

and we observe that $\mathrm{d}^2 W$ is zero at the q deviatoric stress peak. This peak corresponds also to an unstable state. $\mathrm{d}^2 W$ is negative in the descending branch and positive before. Let us now rewrite the constitutive equation in the form

$$\left(\begin{array}{c} \mathrm{d}q \\ \mathrm{d}\epsilon_v \end{array} \right) = \mathbf{B} \left(\begin{array}{c} \mathrm{d}\epsilon_1 \\ \mathrm{d}\sigma_3 \end{array} \right).$$

At q peak for the isochoric condition $\mathrm{d}\epsilon_v = 0$, it follows

$$\left(\begin{array}{c} 0 \\ 0 \end{array} \right) = \mathbf{B} \left(\begin{array}{c} \mathrm{d}\epsilon_1 \\ \mathrm{d}\sigma_3 \end{array} \right).$$

The stress-strain state characterized by the maximum of q leads to an undefined algebraic linearized system. The non-trivial solution is obtained when

$$\det \mathbf{B} = 0, \tag{1.23}$$

which constitutes a bifurcation criterion. The mode of failure, by instability, is given by this non-trivial solution:

$$\mathbf{B} \left(\begin{array}{c} \mathrm{d}\epsilon_1 \\ \mathrm{d}\sigma_3 \end{array} \right) = \left(\begin{array}{c} 0 \\ 0 \end{array} \right).$$

Dense sand and proportional strain paths

In this subsubsection the loading path is now a proportional strain loading path. The case of proportional paths for dense sand was studied in order to generalize the conclusions obtained for loose sand.

Let us consider the following loading program in axisymmetric conditions:

$$\left\{ \begin{array}{l} \mathrm{d}\epsilon_1 = \text{constant}, \\ \mathrm{d}\epsilon_1 + 2\,R\,\mathrm{d}\epsilon_3 = 0 \quad (R \text{ constant}), \end{array} \right.$$

which is equivalent to

$$\left\{ \begin{array}{l} \mathrm{d}\sigma_1 - \dfrac{\mathrm{d}\sigma_3}{R} = \text{constant}, \\ \mathrm{d}\epsilon_1 + 2\,R\,\mathrm{d}\epsilon_3 = 0 \quad (R \text{ constant}), \end{array} \right.$$

because the first pair of variables of previous systems and the second one, respectively, are conjugated quantities according to the work of second-order:

$$\left(\mathrm{d}\sigma_1 - \frac{\mathrm{d}\sigma_3}{R} \right) \mathrm{d}\epsilon_1 + \frac{\mathrm{d}\sigma_3}{R} (\mathrm{d}\epsilon_1 + 2R\,\mathrm{d}\epsilon_3) = \mathrm{d}\sigma_1 \mathrm{d}\epsilon_1 + 2\,\mathrm{d}\sigma_3 \mathrm{d}\epsilon_3. \tag{1.24}$$

For different values of parameter R, which is constant during a given loading path, such as

$$R \in \{0.3, 0.35, 0.40, 0.45, 0.50, 0.60, 0.70, 0.80, 0.90, 1.0\},$$

we obtain the results depicted on Figure 1.11.

One first conclusion is that the maximum of $\sigma_1 - \frac{\sigma_3}{R}$ is an unstable state following Hill's criterion, since the second-order work is vanishing at ($\sigma_1 - \frac{\sigma_3}{R}$) peak for this set of paths according to equation (1.24). Such maxima are reached according to Figure 1.11 for each value of R lower than 0.7. These results can be considered as generalizing the classical results obtained with loose sands and R equal to 1 (the so-called undrained loading).

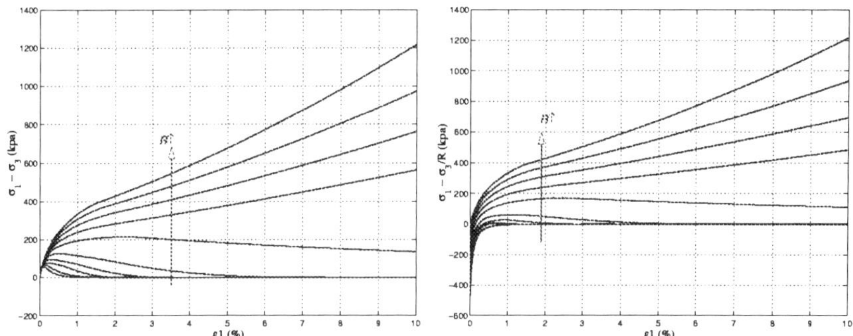

FIGURE 1.11. Deviatoric stress q versus axial strain (ϵ_1) on the left and $\sigma_1 - \dfrac{\sigma_3}{R}$ versus axial strain (ϵ_1) on the right.

Let us note that in this special loading program, which could be described numerically and satisfied locally in some boundary value problems, the constitutive model can be stated as follows:

$$\begin{pmatrix} d\sigma_1 - R^{-1}d\sigma_3 \\ d\epsilon_1 + 2Rd\epsilon_3 \end{pmatrix} = \mathbf{C} \begin{pmatrix} d\epsilon_1 \\ R^{-1}d\sigma_3 \end{pmatrix}$$

and at $\sigma_1 - R^{-1}\sigma_3$ peak it becomes

$$\mathbf{C} \begin{pmatrix} d\epsilon_1 \\ R^{-1}d\sigma_3 \end{pmatrix} = \begin{pmatrix} 0 \\ 0 \end{pmatrix}.$$

Then the bifurcation criterion is given by: det $\mathbf{C} = 0$, and the mode of failure by the non-trivial solution of the above linearized algebraic system.

1.8.2 Instability domains

Instability domain in the Rendulic plane

To determine the instability domain in the stress space, the numerical simulation follows the program below:

1. For a given confining pressure ($\sigma_1 = \sigma_2 = \sigma_3$) we perform either a triaxial compression ($d\epsilon_1 > 0$) or a triaxial extension ($d\epsilon_1 < 0$).

2. Along each triaxial path we check for the first stress state which exhibits one stress direction giving a nil second-order work. Then we obtain one point, in the stress space, of the boundary of the instability domain.

3. We pursue the procedure for different confining pressures. We obtain finally the instability domain in the σ_1-$\sqrt{2}\sigma_2$ plane.

Instability domain for loose sand

The loose sand is an Hostun sand with an initial void ratio equal to 0.92 and a friction angle equal to about 32°. Figure 1.12 gives the instability domain obtained with the non-linear model (o) and the octo-linear model (+). We have also represented the trace of the limit failure surface in the σ_1-$\sqrt{2}\sigma_2$ plane.

We remark on this figure the existence of a large domain of potentially unstable states. Whatever the constitutive model, they exhibit qualitatively the same domain.

Instability domain for dense sand

The dense sand is also Hostun sand with an initial void ratio equal to 0.55 and a friction angle equal to about 41°. The results shown in Figure 1.13 confirm and extend the previous results obtained on loose sand. The instability which has been usually associated to loose sand, and to undrained conditions, is now extended and generalised to all granular materials and to drained conditions. For dense sand the instability domain is smaller quantitatively than in the case of loose sand.

Instability domain in the deviatoric stress plane

Let us now determine the trace of the instability surface in the deviatoric stress plane, obtained by radial stress paths. Figure 1.14 and Figure 1.15 give the numerical results obtained for loose sand and dense sand respectively. The polyhedron represents the Mohr–Coulomb failure surface. These two figures show clearly that unstable states exist strictly inside the limit plastic condition.

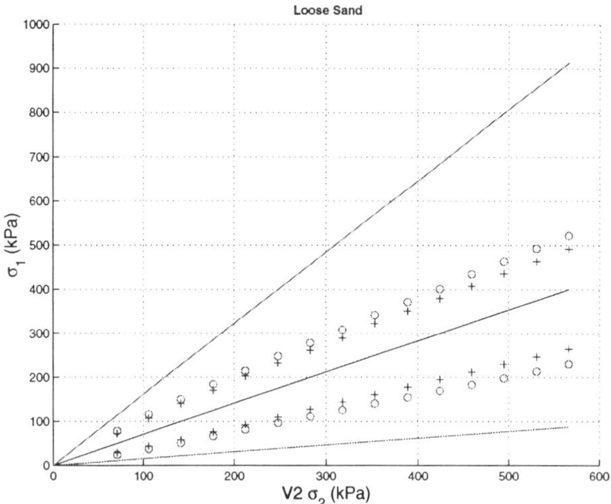

FIGURE 1.12. **Graph of the instability domain for loose sand along the triaxial paths. Representation in the σ_1-$\sqrt{2}\sigma_2$ plane in the case of the octo-linear $(+)$ and the non-linear(o) constitutive models.**

The Mohr–Coulomb failure surface defines in the stress space the physically admissible stress states. In the framework of associated elastoplasticity, unstable states are reached when the stress state joins this failure surface. The proper description of the behaviour of granular materials by incrementally non-linear constitutive relations allows us to state that the failure surface defines only one type of failure which is a plastic flow. There exist mechanical states where 'failure' could be reached before this kind of plastic flow with a non-controllable loading process (Imposimato and Nova [24]). This generalized failure condition is satisfied when the stress-strain state reaches a potentially unstable state and when this unstable mode is activated by an appropriate perturbation or external load.

Instability behaviour in plane strain conditions

This short section gives a compilation of results obtained for different sand densities for plane strain loading conditions. This kinematic condition is observed in many landslides. Figure 1.16 gives the first ratio $\frac{\sigma_1}{\sigma_3}$ leading to the first potentially unstable state, according to Hill's criterion. It must be emphasized that the plastic limit values of $\xi = \frac{\sigma_1}{\sigma_3}$ are around 6.0 for dense Hostun sand and around 4.0 for loose Hostun sand.

From all the previous results it can be concluded that there is a large domain of the stress space which is potentially unstable according to Hill's condition of stability.

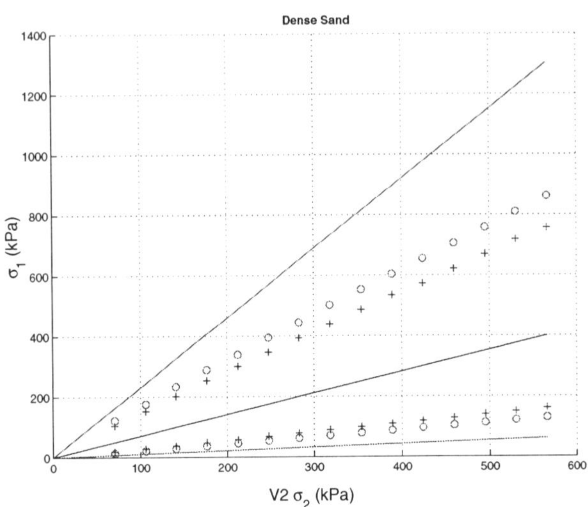

FIGURE 1.13. Graph of the instability domain for dense sand along triaxial paths. Representation in the σ_1-$\sqrt{2}\sigma_2$ plane in the case of the octo-linear (+) and the non-linear (o) constitutive models.

These stability analyses are until now material or local ones and could be applied only to homogeneous stress-strain fields. For boundary value problems the unstable stress directions can be excited or not by the admissible perturbations. The failure will appear only in the first case and if the loading mode allows the failure mechanism to become critical.

1.8.3 Application to slope stability problems

Granular avalanches

The instability criterion has been introduced into the finite elements code developed by Laouafa [27] and Laouafa & Royis [28] and some heuristic sand piles have been simulated.

Of course in order to excite the instabilities we need a perturbation of the boundary conditions. To represent as closely as possible the procedure to induce granular avalanches along sand piles, we have considered locally vanishing mechanical properties for a few elements of the finite element model. For this perturbation, the sign of the second-order work is checked at each stress integration point in the stress direction, precisely excited by the considered perturbation. Figure 1.17 shows some preliminary results coming from that procedure. Some finite elements are perturbed within the mesh. We see in Figure 1.17 the integration points where the unstable states are reached (in the sense recalled just above) and the perturbed area

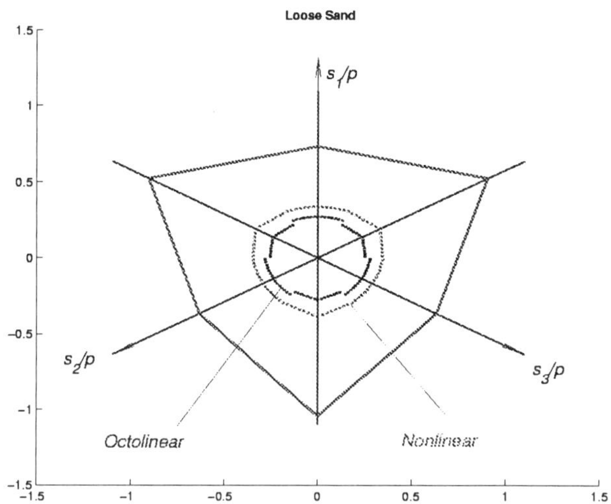

FIGURE 1.14. Trace of the instability domain in the deviatoric stress space for loose sand and both constitutive models.

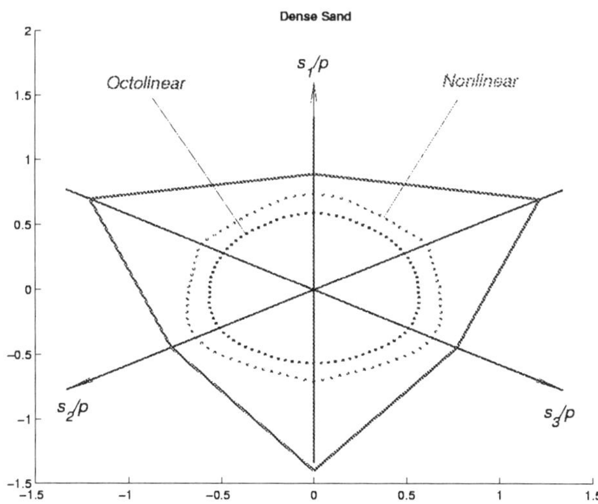

FIGURE 1.15. Trace of the instability domain in the deviatoric stress space for dense sand and both constitutive models.

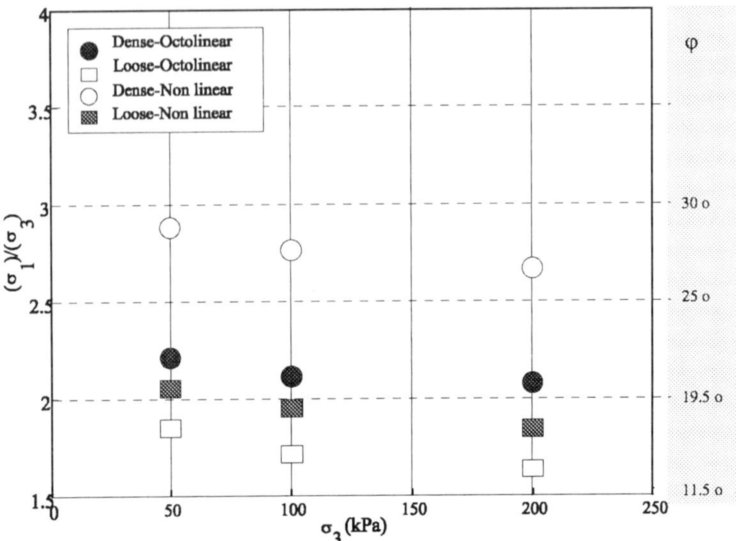

FIGURE 1.16. First states of instability for plane strain conditions for loose sand and dense sand. Two constitutive models have been used.

(in black).

It is interesting to remark that:

- these unstable domains appear for a low slope angle (Figure 1.17),

- the unstable zone is located closely to the sand pile surface, which could indicate a possible new mechanism of failure by avalanches involving only superficial layers.

It is clear that this last mode of failure corresponds to mechanisms which are apparent from experiments on granular avalanches. That mode of failure seems to be not appropriately described by the classical limit analysis leading to failure modes by rigid bodies sliding along shear bands.

Excavation of a soil mass

Finally Figure 1.18 presents a first example of an heuristic slope which has been simulated by an excavation process. The second-order work is checked at each integration point for the last excavated layer. In Figure 1.18 we see that a potentially unstable area is visible close to the slope surface. This new method (see also di Prisco *et al.* [18, 17]) to analyze slope stability problems seems to be particularly attractive in the case of diffuse rupture.

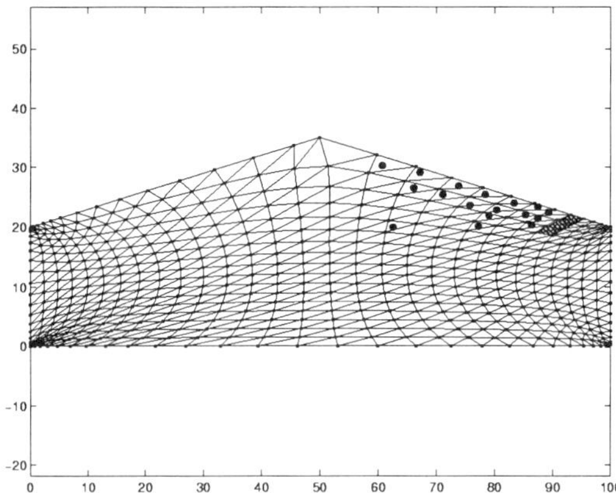

FIGURE 1.17. Modelling of a sand pile with a low slope angle. The perturbation and the induced unstable elements are indicated.

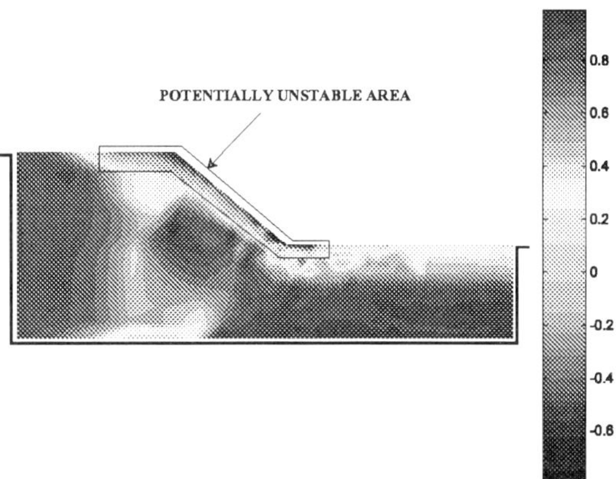

FIGURE 1.18. Finite element modelling of an heuristic slope. The unstable area has been determined for the last excavated sand layer.

Acknowledgments: This work has been performed with the help of the french Ministry of Education and Research ('Instabilités dans les milieux granulaires et applications aux glissements de terrain'), whose support is gratefully acknowledged.

References

[1] D. AUBRY, J.C. HUJEUX, F. LASSOUDIRE, Y. MEIMON: A double memory model with multiple mechanisms for cyclic soil behaviour, in: R. DUNGAR, G.N. PANDE, J. STUDER (eds.): *Proc. Int. Symp. Numer. Models Geomech.* **1**, Balkema, Rotterdam, 1982, pp. 3–13.

[2] Z.P. BAZANT: Endochronic inelasticity and incremental plasticity, *Int. J. Solids Struct.* **14**, pp. 691–714, 1978.

[3] R. CHAMBON, J. DESRUES, W. HAMMAD, R. CHARLIER: CLoE, a new rate-type constitutive model for geomaterials. Theoretical basis and implementation, *Int. J. Num. Anal. Meth. Geomech.* **18**, pp. 253–278, 1994.

[4] Y.F. DAFALIAS: Bounding surface plasticity. I: mathematical foundation and hypoelasticity, *J. Eng. Mech.* **112 (9)**, pp. 966–987, 1986.

[5] Y.F. DAFALIAS, L.R. HERMANN: A bounding surface soil plasticity model, in: G.N. PANDE, O.C. ZIENKIEWICZ (eds.): *Proc. Symp. Soils under Cyclic and Transient Loading*, Balkema, Rotterdam, 1980, pp. 335–345.

[6] F. DARVE: Une loi rhéologique incrémentale non-linéaire pour les solides, *Mech. Res. Comm.* **7 (4)**, pp. 295–212, 1980.

[7] F. DARVE: An incrementally non-linear constitutive law of second-order and its application to strain localization, in: C.S. DESAI, R.H. GALLAGHER (eds.): *Mechanics of Engineering Materials*, John Wiley and Sons, London, 1984, pp. 179–196.

[8] F. DARVE: Incrementally non-linear constitutive relationships, in: F. DARVE (ed.): *Geomaterials Constitutive Equations and Modelling*, Elsevier Applied Science, Amsterdam, 1990, pp. 213–238.

[9] F. DARVE: Liquefaction phenomenon of granular materials and constitutive instability, *Int. J. Eng. Comp.* **7**, pp. 5–28, 1996.

[10] F. DARVE, B. CHAU: Constitutive instabilities in incrementally non-linear modelling, in: C.S. DESAI (ed.): *Constitutive Laws for Engineering Materials*, Elsevier Applied Science, Amsterdam, 1987, pp. 301–310.

[11] F. DARVE, H. DENDANI: An incrementally non-linear constitutive relation and its predictions, in: A.S. SAADA, G.F. BIANCHINI (eds.): *Constitutive Equations for Granular Non-Cohesive Soils*, Balkema, Rotterdam, 1989, pp. 237–254.

[12] F. DARVE, S. LABANIEH: Incremental constitutive law for sands and clays, simulation of monotonic and cyclic tests, *Int. J. Num. Anal. Meth. Geomech.* **6**, pp. 243–273, 1982.

[13] F. DARVE, F. LAOUAFA: Instabilities in granular materials and application to landslides, *Mech. Coh.-Frict. Mater.* **5 (8)**, pp. 627–652, 2000.

[14] F. DARVE, X. ROGUIEZ: Instabilities in granular materials, in: D.R.J. OWEN, E. ONATE, E. HINTON (eds.): *Computational Plasticity - Fundamentals and Applications*, CIMNE, Barcelona, 1997, pp. 720–727.

[15] R.D. DAVIS, G. MULLENGER: A rate-type constitutive model for soils with critical state, *Int. J. Num. Anal. Meth. Geomech.* **2**, pp. 255–282, 1978.

[16] H. DI BENEDETTO, F. DARVE: Comparaison de lois rhéologique en cinématique rotationnelle, *J. Mécan. Théor. Appl.* **2 (5)**, pp. 769–798, 1983.

[17] C. DI PRISCO, R. NOVA: Stability problems related to static liquefaction of loose sand, in: J. DESRUES, R. CHAMBON, I. VARDOULAKIS (eds.): *Localisation and Bifurcation Theory for Soils and Rocks*, Balkema, Rotterdam, 1994, pp. 59–70.

[18] C. DI PRISCO, R. MATIOTTI, R. NOVA: Theoretical investigation of undrained stability of shallow submerged slopes, *Géotechnique* **45**, pp. 479–496, 1995.

[19] J.M. DUNCAN, C.Y. CHANG: Non-linear analyses of stress and stress and strain in soils, *J. Soil. Mech. and Found. Div., ASCE* **95 (SM5)**, pp. 1629–1653, 1970.

[20] G. GUDEHUS: A comparison of some constitutive laws for soils under radially loading symmetric loading and unloading, in: W. WHITTKE (ed.): *Proc. 3rd Int. Conf. Num. Meth. Geomech.*, Balkema, Rotterdam, **4**, pp. 1309–1324, 1979.

[21] P. GUÉLIN: Note sur l'hystérisis mécanique, *J. Mécan.* **19 (2)**, pp. 217–247, 1980.

[22] P.Y. HICHER: *Comportement Mécanique des Argiles Saturées sur Divers Chemins de Sollicitations Monotones et Cycliques: Application à une Modélisation Élasto-Plastique et Visco-Élastique*, Ph.D. thesis, Ecole Centrale de Paris, 1985.

[23] R. HILL: A general theory of uniqueness and stability in elastic-plastic solids, *J. Mech. Phys. Solids* **6**, pp. 239–249, 1958.

[24] S. IMPOSIMATO, R. NOVA: An investigation on the uniqueness of the incremental response of elastoplastic models for virgin sand, *Mech. Coh.-Frict. Mater.* **3**, pp. 65–87, 1998.

[25] D. KOLYMBAS: A rate dependent constitutive equation for soils, *Mech. Res. Comm.* **4(6)**, pp. 367–372, 1977.

[26] P.V. LADE: Elasto-plastic stress theory for cohesionless soils with curved yield surfaces, *Int. J. Solids Struct.* **13**, pp. 1019–1035, 1977.

[27] F. LAOUAFA: *Analyse et Développement de Méthodes d'Éléments Finis. Applications aux Modèles Incrémentaux de Type Interpolation*, Ph.D. thesis, INSA, Lyon, 1996.

[28] F. LAOUAFA, P. ROYIS: Eléments finis déplacements. Nouvelle forme du problème algébrique et algorithmes adaptés, *C. R. Acad. Sc. Paris, Série IIb* **325**, pp. 347–352, 1997.

[29] B. LORET: *Formulation d'Une Loi de Comportement Élasto-Plastique des Milieux Granulaires*, D.I. thesis, Ecole Polytechnique, 1981.

[30] M. MÉGHACHOU: *Stabilité des Sables Lâches. Essais et Modélisations*, Ph.D. thesis, Université Joseph Fourier, Grenoble, 1993.

[31] Z. MROZ: On the description of anisotropic work hardening, *J. Mech. Phys. Sci.* **15**, pp. 163–175, 1967.

[32] R. NOVA, D.M. WOOD: A constitutive model for sand in triaxial compression, *Int. J. Num. Anal. Meth. Geomech.* **3**, pp. 255–278, 1979.

[33] D.R. OWEN, W.O. WILLIAMS: On the time derivatives of equilibrated response functions, *Arch. Rat. Mech. Anal.* **33 (4)**, pp. 288–306, 1969.

[34] J.H. PREVOST: Plasticity theory for soil stress-strain behaviour, *J. Eng. Mech. Div., ASCE* **104 (EM5)**, pp. 1177–1194, 1978.

[35] C. TRUESDELL, W. NOLL: The non-linear field theories of mechanics, in: S. FLÜGGE (ed.): *Handbuch der Physik*, Springer-Verlag, Berlin **III/3**, 1965.

[36] K.C. VALANIS: A theory of viscoplasticity without a yield surface, *Archives of Mechanics* **23**, pp. 517–551, 1971.

[37] I. VARDOULAKIS, M. GOLDSCHEIDER, G. GUDEHUS: Formation of shear bands in sand bodies as a bifurcation problem, *Int. J. Num. Anal. Meth. Geomech.* **2**, pp. 99–128, 1978.

[38] P. VERMEER: A double hardening model for sand, *Geotechnique* **28** (4), pp. 413–433, 1978.

FÉLIX DARVE and FARID LAOUAFA
Institut National Polytechnique de Grenoble
Laboratoire Sols Solides Structures
INPG-UJF-CNRS, Alert Geomaterials
Domaine Universitaire - BP 53
F-38041 Grenoble, FRANCE

E-mail: Felix.Darve@hmg.inpg.fr

http://www.3S.hmg.inpg.fr

Chapter 2

Micromechanical Modeling of Granular Materials

James T. Jenkins and Luigi La Ragione

ABSTRACT We briefly review attempts to predict the mechanical response of granular materials based on a simple characterization of the interparticle force and the statistical geometry of their packing and interactions. The results of two such theories are presented. In the first, the particle displacements are assumed to be determined by the average strain. Here, features of the observed behavior are reproduced but the response of the aggregate is too stiff. In the second, a limited number of additional degrees of freedom are given to pairs of particles, the motion is constrained by force and moment equilibrium and the stiffness is reduced. Both approaches require a choice of state variables, whose values determine the incremental response and which, in turn, change their values with the deformation. Examples of state variables and their evolution are provided.

2.1 Introduction

The goal of micromechanical modeling of granular materials is to derive accurate mathematical descriptions of their mechanical behavior from a consideration of the interaction between individual particles and statistical characterizations of the contact forces and the local geometry of the packing. Here, as an example of this activity, we derive an incremental, rate-independent continuum theory for an idealized granular material consisting of spherical particles from a consideration of force and moment equilibrium of the individual grains and simple but appropriate assumptions regarding the force of contact between the grains and the statistical geometry of their packing. The aggregate is assumed to be subjected to deviatoric strains that are so small compared to the isotropic strains associated with its compression that the elasticity of the contacts can not be ignored. That is, we consider the regime of frictional, elastic deformation that precedes localization and focus on the prediction of the wave speeds, the stress-strain behavior, the volume change, the plastic strain and the onset of localization.

For frictional contacts, the tangential component of the contact force depends upon the history of loading, so an incremental formulation is employed. Then, because the incremental relation between the contact force and the displacement of a contact is independent of rate, the macroscopic relation between the average stress and the average strain is also incremental and rate-independent. Another advantage of an incremental formulation is that the incremental forms of the balance of force and moment for individual particles are linear in the increments of particle displacement and rotation.

In an aggregate that is small enough to be considered to be deformed homogeneously, there is a distribution of the magnitude of the normal component of the contact forces even when it is isotropically compressed. Because the stiffness of a contact depends upon the normal component of the contact force, the stiffness of even similarly oriented contacts can be very different. At any stage of the deformation, scalar and tensor measures of the distribution of contacts and contact stiffness influence the relation between the increment of average stress and average strain. An example of a scalar measure is the average number of contacts per particle, the coordination number k; examples of tensor measures are the anisotropy in the orientation of nearest neighbors, the geometric "fabric" [16] and the anisotropy in the stiffness of contacts with nearest neighbors [10]. Such quantities characterize the present state of the homogeneously deformed aggregate and, by virtue of their dependence on the path of the deformation, they endow it with a memory of its prior history. For this reason, we call such measures state variables. The challenge is first, to identify these state variables; second, to determine how these state variables enter into the incremental stress- strain relation; and third, to establish how the state variables change with the deformation.

Here, we restrict our attention to the behavior of an idealized granular material consisting of identical spheres that are first isotropically compressed and then sheared, perhaps with simultaneous changes in confining pressure, until the onset of strain localization. Because our primary interest is in problems related to soil mechanics, we restrict our attention to a relatively limited range of loading in which the shear strain varies from zero to a substantial fraction of the volume strain associated with the isotropic compression. In this range, we anticipate that both the elasticity and the friction of the contacts will be important, but that changes in the geometry of the packing may be small and are likely to be dominated by the effective deletion of contacts as contact forces relax to zero.

The desired continuum theory relates increments in the average stress $\mathbf{T}$, the state variables $\boldsymbol{\Sigma}$ and the average distortion $\mathbf{L}$, where the average distortion involves both the average strain and an objective measure of the average rotations. These relations are expected to have the structure

$$\dot{T}_{ij} = F_{ijkl}(\boldsymbol{\Sigma}, \mathbf{L})\dot{L}_{kl}$$

and

$$\dot{\Sigma}_{ij} = G_{ijkl}(\mathbf{\Sigma}, \mathbf{L})\dot{L}_{kl},$$

where the overdot denotes an objective increment. In general, we expect such evolution equations to be stiff, so that they provide a general structure that can include behavior that might be interpreted as yielding. Such a theoretical structure is somewhat more general than the plasticity theories for compressible, frictional materials that have so far been employed to interpret physical experiments (*e.g.*, [17, 19]).

2.2   Theory

We focus our attention on a pair of contacting spheres, label them A and B, and denote the vector from the center of A to the center of B by $\mathbf{d}^{(BA)}$. We write the increment $\dot{\mathbf{F}}^{(BA)}$ in the contact force exerted by particle B on particle A in terms of the increment $\dot{\mathbf{u}}^{(BA)}$ in the relative displacement of the points of contact:

$$\dot{F}_i^{(BA)} = K_{ij}^{(BA)}\dot{u}_j^{(BA)},$$

where $\mathbf{K}^{(BA)}$ is the contact stiffness.

In this paper, we assume that the contact stiffness is given in terms of the unit vector $\widehat{\mathbf{d}}^{(BA)}$ in the direction of $\mathbf{d}^{(BA)}$ by

$$K_{ij}^{(BA)} = K_N^{(BA)}\widehat{d}_i^{(BA)}\widehat{d}_j^{(BA)} + K_T^{(BA)}(\delta_{ij} - \widehat{d}_i^{(BA)}\widehat{d}_j^{(BA)}). \qquad (2.1)$$

Here $K_N^{(BA)}$ and $K_T^{(BA)}$ are the normal and tangential contact stiffness, assumed to be given here in terms of the normal component δ of the compressive displacement of the centers of the particles, the diameter d of the spheres, and their material properties by

$$K_N^{(BA)} = \frac{\mu d^{\frac{1}{2}}}{(1-\nu)}\delta^{\frac{1}{2}},$$

where μ and ν are, respectively, the shear modulus and Poisson ratio of the material of the spheres, and

$$K_T^{(BA)} = \frac{2\mu d^{\frac{1}{2}}}{(2-\nu)}\delta^{\frac{1}{2}}. \qquad (2.2)$$

The relations (2.1) and (2.2) are simpler than those employed, for example, by Thornton & Randall [18] in that the stiffness matrix is diagonal, only the elastic contribution to the tangential stiffness is included and frictional sliding is ignored. Frictional sliding could be incorporated; in which case, the sliding contacts must be identified and their stiffnesses set equal to zero (*e.g.* [9]). In ignoring sliding, we overemphasize the importance of the tangential component of the contact force; however, we are able to introduce algebraic evolution equations for the state variables. These incorporate anisotropy associated with the orientation of contacts with respect to the mean strain that enters through the contact displacement.

The increment $\dot{\mathbf{u}}^{(BA)}$ in contact displacement may be written in terms of the increments $\dot{\mathbf{c}}^{(B)}$ and $\dot{\mathbf{c}}^{(A)}$ in the translations of the centers of the two spheres and the increments $\dot{\omega}^{(B)}$ and $\dot{\omega}^{(A)}$ in their rotations about their centers as

$$\dot{u}_i^{(BA)} = \dot{c}_i^{(B)} - \dot{c}_i^{(A)} - \frac{1}{2}\varepsilon_{ijk}\left(\dot{\omega}_j^{(B)} + \dot{\omega}_j^{(A)}\right) d_k^{(BA)}.$$

Alternatively, the relative displacement of the two contacting points may be written in terms of the increments in the averages of quantities and their fluctuations as

$$\dot{u}_i^{(BA)} = \left(\dot{E}_{ij} + \dot{W}_{ij}\right) d_j^{(BA)} + \dot{\Delta}_i^{(BA)} \tag{2.3}$$
$$-\varepsilon_{ijk}\dot{\Omega}_j d_k^{(BA)} - \frac{1}{2}\varepsilon_{ijk}\dot{S}_j^{(BA)} d_k^{(BA)},$$

where $\mathbf{E}$ and $\mathbf{W}$ are, respectively, the average strain and the average rotation based on the positions of the particle centers, $\mathbf{\Omega}$ is the average rotation of the particles about their centers, $\mathbf{\Delta}^{(BA)}$ is the difference between the fluctuation in the translations of the two centers and $\mathbf{S}^{(BA)}$ is the sum of the fluctuations in the rotations about their centers. In addition, we will require $\mathbf{\Sigma}^{(BA)}$, the sum of the fluctuations in the translations of the two centers and $\mathbf{D}^{(BA)}$, to be the difference between the fluctuations in the rotations about their centers.

It is necessary to distinguish between the average rotation based upon the inhomogeneous displacements of the particle centers and the average spin about the centers, because if the initial state is anisotropic or as anisotropy develops in the state of the material, these need not be equal (see [11], [6] and [8]). Their difference is then determined by the requirement that the stress be symmetric.

Given $\dot{\mathbf{F}}$, the incremental stress $\dot{\mathbf{T}}$ may be written as the average over all N particles in a region of homogeneous distortion that is identified with the continuum point as

$$\dot{T}_{ij} = \left\langle \frac{1}{V^{(A)}} \sum_{n=1}^{N^{(A)}} \dot{F}_i^{(nA)} d_j^{(nA)} \right\rangle \equiv \frac{1}{2N}\sum_{A=1}^{N} \frac{1}{V^{(A)}} \sum_{n=1}^{N^{(A)}} \dot{F}_i^{(nA)} d_j^{(nA)}, \tag{2.4}$$

where $N^{(A)}$ is the number of particles in contact with particle A and $V^{(A)}$ is the volume occupied by particle A and its nearest neighbors, including the space between the particles.

Given the increments $\dot{\mathbf{E}}$ and $\dot{\mathbf{W}}$ in average strain and rotation, the calculation of the increment in stress requires the determination of the fluctuations $\dot{\boldsymbol{\Delta}}$ and $\dot{\mathbf{S}}$ for all pairs of particles in the region. As emphasized by Koenders [10, 12, 13], these should be obtained as parts of the solutions of the equations of balance of force and moment for each of the N particles. Then $\dot{\boldsymbol{\Omega}}$ follows from the condition that the increment in stress be symmetric.

However, some important qualitative features of the behavior of the aggregate and, sometimes, some quantitative predictions can be obtained by taking a far simpler approach and ignoring the fluctuations entirely. Such a procedure seems to have a physical justification only when the shear strains are a small fraction of the volume strains associated with the isotropic compression.

2.2.1   Mean field theory

In this case, the increment in stress is given by

$$\dot{T}_{ij} = C_{ijkl}\dot{L}_{kl},$$

where

$$C_{ijkl} \equiv \left\langle \frac{1}{V^{(A)}} \sum_{n=1}^{N^{(A)}} K_{ik}^{(nA)} d_j^{(nA)} d_l^{(nA)} \right\rangle$$

and

$$\dot{L}_{kl} \equiv \dot{E}_{kl} + \dot{W}_{kl} - \varepsilon_{kml}\dot{\Omega}_m.$$

An analytical expression for C may be obtained by employing the continuous analog of equation (2.4). This is phrased in terms of the number of particles per unit volume n and a contact distribution function $f(\hat{\mathbf{d}})$ defined so that $f(\hat{\mathbf{d}})d\alpha$ is the number of contacts in the element of solid angle $d\alpha$ centered at $\hat{\mathbf{d}}$. For an isotropic distribution of contacts, $f = \frac{k}{4\pi}$, where k is the coordination number. Then

$$C_{ijkl} \equiv \frac{n}{2} \int \int f(\hat{\mathbf{d}}) K_{ik} d_j d_l d\alpha,$$

where the integration is over all solid angles. Using such a formula, Digby [3] and Walton [20] consider isotropic aggregates and calculate the effective shear and bulk moduli. Their expression for the effective shear modulus $\bar{\mu}$ is, for example,

$$\bar{\mu} = \frac{1}{15}nkd^3 \frac{\mu}{(1-\nu)} \frac{(5-4\nu)}{(2-\nu)} \left[\frac{9}{8} \frac{(1-\nu)}{\mu} \frac{p}{nkd^3} \right]^{\frac{1}{3}},$$

where p is the confining pressure.

Using $k = 5.36$ as determined in their numerical simulations, Cundall, Jenkins and Ishibashi [2, 7] find the predicted shear modulus to be three times that measured in their experiments and numerical simulations. On the other hand, Norris and Johnson [15] adopt a value of $k = 9$, that may be appropriate to their much higher confining pressures, and find the predicted values of wave speeds based on these to be in reasonable agreement with those measured in experiments by Domenico [4]. These differing results highlight the importance of the coordination number in the mechanics of these materials and indicate its range of variation when friction is present. Makse, Gland, Johnson and Schwartz [14] use numerical simulations to make clear how the coordination number increases with increasing confining pressure for frictionless and frictional contacts.

The corresponding incremental relation between the pressure and the volumetric strain can be integrated to obtain the elastic volume strain e corresponding to the confining pressure (*e.g.* [5]):

$$ e = - \left[27\sqrt{3}\frac{1}{nkd^3}(1 - \nu)\frac{p}{\mu} \right]^{\frac{2}{3}}. $$

The magnitude of this volume strain can be used to scale the shear strains to provide an appropriate measure of their strength. The shear stress should be scaled by the pressure in a similar way. It is only by employing such scaled variables that the results of physical experiments carried out on different materials at different confining pressures can be compared. In typical experiments relevant to the behavior of soils, both the scaled shear stress and the scaled shear strain range from zero to around one. Jenkins [5] also calculates the relationship between the scaled volume change and the scaled shear strain and shows that at least for triaxial compression and extension experiments carried out at constant pressure [1], the agreement between the predictions and the observations can be made to agree if there is an initial anisotropy due to the deposition of the particles.

Jenkins and Strack [9] take the mean field model somewhat further. Using Hertz elasticity for the normal component of the contact force and a crude model of linear elasticity followed by frictional sliding for the tangential component, they focus on triaxial compression. They predict the behavior of contacts as a function of their orientation with respect to the axis of compression and the compressive strain. When, after an initial isotropic compression, deviatoric straining commences, all contacts first deform elastically. As the compression proceeds, contacts oriented at a definite angle first begin to slide; the sliding then spreads to other contacts oriented further away from the axis. After sliding has reached the contacts perpendicular to the axis, continued deformation results in the reduction of their normal force to zero. Such contacts are considered to be deleted. Further compression results in an increasing region of deleted contacts near the perpendicular. The sum of such contact behavior is reflected in the relation between shear stress and shear strain: there is an initial region of stiff elastic response, a region of diminishing stiffness as contacts begin to slide, followed by a region of dramatically reduced stiffness as contacts are deleted.

This simple picture of the mean field behavior of contacts goes far in describing the observed features of the stress strain curve for loading. However, over a range of scaled shear stress between zero and one, the shear stresses predicted by such a theory are about three times larger than those measured in numerical simulations and experiments. The stresses are over-predicted because the relative displacement of the contacts are not so constrained; the particles are free to displace and rotate in accord with stress and moment equilibrium.

2.2.2 Pair fluctuations

Motivated by the over-prediction of the shear stress by the mean-field theory and by the apparent need to incorporate additional internal degrees of freedom, we next analyze the simplest possible situation in which two contacting particles, A and B, have sufficient translational and rotational freedom to satisfy force and moment equilibrium. In order that the equilibrium equations for the two particles determine these translations and rotations, we assume that the other particles in contact with the pair translate and rotate with the average deformation.

We denote the increment in translation of the center of the nth neighbor of particle A by $\dot{\mathbf{c}}^{(n)}$ and the increment in its rotation about its center by $\dot{\omega}^{(n)}$. Then

$$\dot{u}_i^{(nA)} = \dot{c}_i^{(n)} - \dot{c}_i^{(A)} - \frac{1}{2}\varepsilon_{ijk}(\dot{\omega}_j^{(n)} + \dot{\omega}_j^{(A)})d_k^{(nA)}.$$

Because we assume that for $n \neq B$ only the fluctuations in the translation and rotation of particle A occur, for these pairs we may write

$$\dot{c}_i^{(n)} - \dot{c}_i^{(A)} = (\dot{E}_{ij} + \dot{W}_{ij})d_j^{(nA)} + \frac{1}{2}\dot{\Delta}_i^{(BA)} - \frac{1}{2}\dot{\Sigma}_i^{(BA)}$$

and

$$\dot{\omega}_i^{(n)} + \dot{\omega}_i^{(A)} = 2\dot{\Omega}_i - \frac{1}{2}\dot{D}_i^{(BA)} + \frac{1}{2}\dot{S}_i^{(BA)}.$$

When $n = B$,

$$\dot{c}_i^{(B)} - \dot{c}_i^{(A)} = (\dot{E}_{ij} + \dot{W}_{ij})d_j^{(BA)} + \dot{\Delta}_i^{(BA)}$$

and

$$\dot{\omega}_i^{(B)} + \dot{\omega}_i^{(A)} = 2\dot{\Omega}_i + \dot{S}_i^{(BA)}.$$

52   J.T. Jenkins and L. La Ragione

The equations of force equilibrium for particle A are, then,

$$0 = K_{ij}^{(BA)} \left[\left(\dot{E}_{jk} + \dot{W}_{jk} \right) d_k^{(BA)} + \dot{\Delta}_j^{(BA)} \right]$$

$$+ \sum_{n \neq B}^{N^{(A)}} K_{ij}^{(nA)} \left[\left(\dot{E}_{jk} + \dot{W}_{jk} \right) d_k^{(nA)} + \frac{1}{2} \dot{\Delta}_j^{(BA)} - \frac{1}{2} \dot{\Sigma}_j^{(BA)} \right]$$

$$- \frac{1}{2} \varepsilon_{jkl} K_{ij}^{(BA)} \left(2\dot{\Omega}_k + \dot{S}_k^{(BA)} \right) d_l^{(BA)}$$

$$- \frac{1}{2} \varepsilon_{jkl} \sum_{n \neq B}^{N^{(A)}} K_{ij}^{(nA)} \left(2\dot{\Omega}_k - \frac{1}{2} \dot{D}_k^{(BA)} + \frac{1}{2} \dot{S}_k^{(BA)} \right) d_l^{(nA)};$$

while those for moment equilibrium are

$$0 = \varepsilon_{pqr} d_q^{(BA)} K_{ri}^{(BA)} \left[\left(\dot{E}_{ij} + \dot{W}_{ij} \right) d_j^{(BA)} + \dot{\Delta}_i^{(BA)} \right]$$

$$+ \varepsilon_{pqr} \sum_{n \neq B}^{N^{(A)}} d_q^{(nA)} K_{ri}^{(nA)} \left[\left(\dot{E}_{ij} + \dot{W}_{ij} \right) d_j^{(nA)} + \frac{1}{2} \dot{\Delta}_i^{(BA)} - \frac{1}{2} \dot{\Sigma}_i^{(BA)} \right]$$

$$- \frac{1}{2} \varepsilon_{pqr} \varepsilon_{ijk} d_q^{(BA)} K_{ri}^{(BA)} \left(2\dot{\Omega}_j + \dot{S}_j^{(BA)} \right) d_k^{(BA)}$$

$$- \frac{1}{2} \varepsilon_{pqr} \varepsilon_{ijk} \sum_{n \neq B}^{N^{(A)}} d_q^{(nA)} K_{ri}^{(nA)} \left(2\dot{\Omega}_j - \frac{1}{2} \dot{D}_j^{(BA)} + \frac{1}{2} \dot{S}_j^{(BA)} \right) d_k^{(nA)}.$$

The corresponding equilibrium equations for particle B are obtained by interchanging A and B, keeping in mind that $\mathbf{d}^{(AB)} = -\mathbf{d}^{(BA)}$.

We wish to phrase the equilibrium equations in terms of the typical neighborhood of a single particle rather than the typical neighborhood of a pair. Consequently, we write, for example,

$$\sum_{n \neq B}^{N^{(A)}} K_{ij}^{(nA)} d_l^{(nA)} = \sum_{n=1}^{N^{(A)}} K_{ij}^{(nA)} d_l^{(nA)} - K_{ij}^{(BA)} d_l^{(BA)}.$$

Then we identify the sums over the nearest neighbors of particles A and B that are tensors of even rank with their average values over the aggregate and introduce

$$A_{ij} \equiv \left\langle \sum_{n=1}^{N^{(A)}} K_{ij}^{(nA)} \right\rangle,$$

$$C_{pij} \equiv \varepsilon_{pqm} \left\langle \sum_{n=1}^{N^{(A)}} K_{mi}^{(nA)} d_q^{(nA)} d_j^{(nA)} \right\rangle,$$

and

$$B_{pk} \equiv \varepsilon_{ijk} C_{pij} = \varepsilon_{ijk} \varepsilon_{pqi} \left\langle \sum_{n=1}^{N^{(A)}} K_T^{(nA)} d_q^{(nA)} d_j^{(nA)} \right\rangle = B_{kp}.$$

Because sums that are tensors of odd rank have zero average over the aggregate, we introduce an average over all pairs with direction within a small element of angle about $\widehat{\mathbf{d}}^{(BA)}$ and write

$$J_{ijk}^{(BA)} \equiv \left\langle \sum_{n=1}^{N^{(A)}} K_{ij}^{(nA)} d_k^{(nA)} \right\rangle_{\mathbf{d}^{(BA)}} = -J_{ijk}^{(AB)}.$$

The quantity $\mathbf{J}^{(BA)}$ is a key element in the description of the neighborhood of the pair. Although its average over the aggregate is zero, its average is typically non–zero for pairs near a given orientation. Such a term contributes in an important way to the solution for the fluctuations and, consequently, to the contact force. It influences the stress through the contact force and may result in a substantial reduction in the stiffness of the assembly (*e.g.* [10]).

In this elementary treatment of the particle fluctuations and equilibrium, the quantities $\mathbf{A}$, $\mathbf{B}$, $\mathbf{C}$ and $\mathbf{J}^{(BA)}$, are the state variables of the aggregate. That is, as we shall see, knowledge of their values is sufficient to determine the values of the fluctuations and the contact force for a pair with arbitrary orientation; and, upon summing over all pair orientations, they determine the increment in stress in terms of the increments in the average strain and rotations. Phrased in terms of the state variables, the balance of force and moment for particle A are

$$2A_{ij}\left(\dot{\Delta}_j^{(BA)} - \dot{\Sigma}_j^{(BA)}\right) + \varepsilon_{jkl} J_{ijl}^{(BA)}\left(\dot{D}_k^{(BA)} - \dot{S}_k^{(BA)}\right)$$

$$+2K_{ij}^{(BA)}\left(\dot{\Delta}_j^{(BA)} + \dot{\Sigma}_j^{(BA)}\right)$$

$$-\varepsilon_{jkl} K_{ij}^{(BA)}\left(\dot{D}_k^{(BA)} + \dot{S}_k^{(BA)}\right) d_l^{(BA)} = -4J_{ijk}^{(BA)}\dot{L}_{jk}$$

and

$$2\,\varepsilon_{pqr} J_{riq}^{(BA)}\left(\dot{\Delta}_i^{(BA)} - \dot{\Sigma}_i^{(BA)}\right) - 2B_{pj}\left(\dot{D}_j^{(BA)} - \dot{S}_j^{(BA)}\right)$$

$$+2\,\varepsilon_{pqr} d_q^{(BA)} K_{ri}^{(BA)}\left(\dot{\Delta}_i^{(BA)} + \dot{\Sigma}_i^{(BA)}\right)$$

$$-\varepsilon_{pqr}\varepsilon_{ijk} d_q^{(BA)} K_{ri}^{(BA)}\left(\dot{D}_j^{(BA)} + \dot{S}_j^{(BA)}\right) d_k^{(BA)} = -4C_{pij}\dot{L}_{ij}.$$

The increments in the fluctuations that contribute to the contact force are $\dot{\mathbf{\Delta}}^{(BA)}$ and $\dot{\mathbf{S}}^{(BA)}$. Two vector equations involving only these quantities may be obtained by taking the difference between the force equations and the sum of the moment equations. When the resulting equations are written in matrix notation for the unknown vector $\dot{\mathbf{x}}$:

$$\dot{\mathbf{x}} \equiv \left[\begin{array}{c} \dot{\Delta}_j \\ \dot{S}_k \end{array}\right],$$

they are

$$(\mathcal{A} + \mathcal{P})\,\dot{\mathbf{x}} = \dot{\mathbf{b}},$$

where

$$\mathcal{A} \equiv \left[\begin{array}{cc} 2A_{ij} & G_{ik}^{(BA)} \\ -G_{rj}^{(BA)} & \frac{1}{2}B_{rk} \end{array} \right],$$

with

$$G_{iq}^{(BA)} \equiv \varepsilon_{jkq} J_{ijk}^{(BA)} = -G_{qi}^{(BA)};$$

$$\mathcal{P} \equiv \left[\begin{array}{cc} 2K_{ij}^{(BA)} & -\varepsilon_{jkl}K_{ij}^{(BA)}d_{l}^{(BA)} \\ \varepsilon_{rsi}K_{ij}^{(BA)}d_{s}^{(BA)} & -\frac{1}{2}\varepsilon_{rsi}\varepsilon_{jkl}K_{ij}^{(BA)}d_{s}^{(BA)}d_{l}^{(BA)} \end{array} \right];$$

and

$$\dot{\mathbf{b}} = \left[\begin{array}{c} -4J_{imn}^{(BA)}\dot{L}_{mn} \\ -2C_{rmn}\dot{L}_{mn} \end{array} \right].$$

We note that the terms in the matrix $\mathcal{A}$ each involve the average number of contacts k per particle while those in the matrix $\mathcal{P}$ each involve a single contact. We exploit this by first writing the matrix equation as

$$\mathcal{A}\left(\mathcal{I} + \mathcal{A}^{-1}\mathcal{P}\right)\dot{\mathbf{x}} = \dot{\mathbf{b}},$$

where $\mathcal{I}$ is the unit matrix and

$$\mathcal{A}^{-1} = \left[\begin{array}{cc} \frac{1}{2}\mathbf{A}^{-1} + \frac{1}{4}\mathbf{A}^{-1}\mathbf{G}^{(BA)}\mathbf{H}\mathbf{G}^{(BA)}\mathbf{A}^{-1} & \frac{1}{2}\mathbf{A}^{-1}\mathbf{G}^{(BA)}\mathbf{H} \\ -\frac{1}{2}\mathbf{H}\mathbf{G}^{(BA)}\mathbf{A}^{-1} & -\mathbf{H} \end{array} \right],$$

with

$$\mathbf{H} \equiv -2(\mathbf{B} + \mathbf{G}^{(BA)}\mathbf{A}^{-1}\mathbf{G}^{(BA)})^{-1}.$$

Then, upon solving this equation for $\mathbf{x}$ using the approximate inverse obtained by ignoring terms of order $\frac{1}{k^2}$, we obtain

$$\dot{\mathbf{x}} = \left(\mathcal{I} - \mathcal{A}^{-1}\mathcal{P}\right)\mathcal{A}^{-1}\dot{\mathbf{b}}.$$

Here, for simplicity, we evaluate the resulting expressions for $\dot{\boldsymbol{\Delta}}^{(BA)}$ and $\dot{\mathbf{S}}^{(BA)}$ at lowest order and neglect the terms of order $\frac{1}{k}$:

$$\dot{\Delta}_{i}^{(BA)} = -\left(2A_{ij}^{-1} + A_{ik}^{-1}G_{kl}^{(BA)}H_{lp}G_{pq}^{(BA)}A_{qj}^{-1}\right)J_{jmn}^{(BA)}\dot{L}_{mn}$$
$$-A_{ik}^{-1}G_{kl}^{(BA)}H_{lj}C_{jpq}\dot{L}_{pq}$$

and

$$\dot{S}_{r}^{(BA)} = 2H_{rl}G_{lp}^{(BA)}A_{pq}^{-1}J_{qmn}^{(BA)}\dot{L}_{mn} + 2H_{rl}C_{lmn}\dot{L}_{mn}.$$

Using these solutions in the expression for the incremental contact force yields

$$
\begin{aligned}
\dot{F}_s^{(BA)} \;=\;\; & K_{si}^{(BA)} d_j^{(BA)} \dot{L}_{ij} - K_{si}^{(BA)} 2 A_{ik}^{-1} J_{kmn}^{(BA)} \dot{L}_{mn} \\
& - K_{si}^{(BA)} A_{ik}^{-1} G_{kl}^{(BA)} H_{lp} G_{pq}^{(BA)} A_{qj}^{-1} J_{jmn}^{(BA)} \dot{L}_{mn} \\
& - K_{si}^{(BA)} A_{ik}^{-1} G_{kl}^{(BA)} H_{lj} C_{jpq} \dot{L}_{pq} \\
& - \varepsilon_{irj} K_{si}^{(BA)} d_j^{(BA)} H_{kl} G_{lp}^{(BA)} A_{pq}^{-1} J_{qmn}^{(BA)} \dot{L}_{mn} \\
& - \varepsilon_{irj} K_{si}^{(BA)} d_j^{(BA)} H_{kl} C_{lmn} \dot{L}_{mn}.
\end{aligned}
$$

To make any further progress in calculating the incremental stress, we must say something about the state variables and, in particular, the structure of $\mathbf{J}^{(BA)}$.

2.2.3 State variables

The state variables can be evaluated in first approximation using the mean strain. To illustrate how this may be done, we consider only the contribution to the state variables of the normal component of the contact force. That is, when evaluating the state variables, we take $K_T^{(nA)} = 0$. In this case, $\mathbf{C} = \mathbf{0}$, and $\mathbf{J}^{(BA)}$ is symmetric in all of its indices; so $\mathbf{G} = \mathbf{0}$ and $\mathbf{H} = -2\mathbf{B}^{-1} = \mathbf{0}$. Then, using the continuous form of the discrete averages and supposing that the initial state is isotropic, the expression for the incremental stress is

$$
\dot{T}_{ij} \;=\; \frac{n}{2}\frac{k}{4\pi}\left[d^2 \int\!\!\int K_{im}^{(BA)} \widehat{d}_n^{(BA)} \widehat{d}_j^{(BA)} \, d\alpha \right.
$$
$$
\left. - 2 A_{lk}^{-1} d \int\!\!\int K_{il}^{(BA)} J_{kmn}^{(BA)} \widehat{d}_j^{(BA)} \, d\alpha \right] \dot{L}_{mn}, \qquad (2.5)
$$

and the requirement that the incremental in stress be symmetric:

$$
\varepsilon_{pij}\dot{T}_{ij} \equiv 0
$$

may be written as

$$
0 \;\equiv\; \left[\varepsilon_{pmj} d^2 \int\!\!\int K_T^{(BA)} \widehat{d}_n^{(BA)} \widehat{d}_j^{(BA)} \, d\alpha \right.
$$
$$
\left. - 2\varepsilon_{plj} A_{lk}^{-1} d \int\!\!\int K_T^{(BA)} J_{kmn}^{(BA)} \widehat{d}_j^{(BA)} \, d\alpha \right] \dot{L}_{mn}.
$$

We suppose that the displacement of the centers is given by the average strain; in which case, the normal component δ is given by

$$
\delta = -\widehat{d}_i^{(BA)} E_{ij} d_j^{(BA)}.
$$

We then write the average strain as the sum of its isotropic and deviatoric parts,

$$E_{ij} = \frac{e}{3}\delta_{ij} + \widehat{E}_{ij},$$

assume that the deviatoric part is small compared to the isotropic part and expand the term in the stiffness that involves the strain :

$$\left(-\widehat{d}_i^{(BA)} E_{ij} \widehat{d}_j^{(BA)}\right)^{\frac{1}{2}} \doteq \left(-\frac{e}{3}\right)^{\frac{1}{2}} - \frac{1}{2}\left(-\frac{e}{3}\right)^{-\frac{1}{2}} \widehat{d}_i^{(BA)} \widehat{E}_{ij} \widehat{d}_j^{(BA)}.$$

So

$$K_N^{(BA)} \doteq \frac{\mu d}{(1-\nu)}\left(-\frac{e}{3}\right)^{\frac{1}{2}}\left(1 + \frac{3}{2}e^{-1}\widehat{d}_k^{(BA)}\widehat{d}_l^{(BA)}\widehat{E}_{kl}\right).$$

Also, upon using the continuous form of the discrete averages and supposing that the initial state is isotropic, we have

$$
\begin{aligned}
A_{ij} \doteq\ & \frac{k}{4\pi}\frac{\mu d}{(1-\nu)}\left(-\frac{e}{3}\right)^{\frac{1}{2}}\left[\int\int \widehat{d}_i^{(BA)}\widehat{d}_j^{(BA)}\,d\alpha \right. \\
& \left. + \frac{3}{2}e^{-1}\widehat{E}_{kl}\int\int \widehat{d}_i^{(BA)}\widehat{d}_j^{(BA)}\widehat{d}_k^{(BA)}\widehat{d}_l^{(BA)}\,d\alpha\right].
\end{aligned}
\tag{2.6}
$$

Then, upon applying the divergence theorem to the tensor products on the unit sphere,

$$\int\int \widehat{d}_i^{(BA)}\widehat{d}_j^{(BA)}\,d\alpha = \int\int\int r_{i,j}\,dv = \frac{4\pi}{3}\delta_{ij}$$

and

$$\int\int \widehat{d}_i^{(BA)}\widehat{d}_j^{(BA)}\widehat{d}_k^{(BA)}\widehat{d}_l^{(BA)}\,d\alpha = \int\int\int (r_i r_j r_k)_{,l}\,dv = \frac{4\pi}{15}X_{ijkl},$$

where

$$X_{ijkl} \equiv \delta_{il}\delta_{jk} + \delta_{jl}\delta_{ki} + \delta_{kl}\delta_{ij},$$

equation (2.6) becomes

$$A_{ij} \doteq \frac{\mu d}{(1-\nu)}\frac{k}{3}\left(-\frac{e}{3}\right)^{\frac{1}{2}}\left(\delta_{ij} + \frac{6}{10}e^{-1}\widehat{E}_{ij}\right).$$

So, with an error of the same order made elsewhere,

$$A_{ij}^{-1} \doteq \left[\frac{\mu d}{(1-\nu)}\frac{k}{3}\right]^{-1}\left(-\frac{e}{3}\right)^{-\frac{1}{2}}\left(\delta_{ij} - \frac{6}{10}e^{-1}\widehat{E}_{ij}\right).
\tag{2.7}$$

Because the average of $\mathbf{J}^{(BA)}$ over all orientations is zero, its evaluation is somewhat different. We first employ the mean strain assumption in its definition and expand as before to obtain

$$J_{ijk}^{(BA)} \doteq \frac{\mu d^2}{(1-\nu)}\left(-\frac{e}{3}\right)^{\frac{1}{2}}\left(P_{ijk}^{(BA)} + \frac{3}{2}e^{-1}\widehat{E}_{mn}Q_{ijkmn}^{(BA)}\right),
\tag{2.8}$$

where

$$P_{ijk}^{(BA)} \equiv \left\langle \sum_{n=1}^{N^{(A)}} \widehat{d}_i^{(nA)} \widehat{d}_j^{(nA)} \widehat{d}_k^{(nA)} \right\rangle_{\mathbf{d}^{(BA)}}$$

and

$$Q_{ijkmn}^{(BA)} \equiv \left\langle \sum_{n=1}^{N^{(A)}} \widehat{d}_i^{(nA)} \widehat{d}_j^{(nA)} \widehat{d}_k^{(nA)} \widehat{d}_m^{(nA)} \widehat{d}_n^{(nA)} \right\rangle_{\mathbf{d}^{(BA)}} .$$

If we assume that the geometric arrangement of contacts is unchanged when the deviatoric strain is applied, the averages may be evaluated in the isotropically compressed state. However, given that the initial state is isotropic, the most that we can say about these averages is that they are isotropic functions of the vector $\widehat{\mathbf{d}}^{(BA)}$.

General representations for the isotropic functions $\mathbf{P}^{(BA)}$ and $\mathbf{Q}^{(BA)}$ are

$$P_{ijk}^{(BA)} = \epsilon_1 k \widehat{d}_{(i}^{(BA)} \delta_{jk)} + \epsilon_2 k \widehat{d}_i^{(BA)} \widehat{d}_j^{(BA)} \widehat{d}_k^{(BA)}, \tag{2.9}$$

where indices contained within parentheses are meant to be made completely symmetric, and

$$\begin{aligned} Q_{ijkmn}^{(BA)} &= \epsilon_3 k \widehat{d}_{(i}^{(BA)} \delta_{jk} \delta_{mn)} + \epsilon_4 k \widehat{d}_{(i}^{(BA)} \widehat{d}_j^{(BA)} \widehat{d}_k^{(BA)} \delta_{mn)} \\ &\quad + \epsilon_5 k \widehat{d}_i^{(BA)} \widehat{d}_j^{(BA)} \widehat{d}_k^{(BA)} \widehat{d}_m^{(BA)} \widehat{d}_n^{(BA)} . \end{aligned}$$

When these representations are used in (2.8), it provide the general expression for $\mathbf{J}^{(BA)}$. The constant coefficients ϵ_1 through ϵ_5 must be determined through computer simulation or physical experiment. Here, instead of working with greatest generality, we are content to illustrate the method and we suppose that only ϵ_2 and ϵ_5 are non–zero.

We use the resulting form of (2.8), the expression (2.7) for $\mathbf{A}^{-1}$ and the approximation to the complete stiffness matrix:

$$\begin{aligned} K_{ij}^{(BA)} &\doteq \frac{\mu d}{(1-\nu)} \left(-\frac{e}{3}\right)^{\frac{1}{2}} \left(1 + \frac{3}{2} e^{-1} \widehat{d}_k^{(BA)} \widehat{d}_l^{(BA)} \widehat{E}_{kl}\right) \widehat{d}_i^{(BA)} \widehat{d}_j^{(BA)} \\ &\quad + \frac{2\mu d}{(2-\nu)} \left(-\frac{e}{3}\right)^{\frac{1}{2}} \left(1 + \frac{3}{2} e^{-1} \widehat{d}_k^{(BA)} \widehat{d}_l^{(BA)} \widehat{E}_{kl}\right) \left(\delta_{ij} - \widehat{d}_i^{(BA)} \widehat{d}_j^{(BA)}\right), \end{aligned}$$

in the expression for the stress increment and carry out the integrations with the assistance of the additional identity

$$\int \int \widehat{d}_i^{(nA)} \widehat{d}_j^{(nA)} \widehat{d}_k^{(nA)} \widehat{d}_l^{(nA)} \widehat{d}_m^{(nA)} \widehat{d}_n^{(nA)} d\alpha$$

$$= \int \int \int (r_i r_j r_k r_l r_m)_{,n} \, dv = \frac{4\pi}{105} Y_{ijklmn},$$

where

$$Y_{ijklmn} \equiv \delta_{in} X_{jklm} + \delta_{jn} X_{klmi} + \delta_{kn} X_{lmij} + \delta_{ln} X_{mijk} + \delta_{mn} X_{ijkl}.$$

We indicate the intermediate results

$$\int\int K_{im}^{(BA)}\,\widehat{d}_n^{(BA)}\,\widehat{d}_j^{(BA)}\,d\alpha$$

$$=\ \frac{\mu d}{(1-\nu)}\left(-\frac{e}{3}\right)^{\frac{1}{2}}\left[\int\int \widehat{d}_i^{(BA)}\,\widehat{d}_m^{(BA)}\,\widehat{d}_n^{(BA)}\,\widehat{d}_j^{(BA)}\,d\alpha\right.$$

$$\left.+\frac{3}{2}e^{-1}\widehat{E}_{kl}\int\int \widehat{d}_i^{(BA)}\,\widehat{d}_m^{(BA)}\,\widehat{d}_k^{(BA)}\,\widehat{d}_l^{(BA)}\,\widehat{d}_n^{(BA)}\,\widehat{d}_j^{(BA)}\,d\alpha\right]$$

$$+\frac{2\mu d}{(2-\nu)}\left(-\frac{e}{3}\right)^{\frac{1}{2}}\left[\int\int\left(\delta_{im}-\widehat{d}_i^{(BA)}\,\widehat{d}_m^{(BA)}\right)\widehat{d}_n^{(BA)}\,\widehat{d}_j^{(BA)}\,d\alpha\right.$$

$$\left.+\frac{3}{2}e^{-1}\widehat{E}_{kl}\int\int\left(\delta_{im}-\widehat{d}_i^{(BA)}\,\widehat{d}_m^{(BA)}\right)\widehat{d}_k^{(BA)}\,\widehat{d}_l^{(BA)}\,\widehat{d}_n^{(BA)}\,\widehat{d}_j^{(BA)}\,d\alpha\right]$$

and

$$A_{lk}^{-1}\int\int K_{il}^{(BA)}\,J_{kmn}^{(BA)}\,\widehat{d}_j^{(BA)}\,d\alpha$$

$$=\ \frac{3}{k}\frac{\mu d^2}{(1-\nu)}\left(-\frac{e}{3}\right)^{\frac{1}{2}}\left\{\int\int \widehat{d}_i^{(BA)}\,\widehat{d}_k^{(BA)}\,P_{kmn}^{(BA)}\,\widehat{d}_j^{(BA)}\,d\alpha\right.$$

$$-\frac{3}{5}e^{-1}\widehat{E}_{lk}\int\int \widehat{d}_i^{(BA)}\,\widehat{d}_l^{(BA)}\,P_{kmn}^{(BA)}\,\widehat{d}_j^{(BA)}\,d\alpha$$

$$+\frac{3}{2}e^{-1}\widehat{E}_{pq}\int\int \widehat{d}_p^{(BA)}\,\widehat{d}_q^{(BA)}\,\widehat{d}_i^{(BA)}\,\widehat{d}_k^{(BA)}\,P_{kmn}^{(BA)}\,\widehat{d}_j^{(BA)}\,d\alpha$$

$$\left.+\frac{3}{2}e^{-1}\widehat{E}_{pq}\int\int \widehat{d}_i^{(BA)}\,\widehat{d}_k^{(BA)}\,Q_{kmnpq}^{(BA)}\,\widehat{d}_j^{(BA)}\,d\alpha\right\}$$

$$+\frac{3}{k}\frac{2\mu d^2}{(2-\nu)}\left(-\frac{e}{3}\right)^{\frac{1}{2}}\left[\int\int\left(\delta_{ik}-\widehat{d}_i^{(BA)}\,\widehat{d}_k^{(BA)}\right)P_{kmn}^{(BA)}\,\widehat{d}_j^{(BA)}\,d\alpha\right.$$

$$-\frac{3}{5}e^{-1}\widehat{E}_{lk}\int\int\left(\delta_{il}-\widehat{d}_i^{(BA)}\,\widehat{d}_l^{(BA)}\right)P_{kmn}^{(BA)}\,\widehat{d}_j^{(BA)}\,d\alpha$$

$$+\frac{3}{2}e^{-1}\widehat{E}_{pq}\int\int \widehat{d}_p^{(BA)}\,\widehat{d}_q^{(BA)}\left(\delta_{ik}-\widehat{d}_i^{(BA)}\,\widehat{d}_k^{(BA)}\right)P_{kmn}^{(BA)}\,\widehat{d}_j^{(BA)}\,d\alpha$$

$$\left.+\frac{3}{2}e^{-1}\widehat{E}_{pq}\int\int\left(\delta_{ik}-\widehat{d}_i^{(BA)}\,\widehat{d}_k^{(BA)}\right)Q_{kmnpq}^{(BA)}\,\widehat{d}_j^{(BA)}\,d\alpha\right],$$

which, in the event that $\epsilon_1=\epsilon_3=\epsilon_4=0$, simplify to

$$\int\int K_{im}^{(BA)}\,\widehat{d}_n^{(BA)}\,\widehat{d}_j^{(BA)}\,d\alpha$$

$$=\ \frac{4\pi}{3}\frac{\mu d}{(1-\nu)}\left(-\frac{e}{3}\right)^{\frac{1}{2}}\frac{1}{5}\left(X_{imnj}+\frac{3}{14}e^{-1}\widehat{E}_{kl}Y_{imklnj}\right)$$

$$+\frac{4\pi}{3}\frac{2\mu d}{(2-\nu)}\left(-\frac{e}{3}\right)^{\frac{1}{2}}\left[\delta_{im}\delta_{nj}-\frac{1}{5}X_{imnj}\right.$$

$$\left.+\frac{3}{10}e^{-1}\widehat{E}_{kl}\left(\delta_{im}X_{klnj}-\frac{1}{7}Y_{imklnj}\right)\right]$$

and

$$A_{lk}^{-1} \int \int K_{il}^{(BA)} J_{kmn}^{(BA)} \widehat{d}_{j}^{(BA)} d\alpha$$

$$= \frac{\mu d^2}{(1-\nu)} \left(-\frac{e}{3}\right)^{\frac{1}{2}} \frac{4\pi}{5} \left[\epsilon_2 X_{ijmn} - \frac{3}{35}\epsilon_2 e^{-1} \widehat{E}_{lk} Y_{injklm} \right.$$

$$\left. + \frac{3}{14}(\epsilon_2 + \epsilon_5)e^{-1} \widehat{E}_{pq} Y_{pqimnj} \right]$$

$$- \frac{2\mu d^2}{(2-\nu)} \left(-\frac{e}{3}\right)^{\frac{1}{2}} \frac{4\pi}{25} \epsilon_2 e^{-1} \widehat{E}_{lk} \left(\delta_{il} X_{kmnj} - \frac{1}{7} Y_{ijlkmn}\right).$$

Using these in equation (2.5) provides an expression between the increments in stress, strain, and the rotations that incorporates the additional degrees of freedom of the pairs of interacting particles. Then the requirement that the stress increment be symmetric involves only the tangential component of the contact force and reduces to

$$\varepsilon_{pij} \left[\frac{1}{3}\left(\delta_{im}\delta_{nj} + \frac{3}{10}e^{-1}\widehat{E}_{kl}\delta_{im}X_{klnj}\right)\left(\dot{E}_{mn} + \dot{W}_{mn} - \varepsilon_{mqn}\dot{\Omega}_q\right)\right.$$

$$\left. - \frac{1}{25}\epsilon_2 e^{-1}\widehat{E}_{lk}\delta_{il}X_{kmnj}\dot{E}_{mn}\right] = 0,$$

$$\varepsilon_{ipj}\dot{\Omega}_p = \dot{W}_{ij} - \frac{1}{25}\epsilon_2 e^{-1}\varepsilon_{pij}\left(\widehat{E}_{ik}\dot{E}_{jk} - \widehat{E}_{jk}\dot{E}_{ik}\right) + \frac{3}{5}e^{-1}\left(\widehat{E}_{nj}\dot{E}_{in} - \widehat{E}_{ni}\dot{E}_{jn}\right).$$

This determines the increment $\dot{\boldsymbol{\Omega}}$ in average particle rotation in terms of the increments of the average strain and rotation of the aggregate.

2.3   Conclusion

We have outlined a general continuum mechanical description for granular materials that involves incremental constitutive relations for the stress and internal state of the material that are based upon the mechanics of grain contact and a statistical characterization of the grain interactions.

The simplest such theory, based upon the average strain and rotation, captures some important features of the behavior of the aggregate. Formulas for effective moduli based upon the mean field kinematics highlight the importance of the coordination number, the average number of contacts per grain, as an independent internal variable. Calculation of the incremental stress also leads to an expression for the elastic volume strain associated with the confining pressure that provides a natural scale for shear strains. In the context of this theory, the succession of contact mechanisms can be identified that determine the shape of the stress–strain curve.

The hope is that by incorporating the translational and rotational degrees of freedom of pairs of particles at various orientations, the predicted magnitude of the shear stress may be reduced to values more in accord with experiments and numerical simulations. Here, we considered the fluctuations of a pair of particles, each in an average anisotropic neighborhood. In a first attempt, we focused only on the elastic contribution to the tangential stiffness. In this case, a near exact analysis is possible and the reduction in the incremental stiffness and the average rotations of the particles necessary for a symmetric stress can be calculated explicitly. It remains to test these predictions against the results of numerical simulations.

References

[1] Y.C. CHEN, I. ISHIBASHI, J.T. JENKINS: Dynamic shear modulus and fabric. Part I. Depositional and induced fabric, *Geotechnique* **38**, pp. 25–32, 1988.

[2] P.A. CUNDALL, J.T. JENKINS, I. ISHIBASHI: Micromechanical modeling of granular materials with the assistance of experiments and numerical simulations, in J. BIAREZ, R. GOURVÈS (eds.): *Powders and Grains*, Balkema, Rotterdam, 1989, pp. 319–322.

[3] P.J. DIGBY: The effective elastic moduli of porous granular rocks, *J. Appl. Mech.* **48**, pp. 803–808, 1981.

[4] S.N. DOMENICO: Elastic properties of unconsolidated sand reservoirs, *Geophysics* **42**, pp. 1339–1368, 1977.

[5] J.T. JENKINS: Volume change in small strain axisymmetric deformations of a granular material, in M. SATAKE, J.T. JENKINS (eds.): *Micromechanics of Granular Materials*, Elsevier Science Publishers, Amsterdam, 1988, pp. 245–252.

[6] J.T. JENKINS: Anisotropic elasticity for random arrays of identical spheres, in J. WU, T.C.T. TING, D.M. BARNETT (eds.): *Modern Theory of Anisotropic Elasticity and Its Applications*, Society for Industrial and Applied Mathematics, 1991, pp. 368–377.

[7] J.T. JENKINS, P.A. CUNDALL, I. ISHIBASHI: Micromechanical modeling of granular materials with the assistance of experiments and numerical simulations, in J. BIAREZ, R. GOURVÈS (eds.): *Powders and Grains*, Balkema, Rotterdam, 1989, pp. 257–264.

[8] J.T. JENKINS, L. LA RAGIONE: Particle spin in anisotropic granular materials, *Int. J. Solids Struct.* **38**, pp. 1063–1069, 2001.

[9] J.T. JENKINS, O.D.L. STRACK: Mean–field inelastic behavior of random arrays of identical spheres, *Mechanics of Materials* **16**, pp. 25–33, 1993.

[10] M.A. KOENDERS: The incremental stiffness of an assembly of particles, *Acta Mechanica* **70**, pp. 31–49, 1987.

[11] M.A. KOENDERS: Localized deformation using higher order stress/strain theory, *J. Energy Resour. Techn.* **112**, pp. 51–53, 1990.

[12] M.A. KOENDERS: Analytical estimates for constitutive relations of assemblies of particles in elasto– frictional contact, in C. THORNTON (ed.): *Powders and Grains 93*, Balkema, Rotterdam, 1993, pp. 111–116.

[13] M.A. KOENDERS: Least squares method for the mechanics of nonhomogeneous granular assemblies, *Acta Mechanica* **106**, pp. 23–40, 1994.

[14] H.A. MAKSE, N. GLAND, D.L. JOHNSON, L.M. SCHWARTZ: Why effective medium theory fails in granular materials, *Physical Review Letters* **83**, pp. 5070–5074, 1999.

[15] A.N. NORRIS, D.L. JOHNSON: Nonlinear elasticity of granular media, *J. Appl. Mech.* **64**, pp. 39–49, 1997.

[16] M. ODA: *Fabrics and their Effects on the Deformation Behaviours of Sand*, Department of Foundation Engineering, Saitama University, 1976.

[17] F. TATSUOKA, K. ISHIHARA: Yielding of sand in triaxial compression, *Soils and Foundations* **14**, pp. 63–76, 1974.

[18] C. THORNTON, C.W. RANDALL: Application of theoretical contact mechanics to solid particle system simulation, in M. SATAKE, J.T. JENKINS (eds.): *Micromechanics of Granular Materials*, Elsevier Science Publishers, Amsterdam, 1988, pp. 133–142.

[19] G.Z. VOYIADJIS, G. THIAGARAJAN, E. PETRAKIS: Constitutive modeling for granular media using an anisotropic distortional yield model, *Acta Mechanica* **110**, pp. 151–171, 1995.

[20] K. WALTON: The effective elastic moduli of a random packing of spheres, *J. Mech. Phys. Solids* **35**, pp. 213–226, 1987.

JAMES T. JENKINS
Department of Theoretical
and Applied Mechanics
Cornell University, Ithaca
14853 New York, USA

E-mail: Jtj2@cornell.edu

LUIGI LA RAGIONE
Dipartimento di Ingegneria
Civile e Ambientale
Politecnico di Bari
Via Re David, 200
I-70125 Bari, ITALY

E-mail: L.Laragione@poliba.it

Chapter 3

Thermodynamic Modeling of Granular Continua Exhibiting Quasi-Static Frictional Behaviour with Abrasion

Nina P. Kirchner and Kolumban Hutter

ABSTRACT The deformation of a granular heap under the presence of external forces may be described by constitutive relations known as hypoplastic material equations. They describe internal frictional behaviour but ignore abrasion due to interparticle rubbing which is reported to be partly responsible for the formation of localized shear bands and associated instability. We present a thermodynamic continuum theory for granular materials under quasi-static and dynamic loading, treating hypoplasticity by a symmetric second–order tensorial variable and abrasion by a scalar variable for which four versions of an internal balance law are postulated. We present results obtained by employing the entropy principle of [11, 12] and demonstrate that the explicit forms of the constitutive relations for the Cauchy stress tensor depend on the postulation of the evolution equation for the variable describing abrasion. We give arguments which demonstrate that localizations due to abrasion are likely to be predicted only with a gradient type formulation. Proofs of our statements are given in [5].

3.1 Introduction

Observations in-situ and triaxial experiments with dry soil specimens under quasi-static and dynamic loads indicate that failure of the probes is often initiated by formation of shear bands. In computational reproductions of such shear banding a local boundary defect is artificially introduced and then either an internal slip surface (for classical theories) or a diffusive region with enhanced shearing (for Cosserat continua) is formed, always starting from the location of the artificial defect. Such defects should be naturally introduced, and the mechanism we analyse is abrasion of fine material from the surfaces of the grains due to the rubbing activity between the particles. This effect may be a polishing of the particles and thus lead to a reduction of the effective internal friction or, if the particles are initially smooth, abrasion may cause the grains to become edgy. This process is inhomogeneously introduced according to the internal state of stress and deformation. We conjecture that this inhomogeneous distribution of the smoothening or roughening activity acts as the natural internal defect accomplishing the mentioned localization.

We follow [15] in describing the frictional response by a thermodynamically motivated form of hypoplasticity. This can be achieved by introducing a symmetric tensorial internal variable for which an evolution equation is postulated. However, whereas this model of hypoplasticity is able to describe the internal frictional behaviour of a soil under various external loads, it does not per se account for the induced change of internal frictional parameters.

This is now introduced by an additional internal variable, the *roughness* a, an internal length that is reminiscent of the roughness of the grains when they move against each other in a motion induced by the stresses that are established. We postulate that the evolution of this variable in a deformation process – static or dynamic – is described by a local balance law, but the novelty of the physical problems has not let us favour one postulate over the other. To be precise, consider the balance law

$$\frac{\mathrm{d}}{\mathrm{d}t} \int_V \Psi \, \mathrm{d}V = - \oint_{\mathrm{d}V} \mathbf{h}^{(\Psi)} \cdot \mathbf{n} \, \mathrm{d}a + \int_V \pi^{(\Psi)} \mathrm{d}V \qquad (3.1)$$

for a physical quantity Ψ, its flux $\mathbf{h}^{(\Psi)}$ and production $\pi^{(\Psi)}$. Three models are suggested simply by choosing the assignments listed in Table 3.1: These all regard the *roughness* as the basic variable and not *abrasion* which is $\dot{a}$, *i.e.* the time rate of change of roughness. In all three cases a is regarded as the independent variable, and constitutive relations will have to be formulated for the flux and production terms. We have in the fourth column of Table 3.1 added references of researchers who used such postulates for the volume fraction of the granular component as independent variables.

TABLE 3.1. Densities for the three proposed modelling approaches.

Ψ	$\mathbf{h}^{(\Psi)}$	$\pi^{(\Psi)}$	reference	abbreviation
a	$\mathbf{h}^{(a)}$	$\pi^{(a)}$	[19]	(W)
a	–	$\pi^{(a)}$	[14]	(SH)
ρa	$\mathbf{h}^{(\rho a)}$	$\rho\pi^{(\rho a)}$	[17]	(V)

Localization of (3.1) yields

$$
\text{(W)} \qquad
\begin{cases}
\dot{a} + a \operatorname{div} \mathbf{v} - \operatorname{div} \mathbf{h}^{(a)} - \pi^{(a)} = 0, \\[2mm]
[\![\, a(\mathbf{v} - \mathbf{w}) \cdot \mathbf{n} \,]\!] - [\![\, \mathbf{h}^{(a)} \cdot \mathbf{n} \,]\!] = 0,
\end{cases}
$$

$$
\text{(SH)} \qquad
\begin{cases}
\dot{a} + a \operatorname{div} \mathbf{v} - \pi^{(a)} = 0, \\[2mm]
[\![\, a(\mathbf{v} - \mathbf{w}) \cdot \mathbf{n} \,]\!] = 0,
\end{cases}
\qquad (3.2)
$$

$$
\text{(V)} \qquad
\begin{cases}
\rho \dot{a} - \operatorname{div} \mathbf{h}^{(\rho a)} - \rho \pi^{(\rho a)} = 0, \\[2mm]
[\![\, \rho a(\mathbf{v} - \mathbf{w}) \cdot \mathbf{n} \,]\!] - [\![\, \mathbf{h}^{(\rho a)} \cdot \mathbf{n} \,]\!] = 0.
\end{cases}
$$

In these equations, ρ, $\mathbf{v}$, $\mathbf{w}$ are the mass density of the granular material, particle velocity and the velocity of a singular surface with unit normal $\mathbf{n}$, and $[\![\, f \,]\!] := f^{+} - f^{-}$ is the jump of the quantity f across the singular surface. The first equation of the pairs (3.2) applies at smooth points of the material, the second at singular surfaces. Notation is as in usual texts of continuum thermodynamics with which the reader is supposed to be familiar. The above balance laws for roughness must in each situation be complemented by the usual balance laws of mass, (linear) momentum, (int.) energy and entropy,

$$
\left.
\begin{aligned}
0 &= \dot{\rho} + \rho \operatorname{div} \mathbf{v} \\
\mathbf{0} &= \rho \dot{\mathbf{v}} - \operatorname{div} \mathbf{T} - \rho \mathbf{b} \\
0 &= \rho \dot{\varepsilon} + \operatorname{div} \mathbf{q} - \mathbf{T} \cdot \mathbf{D} - \rho r \\
\pi &= \rho \dot{\eta} + \operatorname{div} \boldsymbol{\phi} - \rho \sigma
\end{aligned}
\right\}
\qquad (3.3)
$$

$$
\left.
\begin{aligned}
0 &= [\![\, \rho(\mathbf{w} - \mathbf{v}) \cdot \mathbf{n} \,]\!] \\
\mathbf{0} &= [\![\, \mathbf{T}\mathbf{n} \,]\!] - [\![\, \rho \mathbf{v}(\mathbf{w} - \mathbf{v}) \cdot \mathbf{n} \,]\!] \\
0 &= [\![\, (\tfrac{1}{2}\mathbf{v} \cdot \mathbf{v} + \varepsilon)\rho(\mathbf{w} - \mathbf{v}) \cdot \mathbf{n} \,]\!] - [\![\, (\mathbf{q} - \mathbf{T}\mathbf{v}) \cdot \mathbf{n} \,]\!] \\
0 &= [\![\, \rho\eta(\mathbf{w} - \mathbf{v}) \cdot \mathbf{n} \,]\!] - [\![\, \boldsymbol{\phi} \cdot \mathbf{n} \,]\!]
\end{aligned}
\right\}
\qquad (3.4)
$$

in which $\mathbf{T}$ and $\mathbf{D}$ are the Cauchy stress and the stretching tensor, respectively, $\mathbf{b}$, $\mathbf{q}$ and $\boldsymbol{\phi}$ are an external body force (such as, $e.g.$, gravity), the heat flux and the entropy flux, and ε is the internal energy. Finally, r is an external radiation, and η and σ are the entropy and its supply. Complementing equations (3.2), (3.3) by constitutive relations will yield a continuum mechanical model for the material under consideration.

Note that the mass balance law can be written in terms of the void ratio e, which is defined as the ratio of the volume of voids to the volume of the solid particles: assuming the mass density of the solid grains to be constant, $\gamma \approx$ const., the balance of mass can be rewritten as

$$
\dot{e} = (1 + e)\operatorname{tr}\mathbf{D},
$$

or in integrated form (e_r is the void ratio in the reference configuration) as

$$e = (1 + e_r)\sqrt{\det \mathbf{B}} - 1,$$

where $\mathbf{B} = \mathbf{F}\mathbf{F}^{\mathrm{T}}$ is the left Cauchy–Green deformation tensor. γ and e are related to each other via

$$\rho = \frac{\gamma}{1 + e}.$$

It should be noted that we have for simplicity not written down an additional and separate balance law for the solid volume fraction. This is a justifiable avenue whenever $\gamma = \text{const.}$, since in this case volume fraction and density are essentially the same, and balance of mass governs the latter.

There is a fourth possibility of introducing the induced variations of the internal frictional parameters as a variable. It follows Goodman & Cowin's [2] concept of an equilibrated force balance and establishes a balance state for abrasion instead of roughness by choosing $\Psi = \rho k \dot{a}$, $k = \text{const.}$, $\pi^{(\rho k \dot{a})} = \rho f$, and thus yields the local balance equations

$$\text{(GC)} \qquad \begin{cases} \rho k \ddot{a} - \operatorname{div} \mathbf{h}^{(\rho k \dot{a})} - \rho f & = 0, \\[2mm] [\![\, \rho k \dot{a}(\mathbf{w} - \mathbf{v}) \cdot \mathbf{n} \,]\!] - [\![\, \mathbf{h}^{(\rho k \dot{a})} \cdot \mathbf{n} \,]\!] & = 0. \end{cases} \qquad (3.5)$$

Here, k is the coefficient of abrasive inertia, $\mathbf{h}^{(\rho k \dot{a})}$ is the flux of abrasion and ρf its production rate. An obvious difference between the first three and this fourth model is that here abrasion, the time rate of change of roughness, is the basic field variable whereas in (3.2) it is the roughness itself. More significant is the fact that in this approach following [2] kinetic energy and powers of working associated with abrasion are postulated as

$$K^{(\rho k \dot{a})} := \int_V \rho k \frac{\dot{a}^2}{2} \, \mathrm{d}V, \qquad P^{(\rho k \dot{a})} := \oint_{\partial V} \mathbf{h}^{(\rho k \dot{a})} \dot{a} \cdot \mathbf{n} \, \mathrm{d}a.$$

This changes the (localized) form of the energy balance, which now instead of $(3.3)_3$ takes the form

$$\rho \dot{\varepsilon} + \dot{a} \rho f + \operatorname{div} \mathbf{q} - \mathbf{T} \cdot \mathbf{D} - \mathbf{h}^{(\rho k \dot{a})} \cdot \nabla \dot{a} - \rho r = 0,$$
$$[\![\, \rho\big(\tfrac{1}{2}(\mathbf{v} \cdot \mathbf{v} + k \dot{a}^2) + \varepsilon\big)(\mathbf{w} - \mathbf{v}) \cdot \mathbf{n} \,]\!] - [\![\, (\mathbf{q} - \mathbf{T}\mathbf{v} - \dot{a}\mathbf{h}^{(\rho k \dot{a})}) \cdot \mathbf{n} \,]\!] = 0. \qquad (3.6)$$

The above equations (3.2), (3.5) comprise four different suggestive models which possess the potential to describe the effects of induced changes of internal frictional behaviour of the granular body. To all of them we have given suggestive reasons why we believe them to be physically reasonable (see [5]), but there are no strong arguments to rule out a particular one against any other one.

In the following analysis we shall therefore introduce the same constitutive postulates appropriate for a frictional material of hypoplastic type, subject all of them to the same thermodynamic analysis – *i.e.* the entropy principle of Müller – and will only value their adequacy for the question under consideration from the deduced final form of the equations.

In the following Section 3.2, the constitutive class for the dry granular material is introduced so that in Section 3.3, the exploitation of the entropy principle according to Müller and Liu can be carried out, the procedure of which is briefly sketched. Of the numerous results emerging in the course of such an exploitation, a few outstanding ones are singled out and presented for all model approaches so that a direct comparison is possible. In Section 3.3, the results of particular interest are related to the heat flux vector and the entropy flux vector as well as to the so-called *generalized Gibbs equations*, with the help of which the existence of a universal function (the *coldness function*) can be shown for the granular material under consideration if in addition an *ideal-wall postulate* is formulated. In Section 3.4, the focus is on the model specific constitutive response of selected constitutive quantities in thermodynamic equilibrium, results concerning the equilibrium entropy and the equilibrium Cauchy stresses are presented and compared for all models. In particular, it will be observed that localizations due to abrasion are likely to be predicted only with a gradient type formulation – the Helmholtz free energy will for the models (W), (V) and (GC) provide us with a "driving force" that is likely to capture localization phenomena since it accounts for sudden spatial changes in roughness. Finally, Section 3.5 concludes the article with a few additional remarks on specific features of the theory and a short summary.

3.2 Constitutive equations

As mentioned in Section 3.1, equations (3.2), (3.3) have to be complemented by constitutive equations. To do so, we have to specify the state space $\mathbb{S}$, which comprises the independent variables of the theory and which is the domain of the constitutive equations. For the present modelling, the state space is given by[1]

$$\mathbb{S} := \{a_0,\, a,\, \nabla a,\, \dot{a},\, \theta,\, \nabla\theta, \dot{\theta}, \mathbf{D},\, \mathbf{B},\, \mathbf{Z}\},$$

where $\mathbf{B}$ and $\mathbf{D}$ have been defined above, θ is an empirical temperature and $\mathbf{Z}$ is a symmetric internal tensorial (spatial) variable. $\mathbf{Z}$ is postulated to be an objective tensor of rank 2 and is introduced to model the frictional behaviour of the granular material. Roughness a, abrasion $\dot{a}$ and an initial roughness a_0 are postulated to be objective scalar quantities and

---

[1]The choice of the state space $\mathbb{S}$ can in essence be found in [2] and [18].

with this assumption all above quantities fulfill Euclidean objectivity. The inclusion of an initial roughness accounts for the fact that abrasion may either describe smoothing of initially rough particles or roughening due to the development of scratches at the surface of initially smooth particles. A spatial variation of roughness, ∇a, is accounted for, a spatial variation of abrasion $\nabla \dot{a}$ is for reasons of simplicity not considered. For the internal variable $\mathbf{Z}$, an evolution equation is postulated such that its time-derivative, denoted by an overhead circle, is objective:

$$\overset{\circ}{\mathbf{Z}} := \dot{\mathbf{Z}} - [\mathbf{\Omega}, \mathbf{Z}] := \dot{\mathbf{Z}} - \mathbf{\Omega}\mathbf{Z} + \mathbf{Z}\mathbf{\Omega} = \mathbf{\Phi}. \tag{3.7}$$

In (3.7), the objective time derivative $\overset{\circ}{\mathbf{Z}}$ is in fact the Jaumann derivative of $\mathbf{Z}$ if we choose $\mathbf{\Omega} = \mathbf{W}$, where $\mathbf{W} := \mathrm{skw}\,\nabla\mathbf{v}$ is the vorticity tensor, and $\mathbf{\Phi}$ is the constitutive part of (3.7). The constitutive quantities depend on the modeling approach which is pursued and are gathered in the set

$$\mathbb{C}^{(\cdot)} := \{\mathbf{T}, \varepsilon, \mathbf{q}, \eta, \mathbf{h}^{(\cdot)}, \phi, \mathbf{\Phi}, \pi^{(\cdot)}\}.$$

On the right–hand side of this equation, the quantities marked by $(\cdot)$ have to be replaced according to Table 3.1 and equation (3.5) depending on whether $\mathbb{C}^{(\cdot)}$ stands for $\mathbb{C}^{(\mathrm{W})}$, $\mathbb{C}^{(\mathrm{SH})}$, $\mathbb{C}^{(\mathrm{V})}$ or $\mathbb{C}^{(\mathrm{GC})}$. For each $\mathcal{C} \in \mathbb{C}^{(\cdot)}$ we have a functional relation of the form

$$\mathcal{C} = \hat{\mathcal{C}}(\mathbb{S}),$$

assumed to be smooth and differentiable at least once with respect to its variables. Upon inserting the constitutive equations into the balance equations, the latter become the so-called field equations, and any solution to those is called a thermodynamic process.

We note that the independent field variables are the velocity vector $\mathbf{v}$ (three scalar components), the temperature ("hidden" in the internal energy $\rho\varepsilon$) θ (1), the roughness a (1) *or* abrasion $\dot{a}$ (1), respectively, and the symmetric tensor $\mathbf{Z}$ (6), comprising $3 + 1 + 1 + 6 = 11$ scalar independent fields, for which equations $(3.3)_2$, $(3.3)_3$ or (3.6), respectively, and $(3.2)_1$, $(3.2)_3$, $(3.2)_5$ or (3.5), respectively, as well as (3.7) comprise 11 functional differential equations. Balance of mass is not a genuine equation because ρ (and e) can be computed once the motion is known. Neither must moment of momentum be accounted for if $\mathbf{T}$ is a priori assumed to be symmetric.

3.3   The entropy principle

Before the exploitation of the entropy principle is now carried out according to [11] and [8, 9], let us comment briefly upon assumptions that have tacitly been made in the two preceding sections.

As far as entropy η is concerned, it is required to be a scalar–valued, objective and additive quantity such that it gives rise to the balance law $(3.3)_4$. Moreover, it is determined by a general constitutive equation, $\eta = \hat{\eta}(\mathbb{S})$. Likewise, the entropy flux ϕ is an objective vector–valued function given by $\phi = \hat{\phi}(\mathbb{S})$, implying in particular that no a priori assumptions on the relation between the heat flux and the entropy flux have been made. The entropy supply σ is given by a linear combination of the (momentum and energy) supplies.

Following the rational thermodynamics interpretation of the second law of thermodynamics, it is required that all solutions to the field equations must satisfy the second law, or, in other words, the entropy production π must be non-negative for all thermodynamic processes. The second law is thus interpreted as a restriction to the material response, since it implies that the constitutive equations have to be chosen such that the second law of thermodynamics is fulfilled irrespective of the processes a body under consideration may experience.

Within the field of rational thermodynamics, approaches of various degrees of complexity have been undertaken to fulfill the entropy principle; among them, one finds classical irreversible thermodynamics ([10]), the large group of Clausius–Duhem Coleman–Noll approaches (see, *e.g.*, [1]), and the approach proposed by [11] complemented by its method of exploitation as suggested by [8, 9] (for review articles covering all these approaches *cf.*, *e.g.*, [3], [13] or [4]). We will here invoke the entropy principle of Müller and Liu to reduce the complexity of the constitutive functions and to derive the response of constitutive quantities such as, *e.g.*, the Cauchy stress in thermodynamic equilibrium.

The exploitation of the entropy principle starts from a modified entropy inequality in which all balance equations are considered as constraining equations and are thus subtracted from the entropy inequality $(3.3)_4$. Note that the balance of mass does not appear as a constraint since ρ is treated as a function of the independent constitutive field $\mathbf{B}$ via the equation

$$\rho_r = \rho \det\mathbf{F} = \rho\sqrt{\det\mathbf{B}},$$

where ρ_r is the mass density in the reference configuration. We thus start from the following form of the entropy inequality:

$$\pi \;=\; \rho\dot{\eta} + \operatorname{div}\phi + \rho\sigma - \boldsymbol{\lambda}^{\mathrm{mom}} \cdot (\rho\dot{\mathbf{v}} - \operatorname{div}\mathbf{T} + \rho\mathbf{b})$$

$$-\begin{cases} \lambda^a_{\mathrm{W}}\left(\dot{a} + a\operatorname{div}\mathbf{v} - \operatorname{div}\mathbf{h}^{(a)} - \pi^{(a)}\right) & \text{for the model (W)} \\[2mm] \lambda^a_{\mathrm{SH}}\left(\dot{a} + a\operatorname{div}\mathbf{v} - \pi^{(a)}\right) & \text{for the model (SH)} \\[2mm] \lambda^a_{\mathrm{V}}\left(\rho\dot{a} - \operatorname{div}\mathbf{h}^{(\rho a)} - \rho\pi^{(\rho a)}\right) & \text{for the model (V)} \\[2mm] \lambda^a_{\mathrm{GC}}\left(\rho k\ddot{a} - \operatorname{div}\mathbf{h}^{(\rho k\dot{a})} - \rho f\right) & \text{for the model (GC)} \end{cases}$$

$$
- \quad \lambda^\varepsilon \begin{cases} \rho\dot{\varepsilon} + \text{div } \mathbf{q} - \mathbf{T}\cdot\mathbf{D} + \rho r & \text{for models (W), (SH), (V)} \\[4pt] \rho\dot{\varepsilon} + \dot{a}\rho f + \text{div } \mathbf{q} - \mathbf{T}\cdot\mathbf{D} & \\ \quad - \mathbf{h}^{(\rho k \dot{a})}\cdot\nabla\dot{a} + \rho r & \text{for the model (GC)} \end{cases} \tag{3.8}
$$

$$
- \quad \boldsymbol{\Lambda}^{\text{fric}}\cdot\left(\dot{\mathbf{Z}} - [\boldsymbol{\Omega},\mathbf{Z}] - \boldsymbol{\Phi}\right) \quad \geq \quad 0.
$$

Here, λ_{W}^a, λ_{SH}^a, λ_{V}^a, λ_{GC}^a and λ^ε are scalar–valued, $\boldsymbol{\lambda}^{\text{mom}}$ is a vector–valued and $\boldsymbol{\Lambda}^{\text{fric}}$ is a tensor–valued symmetric[2] Lagrange multiplier, *i.e.*, $\boldsymbol{\Lambda}^{\text{fric}} = \boldsymbol{\Lambda}^{\text{fric}\,\text{T}}$. In general, these Lagrange multipliers may depend on all independent variables collected in the set $\mathbb{S}$, and possibly on $\mathbf{v}$, r, $\mathbf{b}$ and σ.

Inequality (3.8) is a requirement aiming at a restriction of constitutive relations. It is physically reasonable to suppose that this constitutive behaviour cannot depend on the source terms. Thus all source terms arising in (3.8) must add up to zero, implying

$$
\sigma = \boldsymbol{\lambda}^{\text{mom}}\cdot\mathbf{b} + \lambda^\varepsilon r,
$$

which says that the entropy source is known in terms of all other sources if the Lagrange multipliers are known.

If the differentiations in inequality (3.8) are carried out with regard to the state space $\mathbb{S}$ using the chain rule of differentiation, one arrives after lengthy but straightforward calculations at an expression for π which reads in principle as

$$
\pi = \sum_\alpha f_\alpha^s(\mathbb{S})s_\alpha + \sum_\beta f_\beta^n(\mathbb{S})n_\beta + f_r \geq 0. \tag{3.9}
$$

Here, f_α^s and f_β^n are coefficient functions preceding variables s_α which are *contained* in $\mathbb{S}$ and variables n_β *exceeding* the state space $\mathbb{S}$, respectively. The latter are often simply called the *higher derivatives*. f_r is a function collecting those terms associated with variables that belong to none of the just described classes; examples are the terms $\lambda_{\text{GC}}^a\rho f$ or $\boldsymbol{\Lambda}^{\text{fric}}\cdot\boldsymbol{\Phi}$. It should be noted that s_α and n_β may be scalar–, vector– or tensor–valued (and that therefore, the same holds true for the coefficient functions), but for simplicity of notation, this is not reflected in the symbols chosen in (3.9).

In [8] Liu has shown that, in order to satisfy the entropy principle, the coefficient functions preceding the higher derivatives n_β are each required to vanish identically. This provides us with two kinds of useful information: First, the vanishing of the coefficient functions f_β^n can be exploited – it gives rise to the so-called *Liu equations* which are at this point, however, not discussed in detail. An important result that is worth mentioning is that irrespective of the modeling approach chosen, the Lagrange multiplier

---

[2]Since $\dot{\mathbf{Z}} - [\boldsymbol{\Omega},\mathbf{Z}] - \boldsymbol{\Phi}$ is symmetric, only the symmetric part of $\boldsymbol{\Lambda}^{\text{fric}}$ is relevant.

associated with the balance of linear momentum vanishes identically and that the one related to the internal friction evolution is known in terms of the constitutive quantities η, ε and the Lagrange multiplier λ^ε:

$$\boldsymbol{\lambda}^{\mathrm{mom}} = \mathbf{0}, \qquad \boldsymbol{\Lambda}^{\mathrm{fric}} = \rho \left(\frac{\partial \eta}{\partial \mathbf{Z}} - \lambda^\varepsilon \frac{\partial \varepsilon}{\partial \mathbf{Z}} \right).$$

Second, inequality (3.9) simplifies considerably due to the vanishing coefficient functions and becomes the so-called *residual inequality* which is in essence of the form

$$\pi = \sum_\alpha f_\alpha^s(\mathbb{S}) s_\alpha + \tilde{f}_r \geq 0, \tag{3.10}$$

where a modified function $\tilde{f}_r$ accounts for possible changes in f_r due to the vanishing coefficient functions f_β^n. This inequality will in Section 3.4 be evaluated in equilibrium, but before this is done it is sketched how additional information, valid also in non-equilibrium, can be deduced from the results obtained so far. The major results emerging from the different modeling approaches (W), (SH), (V) and (GC) will briefly be compared.

We define the vector valued quantity

$$\mathbf{l} := \underbrace{\boldsymbol{\phi} - \lambda^\varepsilon \mathbf{q}}_{=:\mathbf{k}} + \begin{cases} \lambda_{\mathrm{W}}^a\, \mathbf{h}^{(a)} & \text{(W)} \\[2mm] \mathbf{0} & \text{(SH)} \\[2mm] \lambda_{\mathrm{V}}^a\, \mathbf{h}^{(\rho a)} & \text{(V)} \\[2mm] \lambda_{\mathrm{GC}}^a\, \mathbf{h}^{(\rho k \dot{a})} & \text{(GC)} \end{cases}$$

where $\mathbf{k}$ is called the *extra entropy flux*, and explore symmetry group properties to deduce additional constraining conditions. If in particular the conditions of isotropy for the vector–valued quantity $\mathbf{l}$ defined above are considered, one finds in combination with the Liu equations that for three of the proposed models, the entropy flux $\boldsymbol{\phi}$ is *not* collinear to the heat flux: to be precise, we find

$$\boldsymbol{\phi} = \begin{cases} \lambda^\varepsilon \mathbf{q} - \lambda_{\mathrm{W}}^a\, \mathbf{h}^{(a)} & \text{(W)} \\[2mm] \lambda^\varepsilon \mathbf{q} & \text{(SH)} \\[2mm] \lambda^\varepsilon \mathbf{q} - \lambda_{\mathrm{V}}^a\, \mathbf{h}^{(\rho a)} & \text{(V)} \\[2mm] \lambda^\varepsilon \mathbf{q} - \lambda_{\mathrm{GC}}^a\, \mathbf{h}^{(\rho k \dot{a})} & \text{(GC)}, \end{cases} \tag{3.11}$$

such that only for the model proposed by [14] the collinearity of the heat flux and the entropy flux is "recovered" in the sense that this result is postulated to be valid from the outset if a Coleman–Noll Clausius–Duhem approach is pursued. We will see that in thermodynamic equilibrium, $\lambda^\varepsilon = 1/\theta$ holds where θ can be identified with the absolute temperature, so that in equilibrium, $\phi = \mathbf{q}/\theta$ is obtained, which is exactly the form used in the Coleman–Noll approaches.

Moreover, the Liu equations can be cast into a compact form and then yield the so-called *generalized Gibbs relations*

$$
\rho\,\mathrm{d}\eta = \rho\lambda^\varepsilon \mathrm{d}\varepsilon + \mathcal{P}, \quad \mathrm{d}\phi = \lambda^\varepsilon \mathrm{d}\mathbf{q} + \mathcal{F} -
\begin{cases}
\lambda^a_{\mathrm{W}}\,\mathrm{d}\mathbf{h}^{(a)} & \text{(W)} \\
0 & \text{(SH)} \\
\lambda^a_{\mathrm{V}}\,\mathrm{d}\mathbf{h}^{(\rho a)} & \text{(V)} \\
\lambda^a_{\mathrm{GC}}\,\mathrm{d}\mathbf{h}^{(\rho k \dot{a})} & \text{(GC)}
\end{cases}
\tag{3.12}
$$

where the scalar– and vector–valued quantities $\mathcal{P}$ and $\mathcal{F}$ are sums of differentials of variables s_α belonging to the state space $\mathbb{S}$:

$$
\mathcal{P} = \sum_\alpha \mathcal{P}_\alpha(\cdot)\mathrm{d}s_\alpha, \qquad \mathcal{F} = \sum_\alpha \mathcal{F}_\alpha(\cdot)\mathrm{d}s_\alpha.
\tag{3.13}
$$

In the above equations, $\mathcal{P}_\alpha(\cdot)$ and $\mathcal{F}_\alpha(\cdot)$ are coefficient functions of the form

$$
\mathcal{P}_\alpha = \rho\left(\frac{\partial\eta}{\partial\alpha} - \lambda^\varepsilon\frac{\partial\varepsilon}{\partial\alpha}\right), \qquad \mathcal{F}_\alpha(\cdot) = \frac{\partial\phi}{\partial\alpha} - \lambda^\varepsilon\frac{\partial\mathbf{q}}{\partial\alpha} + \lambda^a_{\mathrm{W}}\frac{\partial\mathbf{h}^{(a)}}{\partial\alpha}, \tag{3.14}
$$

and for brevity, only the form which results for the modeling according to [19] is given. The different modeling approaches are in the above equations reflected only in the last term on the right-hand side of $\mathcal{F}_\alpha$, which is consequently not existing in case the model (SH) is used. $(3.12)_2$ can with the help of the above derived results be rewritten in the following form:

$$
\mathcal{F} = \mathbf{q}\,\mathrm{d}\lambda^\varepsilon -
\begin{cases}
\mathbf{h}^{(a)}\,\mathrm{d}\lambda^a_{\mathrm{W}} & \text{(W)} \\
0 & \text{(SH)} \\
\mathbf{h}^{(\rho a)}\,\mathrm{d}\lambda^a_{\mathrm{V}} & \text{(V)} \\
\mathbf{h}^{(\rho k \dot{a})}\,\mathrm{d}\lambda^a_{\mathrm{GC}} & \text{(GC)}
\end{cases}
$$

$$
= \mathcal{F}_a\,\mathrm{d}a + \mathcal{F}_{\dot{a}}\,\mathrm{d}\dot{a} + \mathcal{F}_\theta\,\mathrm{d}\theta + \mathcal{F}_{\dot{\theta}}\,\mathrm{d}\dot{\theta}.
$$

A comparison of coefficients and the assumption that $\mathbf{q}$ and $\mathbf{h}$ are not collinear leads for the models (W), (V) and (GC) to a restriction on the dependencies of the Lagrange multipliers associated with the balances of energy and roughness or abrasion, respectively:

$$
\lambda^\varepsilon = \hat{\lambda}^\varepsilon(a,\dot{a},\theta,\dot{\theta}), \qquad \lambda^a_{\mathrm{W}} = \hat{\lambda}^a_{\mathrm{W}}(a,\dot{a},\theta,\dot{\theta}).
\tag{3.15}
$$

Note that again, for brevity, $(3.15)_2$ has only been given for the model (W) but reads similarly for models (V) and (GC). In case of model (SH),

$(3.15)_1$ is also obtained, but no equation corresponding to $(3.15)_2$ can be deduced, leaving the Lagrange multiplier λ^a_{SH} undetermined. Motivated from classical results, where λ^ε is shown to be a universal function (the so-called *coldness function*, see, *e.g.*, [11]), we wish to reduce the above indicated dependency of λ^ε even further to eventually be able to carry the classical results over to the granular material under consideration. To achieve this, an *ideal wall* condition is postulated in the following form:

There exist ideal material walls between a granular solid and other single non–abrasive constituent continua across which the empirical temperature and the tangential velocity, denoted by $\mathbf{v}_\parallel$, *are continuous:* $[\![\,\theta\,]\!] = 0, [\![\,\mathbf{v}_\parallel\,]\!] = \mathbf{0}$.

The jump conditions $(3.4)_{2,3,4}$ and $(3.5)_2$ then imply that λ^ε is continuous across this ideal wall, $[\![\,\lambda^\varepsilon\,]\!] = 0$, or, in other words, $\lambda^{\varepsilon+} = \lambda^{\varepsilon-}$. We now let the material on the negative side of the wall be given by an *ideal gas*, for which it has been proven that λ^ε is a materially independent function of θ, $\dot\theta$ such that

$$\lambda^{\varepsilon+}(a,\dot a,\theta,\dot\theta) = \lambda^{\varepsilon-}(a,\dot a,\theta,\dot\theta) = \lambda^{\varepsilon-}(\theta,\dot\theta) \quad \Longrightarrow \quad \lambda^{\varepsilon+} = \lambda^{\varepsilon+}(\theta,\dot\theta).$$

We thus find that in all modeling approaches (note that the prerequisites for the ideal wall postulate can be slightly weakened for the (SH) model) λ^ε can only be a function of θ and $\dot\theta$, which allows us to identify λ^ε with a universal function (the *coldness function*) and with the help of which one can define the absolute temperature

$$\Theta = \frac{1}{\lambda^\varepsilon(\theta,0)}. \tag{3.16}$$

For an ideal gas, it has been shown that Θ can be identified with θ, see [11, 12] and [3], and this result in fact carries over to the granular material under consideration, because with the ideal wall condition we have just indicated that it is easily shown that $\hat\lambda^\varepsilon(\theta,\dot\theta)$ is the same function as for ideal gases. Since (3.16) is in fact an equilibrium quantity (the terminology will be defined in Section 3.4), the above comments immediately imply that in equilibrium, $\lambda^\varepsilon = 1/\theta$ holds, an issue already mentioned in the context of (3.11).

3.4   Thermodynamic equilibrium

As already mentioned in Section 3.3 we will now evaluate the residual inequality (3.10) in thermodynamic equilibrium. In *thermodynamic equilibrium*, the thermodynamic processes are time-independent and characterized by a homogeneous temperature and vanishing velocity field: $\nabla\theta = \mathbf{0}$ and $\mathbf{v} = \mathbf{0}$.

The evaluation of a quantity or a functional $\mathcal{C}$ in thermodynamic equilibrium is denoted by a lower index $\big|_E$ as in $\mathcal{C}\big|_E$. To be more precise, we first define the sets

$$\mathbb{E} := \{a_0, a, \nabla a, \theta, \mathbf{B}, \mathbf{Z}\} \subset \mathbb{S} \quad \text{and} \quad \mathbb{N} := \mathbb{S} \setminus \mathbb{E} = \{\dot{a}, \dot{\theta}, \nabla\theta, \mathbf{D}\}.$$

$\mathbb{E}$ contains all independent variables which may be non-vanishing in equilibrium. The elements of these sets are for brevity denoted by $e \in \mathbb{E}$ and $n \in \mathbb{N}$, respectively. With this,

$$\hat{\mathcal{C}}\big|_E := \lim_{\mathbb{N} \to 0} \hat{\mathcal{C}}(\mathbb{E}, \mathbb{N}),$$

where $\mathbb{N} \to 0$ is a shorthand for $\dot{a}, \dot{\theta} \to 0, \nabla\theta, \mathbf{D} \to \mathbf{0}$. Note that the above definition implies that $\boldsymbol{\Phi}\big|_E = \mathbf{0}$ holds, since (see equation (3.7))

$$\overset{\circ}{\mathbf{Z}}\Big|_E = \left(\dot{\mathbf{Z}} - [\boldsymbol{\Omega}, \mathbf{Z}]\right)\Big|_E = \boldsymbol{\Phi}\big|_E,$$

where the left–hand side is an objective time derivative which automatically vanishes in thermodynamic equilibrium.

We also notice that a functional $\hat{\mathcal{C}}(\mathbb{S})$ can be decomposed into an *equilibrium* and a *non-equilibrium part*, $\hat{\mathcal{C}}_E$ and $\hat{\mathcal{C}}_N$, which are defined by

$$\hat{\mathcal{C}}_E := \hat{\mathcal{C}}(a_0, a, \nabla a, \dot{a} = 0, \theta, \dot{\theta} = 0, \nabla\theta = \mathbf{0}, \mathbf{D} = \mathbf{0}, \mathbf{B}, \mathbf{Z}),$$

$$\hat{\mathcal{C}}_N(a_0, a, \nabla a, \dot{a}, \theta, \dot{\theta}, \nabla\theta, \mathbf{D}, \mathbf{B}, \mathbf{Z}) := \hat{\mathcal{C}} - \hat{\mathcal{C}}_E,$$

so that according to this definition, $\hat{\mathcal{C}}(\mathbb{S}) = \hat{\mathcal{C}}_E(\mathbb{E}) + \hat{\mathcal{C}}_N(\mathbb{S})$ holds. This is of importance in the results to follow.

Let us now focus on the residual inequality (3.10), which is for convenience repeated here:

$$\pi = \sum_\alpha f_\alpha^s(\mathbb{S}) s_\alpha + \tilde{f}_r \geq 0.$$

Common to all modeling approaches is the fact that in the sum on the right-hand side, s_α assumes the "values"[3] $\mathbf{D}$, $\dot{\theta}$, $\nabla\theta$ and $\dot{a}$, which all vanish in thermodynamic equilibrium. For the model (SH), s_α assumes no additional values, but for the models (W), (V) and (GC), s_α also assumes the value ∇a, which is non-vanishing in equilibrium. The second term on the right-hand side contains in all four modeling approaches the terms associated with $\boldsymbol{\Phi}$ and those related to the production terms $\pi^{(\Psi)}$, see Table 3.1. As mentioned above, $\boldsymbol{\Phi}\big|_E = \mathbf{0}$ holds, but nothing can be said a priori on the production terms and those associated with ∇a. These terms are referred to as "critical" terms in Table 3.2.

---

[3]This terminology is a little bit sloppy, since the independent variables are in fact field quantities. It is hoped that nonetheless, no confusion arises.

TABLE 3.2. Additional postulates to guarantee the vanishing of the entropy production in equilibrium, $\pi\big|_E = 0$.

model approach	"critical" terms	additional postulates
(W)	$\mathbf{h}^{(a)}\dfrac{\partial \lambda_W^a}{\partial a}\cdot\nabla a,\ \lambda_W^a\pi^{(a)}$	$\dfrac{\partial \lambda_W^a}{\partial a}\Big|_E := 0,\ \pi^{(a)}\big|_E := 0$
(SH)	$\lambda_{SH}^a\pi^{(a)}$	–
(V)	$\mathbf{h}^{(\rho a)}\dfrac{\partial \lambda_V^a}{\partial a}\cdot\nabla a,\ \lambda_V^a\pi^{(\rho a)}$	$\dfrac{\partial \lambda_V^a}{\partial a}\Big|_E := 0,\ \pi^{(\rho a)}\big|_E := 0$
(GC)	$\mathbf{h}^{(\rho k\dot a)}\dfrac{\partial \lambda_{GC}^a}{\partial a}\cdot\nabla a,\ \lambda_{GC}^a\rho f$	$\lambda_{GC}^a\big|_E := 0$

To guarantee that the entropy production assumes its minimum in thermodynamic equilibrium, $\pi\big|_E = 0$, different postulates have thus to be formulated which enforce this equality, that is, which imply the identical vanishing of the critical terms. These postulates are summarized in Table 3.2 and are now subject to a very brief discussion. Let us start with model (SH). At the first glance, the only non-vanishing term in the residual inequality is the expression $\lambda_{SH}^a\pi^{(a)}$, but an evaluation of $(3.2)_3$ (which is the evolution equation for roughness) in equilibrium shows that $\pi^{(a)}\big|_E = 0$ must necessarily hold so that indeed no additional assumptions have to be imposed to satisfy $\pi\big|_E = 0$.

In case of model (GC), a reasonable postulate is $\lambda_{GC}^a\big|_E := 0$. This requirement *implies* that also $\partial \lambda_{GC}^a/\partial a$ vanishes in equilibrium, so that no additional conditions have to be imposed to force the second critical term to be zero in equilibrium.

As far as the models (W) and (V) are concerned, one has different possibilities to enforce the vanishing of the critical terms in equilibrium: on the one hand, one could require $\lambda_W^a\big|_E := 0$ and $\lambda_V^a\big|_E := 0$, respectively, establishing a close formal similarity to the model (GC). On the other hand, one may require that $\pi^a\big|_E := 0$ (and $\pi^{(\rho a)}\big|_E := 0$, respectively) holds and impose simultaneously the condition $(\partial \lambda_W^a/\partial a)\big|_E := 0$ (and $(\partial \lambda_V^a/\partial a)\big|_E := 0$). This choice is advantageous if a comparison to the results emerging from model (SH) is intended: in the latter, the Lagrange multiplier λ_{SH}^a is not restricted at all, so that we prefer imposing a condition on the derivative of the Lagrange multiplier to imposing it on the Lagrange multiplier itself. Physically, it seems also plausible to assume that no roughness is produced in equilibrium, and so the second set of postulates is adopted for the models (W) and (V).

From $(3.3)_4$ and the remarks made in the beginning of Section 3.3 it is clear that the entropy production is a function defined on the state space

$\mathbb{S}$, $\pi = \hat{\pi}(\mathbb{S})$. However, for fixed a_0, a, ∇a, θ, $\mathbf{Z}$ and $\mathbf{B}$ we may interpret π as a function of $\dot{a}$, $\dot{\theta}$, $\nabla\theta$ and $\mathbf{D}$ with parameters $\mathrm{e} \in \mathbb{E}$,

$$\pi = \tilde{\pi}_{\mathbb{E}}(\mathbb{N}).$$

Since π is a non-negative function of its (independent) variables $\dot{a}$, $\dot{\theta}$, $\nabla\theta$ and $\mathbf{D}$ which assumes its minimum in thermodynamic equilibrium, the conditions

$$\left.\frac{\partial\pi}{\partial\dot{\theta}}\right|_{\mathrm{E}} = 0, \quad \left.\frac{\partial\pi}{\partial\nabla\theta}\right|_{\mathrm{E}} = \mathbf{0}, \quad \left.\frac{\partial\pi}{\partial\mathbf{D}}\right|_{\mathrm{E}} = \mathbf{0}, \quad \left.\frac{\partial\pi}{\partial\dot{a}}\right|_{\mathrm{E}} = 0 \qquad (3.17)$$

are necessarily satisfied. The condition that is also sufficient for π to assume its minimum is given by the requirement that the Hessian matrix

$$\begin{pmatrix} \pi_{,\dot{\theta}\,\dot{\theta}}\big|_{\mathrm{E}} & \pi_{,\dot{\theta}\,\nabla\theta}\big|_{\mathrm{E}} & \pi_{,\dot{\theta}\,\mathbf{D}}\big|_{\mathrm{E}} & \pi_{,\dot{\theta}\,\dot{a}}\big|_{\mathrm{E}} \\[2mm] \pi_{,\nabla\theta\,\dot{\theta}}\big|_{\mathrm{E}} & \pi_{,\nabla\theta\,\nabla\theta}\big|_{\mathrm{E}} & \pi_{,\nabla\theta\,\mathbf{D}}\big|_{\mathrm{E}} & \pi_{,\nabla\theta\,\dot{a}}\big|_{\mathrm{E}} \\[2mm] \pi_{,\mathbf{D}\,\dot{\theta}}\big|_{\mathrm{E}} & \pi_{,\mathbf{D}\,\nabla\theta}\big|_{\mathrm{E}} & \pi_{,\mathbf{D}\,\mathbf{D}}\big|_{\mathrm{E}} & \pi_{,\mathbf{D}\,\dot{a}}\big|_{\mathrm{E}} \\[2mm] \pi_{,\dot{a}\,\dot{\theta}}\big|_{\mathrm{E}} & \pi_{,\dot{a}\,\nabla\theta}\big|_{\mathrm{E}} & \pi_{,\dot{a}\,\mathbf{D}}\big|_{\mathrm{E}} & \pi_{,\dot{a}\,\dot{a}}\big|_{\mathrm{E}} \end{pmatrix} \qquad (3.18)$$

is positive semi-definite[4], and $\pi_{,\nabla\theta\dot{\theta}}$ has been introduced as a shorthand for $\partial^2\pi/(\partial\nabla\theta\,\partial\dot{\theta})$ etc. However, the implications of (3.18) will not be considered here and the reader is referred to [5] instead.

As an immediate consequence of $(3.17)_3$ the following equilibrium expressions for the Cauchy stress tensor are obtained:

$$\mathbf{T}_{\mathrm{E}} = \begin{cases} -\theta\left[\left\langle \mathcal{P}_{\mathbf{B}}\big|_{\mathrm{E}}, \mathbf{B}\right\rangle - \left.\frac{\partial\lambda_{\mathrm{W}}^a}{\partial\dot{a}}\right|_{\mathrm{E}} \mathrm{sym}(\nabla a \otimes \mathbf{h}^{(a)})\big|_{\mathrm{E}}\right. \\[3mm] \qquad \left. + \left.\frac{\partial\boldsymbol{\Phi}}{\partial\mathbf{D}}\right|_{\mathrm{E}} \boldsymbol{\Lambda}^{\mathrm{fric}}\big|_{\mathrm{E}} + \lambda_{\mathrm{W}}^a\big|_{\mathrm{E}}\left(\left.\frac{\partial\pi^{(a)}}{\partial\mathbf{D}}\right|_{\mathrm{E}} - a\mathbf{I}\big|_{\mathrm{E}}\right)\right] \quad (\mathrm{W}) \\[6mm] -\theta\left[\left\langle \mathcal{P}_{\mathbf{B}}, \mathbf{B}\right\rangle\big|_{\mathrm{E}} - \lambda_{\mathrm{SH}}^a\big|_{\mathrm{E}}\left(a\mathbf{I}\big|_{\mathrm{E}} - \left.\frac{\partial\pi^{(a)}}{\partial\mathbf{D}}\right|_{\mathrm{E}}\right)\right. \\[3mm] \qquad \left. + \left.\frac{\partial\boldsymbol{\Phi}}{\partial\mathbf{D}}\right|_{\mathrm{E}} \boldsymbol{\Lambda}^{\mathrm{fric}}\big|_{\mathrm{E}}\right] \qquad\qquad\qquad\qquad (\mathrm{SH}) \end{cases}$$

---

[4]In general, to exclude the possibility of the function assuming a saddle-point value, positive definiteness of the Hessian has to be required as a necessary condition for extremal values. Here however, since π is a non-negative function, $\pi = 0$ is a (global) minimum and it is sufficient to require positive semi-definiteness of the Hessian matrix.

$$
\mathbf{T}_{\mathrm{E}} = \begin{cases}
-\theta \left(\langle \mathcal{P}_{\mathbf{B}}|_{\mathrm{E}}, \mathbf{B} \rangle - \left.\dfrac{\partial \lambda_{\mathrm{V}}^{a}}{\partial \dot{a}}\right|_{\mathrm{E}} \mathrm{sym}(\nabla a \otimes \mathbf{h}^{(\rho a)})|_{\mathrm{E}} \right. \\[2ex]
\left. \quad + \left.\dfrac{\partial \mathbf{\Phi}}{\partial \mathbf{D}}\right|_{\mathrm{E}} \mathbf{\Lambda}^{\mathrm{fric}}|_{\mathrm{E}} + \rho \lambda_{\mathrm{V}}^{a}|_{\mathrm{E}} \left.\dfrac{\partial \pi^{(\rho a)}}{\partial \mathbf{D}}\right|_{\mathrm{E}} \right) & \text{(V)} \\[3ex]
-\theta \left[\langle \mathcal{P}_{\mathbf{B}}|_{\mathrm{E}}, \mathbf{B} \rangle + \left(\dfrac{1}{\theta} - \left.\dfrac{\partial \lambda_{\mathrm{GC}}^{a}}{\partial \dot{a}}\right|_{\mathrm{E}} \right) \mathrm{sym}(\nabla a \otimes \mathbf{h}^{\rho k \dot{a}})|_{\mathrm{E}} \right. \\[2ex]
\left. \quad + \left.\dfrac{\partial \mathbf{\Phi}}{\partial \mathbf{D}}\right|_{\mathrm{E}} \mathbf{\Lambda}^{\mathrm{fric}}|_{\mathrm{E}} \right] & \text{(GC)},
\end{cases}
$$

where $\mathcal{P}_{\mathbf{B}}$ has been defined in (3.14) and the pointed brackets $\langle \cdot, \cdot \rangle$ are defined by $\langle \mathcal{P}_{\mathbf{B}}, \mathbf{B} \rangle := \mathcal{P}_{\mathbf{B}} \mathbf{B} + \mathbf{B} \mathcal{P}_{\mathbf{B}}$. Before we comment on these results in more detail, let us define the *Helmholtz free energy* Ψ as follows:

$$
\hat{\Psi} := \hat{\Psi}(\mathbb{S}) := \varepsilon - \theta \eta = \varepsilon_{\mathrm{E}} - \theta \eta_{\mathrm{E}} + \varepsilon_{\mathrm{N}} - \theta \eta_{\mathrm{N}} =: \hat{\psi}_{\mathrm{E}}(\mathbb{E}) + \hat{\psi}_{\mathrm{N}}(\mathbb{S}).
$$

The introduction of Ψ is motivated by the following observations: Consider the expression $\theta \mathcal{P}|_{\mathrm{E}}$. On the one hand, this expression can be evaluated starting from the form of $\mathcal{P}$ as given in (3.13), and the results obtained from (3.17) (which have however not been presented here) enter these calculations. On the other hand, we can evaluate $\theta \mathcal{P}|_{\mathrm{E}}$ starting from (3.12)$_1$, making use of the Helmholtz free energy. Consequently, we arrive at two expressions – each of which is the right–hand side of $\theta \mathcal{P}|_{\mathrm{E}}$ – so that a comparison of their coefficients is possible and provides us with useful information. In particular, we find three types of results, which are of special interest:

1. Irrespective of the chosen modeling approach, one finds the following relation for the Lagrange multiplier $\mathbf{\Lambda}^{\mathrm{fric}}$ associated with the internal frictional behaviour of the granular material and for the coefficient $\mathcal{P}_{\mathbf{B}}$:

$$
\mathbf{\Lambda}^{\mathrm{fric}}|_{\mathrm{E}} = -\frac{\rho}{\theta} \frac{\partial \psi_{\mathrm{E}}}{\partial \mathbf{Z}}, \quad \mathcal{P}_{\mathbf{B}}|_{\mathrm{E}} = -\frac{\rho}{\theta} \frac{\partial \psi_{\mathrm{E}}}{\partial \mathbf{B}}.
$$

This is not surprising, since the modeling of internal friction has not been changed and (3.7) is used for all models under consideration.

2. The second type of results concerns the partial derivative of ψ_{E} with respect to the roughness gradient ∇a. For the different models, we

obtain

$$\rho\frac{\partial\psi_{\mathrm{E}}}{\partial\nabla a} = \begin{cases} -\theta\mathbf{h}^{(a)}\big|_{\mathrm{E}}\dfrac{\partial\lambda_{\mathrm{W}}^{a}}{\partial\dot{a}}\Big|_{\mathrm{E}} & \text{(W)} \\[2ex] 0 & \text{(SH)} \\[2ex] -\theta\mathbf{h}^{(\rho a)}\big|_{\mathrm{E}}\dfrac{\partial\lambda_{\mathrm{V}}^{a}}{\partial\dot{a}}\Big|_{\mathrm{E}} & \text{(V)} \\[2ex] \mathbf{h}^{\rho k\dot{a}}\big|_{\mathrm{E}}\Big(1-\theta\dfrac{\partial\lambda_{\mathrm{GC}}^{a}}{\partial\dot{a}}\Big|_{\mathrm{E}}\Big) & \text{(GC)}. \end{cases}$$

From these equations it is observed that for three models, namely (W), (V) and (GC), thermodynamic considerations provide us with a "driving force" which captures the surface stress vector $\mathbf{h}^{(\cdot)}$ of roughness (for the former two models) or abrasion, (in the latter case), respectively. This driving force $\partial\psi_{\mathrm{E}}/\partial\nabla a$ might be a means to capture localization phenomena, since it accounts for abrupt spatial changes. As seen from the second of the above equations, spatial variations of roughness can for the model (SH) *not* be detected by the Helmholtz free energy. As a consequence, thermodynamics in this model does not provide for a (dynamic) driving force that is essentially generated by the roughness gradient ∇a.

3. As far as the equilibrium parts of the entropy are concerned, we find the following results:

$$\eta_{\mathrm{E}} = \begin{cases} -\dfrac{\partial\psi_{\mathrm{E}}}{\partial\mathbf{Z}}\cdot\dfrac{\partial\boldsymbol{\Phi}}{\partial\dot{\theta}}\Big|_{\mathrm{E}} - \dfrac{\partial\psi_{\mathrm{E}}}{\partial\theta} + \dfrac{\theta}{\rho}\Big(\lambda_{\mathrm{W}}^{a}\big|_{\mathrm{E}}\dfrac{\partial\pi^{(a)}}{\partial\dot{\theta}}\Big|_{\mathrm{E}} \\ \qquad -\mathbf{h}^{(a)}\big|_{\mathrm{E}}\cdot\nabla a\dfrac{\partial^{2}\lambda_{\mathrm{W}}^{a}}{\partial a\,\partial\dot{\theta}}\Big|_{\mathrm{E}}\Big) & \text{(W)} \\[3ex] -\dfrac{\partial\psi_{\mathrm{E}}}{\partial\mathbf{Z}}\cdot\dfrac{\partial\boldsymbol{\Phi}}{\partial\dot{\theta}}\Big|_{\mathrm{E}} - \dfrac{\partial\psi_{\mathrm{E}}}{\partial\theta} + \dfrac{\theta}{\rho}\lambda_{\mathrm{SH}}^{a}\big|_{\mathrm{E}}\dfrac{\partial\pi^{(a)}}{\partial\dot{\theta}}\Big|_{\mathrm{E}} & \text{(SH)} \\[3ex] -\dfrac{\partial\psi_{\mathrm{E}}}{\partial\mathbf{Z}}\cdot\dfrac{\partial\boldsymbol{\Phi}}{\partial\dot{\theta}}\Big|_{\mathrm{E}} - \dfrac{\partial\psi_{\mathrm{E}}}{\partial\theta} + \theta\,\lambda_{\mathrm{V}}^{a}\big|_{\mathrm{E}}\dfrac{\partial\pi^{(\rho a)}}{\partial\dot{\theta}}\Big|_{\mathrm{E}} \\ \qquad -\dfrac{\theta}{\rho}\mathbf{h}^{(\rho a)}\big|_{\mathrm{E}}\cdot\nabla a\dfrac{\partial^{2}\lambda_{\mathrm{V}}^{a}}{\partial a\,\partial\dot{\theta}}\Big|_{\mathrm{E}} & \text{(V)} \\[3ex] -\dfrac{\partial\psi_{\mathrm{E}}}{\partial\mathbf{Z}}\cdot\dfrac{\partial\boldsymbol{\Phi}}{\partial\dot{\theta}}\Big|_{\mathrm{E}} - \dfrac{\partial\psi_{\mathrm{E}}}{\partial\theta} + \theta\dfrac{\partial\lambda_{\mathrm{GC}}^{a}}{\partial\dot{\theta}}\Big|_{\mathrm{E}}f\big|_{\mathrm{E}} \\ \qquad -\dfrac{\theta}{\rho}\mathbf{h}^{(\rho k\dot{a})}\big|_{\mathrm{E}}\cdot\nabla a\dfrac{\partial^{2}\lambda_{\mathrm{GC}}^{a}}{\partial a\,\partial\dot{\theta}}\Big|_{\mathrm{E}} & \text{(GC)}. \end{cases}$$

As was the case for the Lagrange multiplier $\boldsymbol{\Lambda}^{\mathrm{fric}}$, the *frictional contributions* to the equilibrium entropy are in all modeling approaches the same – they are given by the first term on the right–hand side for each model and explictly involve $\mathbf{Z}$ and $\boldsymbol{\Phi}$. In addition, *elastic contributions* associated with the partial derivative of ψ_E with respect to θ – the second term on the right–hand side – remain unchanged irrespective of the model chosen. The different modeling alternatives for roughness and abrasion, respectively, are however clearly detectable in the remaining terms (one in case of model (SH), two for models (W), (V) and (GC)), which will be referred to as *anisotropic contributions*. At the present stage of the investigations, this terminology seems to have but little motivation. However, as soon as the Cauchy stresses are rewritten with the results obtained so far, anisotropic stress contributions become clearly detectable, so that the analogy will become obvious and justify the above terminology. Structurally, the anisotropic contributions to the equilibrium entropy do not, for the models involving a flux term $\mathbf{h}^{(\cdot)}$, differ from one another, a fact that will be different in case of the equilibrium Cauchy stresses, as will soon be seen.

Let us now present the equilibrium Cauchy stresses for all four models proposed:

$$
\mathbf{T}_\mathrm{E} =
\begin{cases}
\rho\dfrac{\partial\boldsymbol{\Phi}}{\partial\mathbf{D}}\Big|_\mathrm{E}\dfrac{\partial\psi_\mathrm{E}}{\partial\mathbf{Z}} + \rho\,\langle\,\dfrac{\partial\psi_\mathrm{E}}{\partial\mathbf{B}},\mathbf{B}\,\rangle + \theta\dfrac{\partial\lambda_\mathrm{W}^a}{\partial\dot{a}}\Big|_\mathrm{E}\mathrm{sym}(\nabla a\otimes\mathbf{h}^{(a)})\big|_\mathrm{E} \\[2ex]
\quad -\theta\lambda_\mathrm{W}^a\big|_\mathrm{E}\big(\dfrac{\partial\pi^{(a)}}{\partial\mathbf{D}}\Big|_\mathrm{E} - a\mathbf{I}\big|_\mathrm{E}\big) & \text{(W)} \\[4ex]
\rho\dfrac{\partial\boldsymbol{\Phi}}{\partial\mathbf{D}}\Big|_\mathrm{E}\dfrac{\partial\psi_\mathrm{E}}{\partial\mathbf{Z}} + \rho\,\langle\,\dfrac{\partial\psi_\mathrm{E}}{\partial\mathbf{B}},\mathbf{B}\,\rangle + \theta\lambda_\mathrm{SH}^a\big|_\mathrm{E}\big(a\mathbf{I}\big|_\mathrm{E} - \dfrac{\partial\pi^{(a)}}{\partial\mathbf{D}}\Big|_\mathrm{E}\big) & \text{(SH)} \\[4ex]
\rho\dfrac{\partial\boldsymbol{\Phi}}{\partial\mathbf{D}}\Big|_\mathrm{E}\dfrac{\partial\psi_\mathrm{E}}{\partial\mathbf{Z}} + \rho\,\langle\,\dfrac{\partial\psi_\mathrm{E}}{\partial\mathbf{B}},\mathbf{B}\,\rangle + \theta\dfrac{\partial\lambda_\mathrm{V}^a}{\partial\dot{a}}\Big|_\mathrm{E}\mathrm{sym}(\nabla a\otimes\mathbf{h}^{(\rho a)})\big|_\mathrm{E} \\[2ex]
\quad -\theta\rho\lambda_\mathrm{V}^a\big|_\mathrm{E}\dfrac{\partial\pi^{(\rho a)}}{\partial\mathbf{D}}\Big|_\mathrm{E} & \text{(V)} \\[4ex]
\rho\dfrac{\partial\boldsymbol{\Phi}}{\partial\mathbf{D}}\Big|_\mathrm{E}\dfrac{\partial\psi_\mathrm{E}}{\partial\mathbf{Z}} + \rho\,\langle\,\dfrac{\partial\psi_\mathrm{E}}{\partial\mathbf{B}},\mathbf{B}\,\rangle \\[2ex]
\quad -\Big(1-\theta\dfrac{\partial\lambda_\mathrm{GC}^a}{\partial\dot{a}}\Big|_\mathrm{E}\Big)\,\mathrm{sym}(\nabla a\otimes\mathbf{h}^{(\rho k\dot{a})}\big|_\mathrm{E}) & \text{(GC)}.
\end{cases}
$$

These refined expressions for $\mathbf{T}_\mathrm{E}$ are obtained with the help of the results given above in 1. and 2., and it is observed that again frictional contributions – first term on the right–hand side – enter $\mathbf{T}_\mathrm{E}$, and that these contributions remain unaffected by the different modeling approaches. Likewise, the second terms on the right–hand side capture the elastic contributions (as was already the case for the equilibrium entropy, see above) and are identical for all models. The remaining terms are referred to as *anisotropic stress contributions*, which is a jargon often used in engineering sciences and by which we mean deviatoric stress contributions. These anisotropic contributions shall now be discussed for each model.

Let us start with the model (GC), in which a balance equation for abrasion $\dot{a}$ has been suggested, see (3.5). Of course, abrasion itself cannot contribute to the equilibrium Cauchy stress since $\dot{a}$ vanishes in equilibrium. Instead of abrasion, we thus find that the roughness gradient ∇a enters $\mathbf{T}_{\mathrm{E}}$ in connection with the abrasive flux $\mathbf{h}^{(\rho k \dot{a})}$.

In contrast to this, the equilibrium Cauchy stress as obtained from model (SH) (in which no flux term entering the evolution of roughness is taken into account) is unaffected by ∇a. If in equilibrium, the production of roughness $\pi^{(a)}$ does not vary with $\mathbf{D}$, the additional stress contribution due to roughness is a purely isotropic one. The models (W) and (V) differ structurally from each other only in the last term. In particular, an isotropic stress contribution which is not present in the model (V) enters the model (W). It is further observed that these models again account for the roughness gradient in combination with the flux $\mathbf{h}^{(\cdot)}$.

With these remarks, this section is drawn to a close. A few additional remarks on specific aspects of the theory are now addressed in Section 3.5.

3.5   Additional remarks and conclusions

Three aspects which have so far not been addressed at all shall now be briefly mentioned. Since a more detailed discussion would exceed the scope of this article, the following comments are to be understood as a reminder of what sort of additional problems can arise in the course of the ideas presented and what has so far been undertaken to resolve them.

First, it should be noted that we have included $\dot{\theta}$ as an independent variable in the state space $\mathbb{S}$. This certainly causes additional computational effort in the course of exploiting the entropy principle, but this is not really considered as an argument against its occurence in $\mathbb{S}$. The reason why it is treated as an independent variable is that with this, the possibility of the linearized heat conduction equation assuming hyperbolic type is no longer excluded (as is the case if $\dot{\theta} \notin \mathbb{S}$). In case of, *e.g.*, the model (GC), a hyperbolic linearized heat conduction equation is achieved by imposing the condition

$$\frac{1}{\rho k}\frac{\partial^2 \varepsilon}{\partial a\, \partial \dot{a}}\bigg|_{\mathrm{E}} \frac{\partial \mathbf{h}^{(\rho k \dot{a})}}{\partial \dot{\theta}}\bigg|_{\mathrm{E}} \cdot \nabla a - \frac{\partial \varepsilon_{\mathrm{E}}}{\partial \mathbf{Z}}\cdot\frac{\partial \mathbf{\Phi}}{\partial \dot{\theta}}\bigg|_{\mathrm{E}} - \frac{1}{\rho k}\frac{\partial \varepsilon}{\partial \dot{a}}\bigg|_{\mathrm{E}}\frac{\partial f}{\partial \dot{\theta}}\bigg|_{\mathrm{E}} \leq \frac{\partial \varepsilon_{\mathrm{E}}}{\partial \theta},$$

which is in fact a restriction on the constitutive response of the internal energy and the frictional behaviour, and an extension of the results derived by [12]. If in addition $(\partial \lambda^{\varepsilon}/\partial \dot{\theta})\big|_{\mathrm{E}} < 0$ holds, the heat conduction equation is indeed of hyperbolic type, which is desirable since otherwise thermal pulses propagate with infinite speed.

Second, $d\eta$ and $d\phi$ have to be considered in more detail. These differentials arise in the generalized Gibbs equations (3.12) and it is a priori unclear whether these are *exact* differentials. This condition however guarantees that, *e.g.*, the entropy η is in fact a potential. In equilibrium, one finds

$$d\eta = \frac{1}{\theta}(d\varepsilon_{\mathrm{E}} - d\psi_{\mathrm{E}}),$$

where $d\varepsilon_{\mathrm{E}}$ and $d\psi_{\mathrm{E}}$ are both exact differentials and hence, so is $d\eta$. In general, the conditions of Poincaré and Frobenius provide the answer to the question whether $d\eta$ and $d\phi$ are exact, but in the present cases, the latter can be made exact (also in non-equilibrium) by imposing constraining conditions on the quantities $\mathcal{F}$, $\mathcal{P}$ and the Lagrange multipliers associated with the evolution of roughness (or abrasion, respectively) and internal energy.

Third and last, the quasi-static frictional behaviour of a dry, rough granular material has been treated by means of a fairly general evolution equation which is of hypoplastic type. To date, the existing hypoplastic constitutive models do not yet take abrasion or an explicit dependence of roughness into account. An extension of the existing hypoplastic model (see [6, 7] and, for a refined version, [20]) has been proposed by [16], where, for model (GC), the effects of an extended hypoplastic constitutive equation when applied to rough granular materials in simple flow configurations (such as, *e.g.*, the Poiseuille flow) have been investigated.

To summarize, we have in this paper presented four different modeling approaches to deal with an internal variable describing the roughness (and – in one case – abrasion, its time rate of change) of a dry granular single constituent continuum. The introduction of such an internal variable is regarded as a necessary amendment to the hypoplastic constitutive equations which account for internal frictional behaviour but ignore the effects of abrasion which are due to a mutual rubbing (and thus polishing) of the particles as a consequence of their differential motion. Localization phenomena such as the formation of shear bands may be triggered by such induced inhomogeneities, and it is believed that with the introduction of the internal variable roughness (and its variations in space and time) a "natural" mechanism has been included and coupled to the remaining governing equations which acts as an internal defect accounting for the mentioned instability phenomena.

The main results emerging from the exploitation of the entropy principle according to Müller and Liu have been presented and compared for all four modeling alternatives. Proofs have not been given here but can be found in [5]. On theoretical grounds, none of the proposed models involving a flux term (models (W), (V) and (GC)) can to date be favoured over any other, and numerical investigations as currently carried out by [16] may be a tool to gather more information as far as the possible acceptance or rejection of a particular model is concerned. In case of model (SH), it is suspected that due to the missing flux term, certain relevant pieces of information cannot be captured. The collinearity of the heat flux and the entropy flux may be seen as an "indicator" for this, since it is known that in many cases dealing with structured granular material the entropy flux has a contribution that is not collinear to the heat flux. However, with soil mechanics or geotechnical applications in mind, where numerical results are needed, none of the above presented models is rejected at this stage.

Acknowledgments: This work was supported by the German Research Foundation (DFG) through the special collaborative research project SFB 298 "Deformation and failure of metallic and granular media".

References

[1] B.D. COLEMAN, W. NOLL: The thermodynamics of elastic materials with heat conduction and viscosity, *Arch. Rational Mech. Anal.* **13**, pp. 167–178, 1963.

[2] M.A. GOODMAN, S.C. COWIN: A continuum theory for granular materials, *Arch. Rational Mech. Anal.* **44**, pp. 249–266, 1972.

[3] K. HUTTER: The foundations of thermodynamics, its basic postulates and implications. A review of modern thermodynamics, *Acta Mechanica* **27**, pp. 1–54, 1977.

[4] K. HUTTER, Y. WANG: Phenomenological thermodynamics and entropy principles, in: A. GREVEN, G. KELLER, G. WARNECKE (eds.): *Entropy*, to appear.

[5] N.P. KIRCHNER: *Thermodynamics of Structured Granular Materials*, Shaker Verlag, Aachen, 2001.

[6] D. KOLYMBAS: A rate-dependent constitutive equation for soils, *Mech. Research Comm.* **4**, pp. 367–372, 1977.

[7] D. KOLYMBAS: *Introduction to Hypoplasticity*, A.A. Balkema, Rotterdam, 1st–edition, 2000.

[8] I-S. LIU: Method of Lagrange multipliers for exploitation of the entropy principle, *Arch. Rational Mech. Anal.* **46**, pp. 131–148, 1972.

[9] I-S. LIU: On the entropy supply in a classical and a relativistic fluid, *Arch. Rational Mech. Anal.* **50**, pp. 111–117, 1973.

[10] J. MEIXNER: Zur Thermodynamik irreversible Prozesse (German) [Thermodynamics of irreversible processes], *Z. Physik. Chem.* **538**, pp. 235–263, 1943.

[11] I. MÜLLER: Die Kältefunktion, eine universelle Funktion in der Thermodynamik viskoser wärmeleitender Flüssigkeiten (German) [The coldness function, a universal function in thermodynamics of viscous, heat conducting fluids], *Arch. Rational Mech. Anal.* **40**, pp. 1–36, 1971.

[12] I. MÜLLER: *Thermodynamics*, Pitman, London, 1st–edition, 1985.

[13] W. MUSCHIK, C. PAPENFUSS, H. EHRENTRAUT: A Sketch of Continuum Thermodynamics, *J. Non-Newtonian Fluid Mech.* **96**, pp. 255–290, 2001.

[14] B. SVENDSEN, K. HUTTER: On the thermodynamics of a mixture of isotropic materials with constraints, *Int. J. Engng. Sciences* **33**, pp. 2021–2054, 1995.

[15] B. SVENDSEN, K. HUTTER, L. LALOUI: Constitutive models for granular materials including quasi-static frictional behaviour: toward a theory of plasticity, *Cont. Mech. Thermodyn.* **11**, pp. 263–275, 1999.

[16] A. TEUFEL: Simple flow configurations in hypoplastic abrasive materials. Diploma thesis, Institute of Mechanics III, Darmstadt University of Technology, Germany, 2001.

[17] A. VILCHINSKI: Personal communication, 2000.

[18] Y. WANG, K. HUTTER: Shearing flows in a Goodman–Cowin type granular material - Theory and numerical results, *J. of Particulate Materials* **17**, pp. 97–124, 1999.

[19] K. WILMAŃSKI: *Mechanics of Continuous Media*, Springer, Heidelberg-Berlin-New York, 1st–edition, 1999.

[20] P.A. VON WOLFFERSDORF: A hypoplastic relation for granular materials with a predefined limit state surface, *Mech. Coh.-Frict. Mat.* **1**, pp. 251–271, 1996.

NINA P. KIRCHNER and KOLUMBAN HUTTER
Institute of Mechanics III
Darmstadt University of Technology
Hochschulstrasse, 1
D-64289 Darmstadt, GERMANY
E-mail: Kirchner@mechanik.tu-darmstadt.de
         Hutter@mechanik.tu-darmstadt.de

Chapter 4

Modeling of Soil Behaviour: from Micro-Mechanical Analysis to Macroscopic Description

Roberto Nova

ABSTRACT The macroscopic behaviour of soil is influenced by its microscopic characteristics. Elementary considerations on friction and grain interlocking lead first to the formulation of an elastic plastic model with isotropic hardening or softening. Micromechanical considerations suggest also that the flow rule should be non-associative. As a consequence, unstable specimen responses, such as static liquefaction and shear banding are possible, even in the hardening regime. In order to model the behaviour in complex tests, an extended model taking induced anisotropy into account is formulated next. Further, we shall show how the time needed for rearranging the internal structure under loading influences the overall response of a specimen. The introduction of a time and a length scale in the macroscopic constitutive model will also help in regularising the numerical response in initial boundary value problems. Finally some features related to the description of soil behaviour at small strains and in unloading-reloading will be briefly discussed.

4.1   Introduction

From an engineering viewpoint, soils are aggregates of mineral grains resulting from the degradation, erosion, transport and sedimentation of the rocks constituting the earth crust. The characteristic size of such grains ranges from several millimeters to less than a micron. Depending on the type of parent rock and the degree of alteration, the grains are made of various minerals, such as quartz, feldspars, mica, carbonates and so on.

Despite such a large difference in size and mineral composition, however, the qualitative behaviour of most soil types is similar. For instance, all soil types behave differently when loaded for the first time in their geological life as they do in unloading-reloading. Furthermore, their behaviour is very sensitive to the level of the confining pressure, and volumetric strains have a remarkable effect on their strength and ductility. As a matter of fact, from an engineering viewpoint, the only macroscopic quantity which depends on the particle size in such a way that soil behaviour is affected not only quantitatively but even qualitatively, is the coefficient of permeability: coarse grained soils are free to drain the water in the pores when compressed, while for fine grained soil, drainage is difficult and consequent consolidation takes a long time.

Such a difference is not relevant in the long term, however: when the consolidation is over, a fine grained soil such as clay does not behave differently, qualitatively, from a coarse grained gravel. Indeed, with the exception of carbonate soils, made of fragile, collapsible, grains, under usual working levels all soils can be considered, as a first approximation, as aggregates of quasi-rigid, rough, particles.

The mechanical properties of the particle surface and the geometry of the assembly have a profound influence on the overall behaviour of a soil specimen. The global strength and stiffness parameters of the soil are affected by the friction and adhesion between the particles influence, as well as by the density and the degree of interlocking of the grains. A qualitative micro-mechanical analysis can be therefore useful for the understanding of soil behaviour. In the following, such an analysis will allow the important phenomena occurring within a soil specimen to be highlighted and we shall use it as a guideline for the formulation of a macroscopic model of soil behaviour.

Elementary considerations on friction and grain interlocking will lead first to the formulation of an elastic plastic model with isotropic hardening or softening. A key role will be played in this respect by the experimental relationship between stress obliquity and dilatancy and by the hardening rule relating the variation of the size of the elastic domain with plastic strains. The micro-mechanical analysis suggests further that the dilation angle is less than the friction angle. The macroscopic counterpart of this phenomenon is the non-associativeness of the plastic flow rule. The experimental data illustrating this feature will be presented and discussed. Non-associativeness of the flow rule has far-reaching consequences. It will be shown in fact that unstable specimen responses are possible, even in the hardening regime. Special cases of such instabilities are the so-called static liquefaction phenomenon in undrained tests on sands and the occurrence of shear banding in plane strain tests.

Although by means of such a basic model it is possible to model the behaviour of soils in many simple tests with a sufficient degree of accuracy, when more complex tests are considered, usually involving unloading and reloading in different directions in the stress space, predictions are far from

being satisfactory. The micro-mechanical analysis will lead us to formulate
an extended model which takes induced anisotropy into account. Further,
we shall show how the time needed for rearranging the internal structure
under loading influences the overall response of a specimen. The introduction of a time scale in the macroscopic constitutive model will also help
in regularising the numerical response in initial boundary value problems.
The issue of the regularisation of the numerical solution is very important
indeed. Whenever bifurcations of the response are possible, as is the case
for soils, either for the non-associativeness of the flow rule or for the dilation
induced softening or for the collapsibility of the grain skeleton, in order to
regularise the solution and to make it objective, it is necessary to introduce a length scale in the macroscopic description. The micro-mechanical
analysis suggests that a convenient way of doing that is to use non-local
relationships. It will be shown that in this way the previously conceived
constitutive model can overcome the difficulties intimately connected with
such unstable responses.

Finally the problems related to the description of soil behaviour at small
strains and in unloading-reloading will be briefly discussed. Once again a
simple micro-mechanical model will be useful for defining the appropriate
way of tackling the problem.

4.2   Elementary considerations

Consider an assembly of quasi-rigid spheroidal grains confined by six rigid
platens, which can be displaced, laterally or up and down one against the
opposite one, at constant speed ("true triaxial apparatus"). Such a displacement is contrasted by the grain assembly and a system of force chains
is created within the specimen.

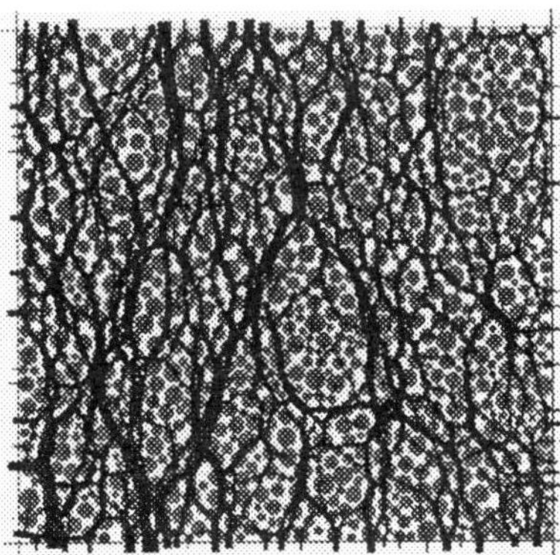

**FIGURE 4.1. Pattern of force chains in a Schneebeli material (after
Calvetti [2]).**

A typical picture of such force chains is depicted in Figure 4.1 (after Calvetti [2]). For the sake of simplicity, here and in the following, a material composed of long rods, with the axis orthogonal to the plane of the figure, is considered. Such a two-dimensional medium, called Schneebeli material (Schneebeli [57]), gives a schematic picture of a section of an actual granular material and is very helpful in illustrating qualitatively the basic mechanisms governing the behaviour of the latter.

The force transmitted at the contact between two grains can be decomposed in two components, one, N, orthogonal and the other one, T, tangential to the plane tangent to both grains at their contact point. Such a contact remains active until the tangential force reaches a limit value, which, in absence of bonds and according to Amonton's law, is directly proportional to the modulus of the normal force:

$$T = N \tan \phi_\mu \,. \tag{4.1}$$

The constant ϕ_μ is known as the grain friction angle. It depends on the grain properties such as surface roughness and mineral hardness.

When equation (4.1) is fulfilled, one grain can slide over the other one, so breaking the contact. The force chain is consequently abruptly interrupted. Since equilibrium must be guaranteed at any time, the load that was carried by the broken chain is instantaneously redistributed to the other force chains, possibly with creation of one or more new ones, made possible by the rearrangement of the grains and the generation of new contacts.

Together with sliding, another possible deformation mechanism exists. Actual grains are not perfectly rigid and rounded so that the contact is not perfectly pointwise. Two grains in contact may therefore exchange also a couple, $\mathcal{M}$. A grain may then rotate with respect to the other. This can occur when the sum of the couples and moments of the forces acting at the other grain contacts with respect to a particular contact reach a critical value.

The larger the size of the grain the larger the lever arm and therefore the easier the achievement of such value. Since the contacts are usually rather flat, such a critical value is rather high, however, and only under particular loading conditions and in localised regions of the specimen grain rotation is the relevant deformation mechanism, as will be clarified later on.

When grains are displaced, they tend to occupy the void space. However, in so doing, new voids are created. The final balance between filled and emptied space may be positive (compression) or negative (dilation), depending on the imposed displacements at the boundaries. The sum of the reactions transmitted by the grains to each platen divided by the platen area can be interpreted as a stress vector acting on a face of a fictitious solid, which occupies entirely the volume delimited by the platens. By knowing the modulus and direction of the stress vector on three mutually orthogonal faces, a fictitious stress tensor can be constructed, which is assumed to act homogeneously within that volume. Similarly, by dividing the relative displacements of two opposite faces by their distance, it is possible to define three principal strains. A fictitious strain tensor can be constructed therefore, equally assumed to characterise homogeneously the strain state of the specimen, that appears as a representative elementary volume of a fictitious solid continuum.

A stress-strain relationship can be established therefore for such a continuum starting from convenient averages operated on forces and displacements acting at the boundaries of the grain assembly. The characteristics of such a relationship will clearly reflect the basic properties which control the deformation pattern of the granular assembly.

4.3   Behaviour in proportional compression tests

Consider first a test in which all platens are displaced at the same speed, isotropically compressing the specimen. The grains are pushed one against the other. If the grains are of spheroidal shape, the average stress state which is generated is also isotropic.

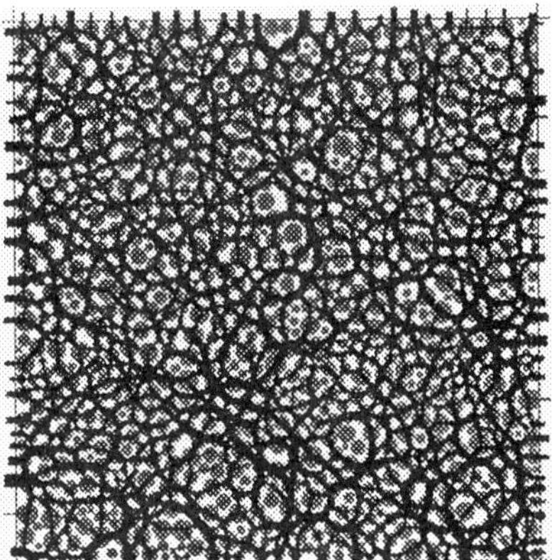

FIGURE 4.2. Pattern of force chains under isotropic compression (after Calvetti [2]).

However since grain displacement is transmitted via the force chains, which are not isotropic, Figure 4.2 after Calvetti [2], some grains can slide anyway over each other giving rise to irreversible changes of the array. On the whole, therefore, irreversible volumetric strains are generated under isotropic loading conditions. If the grains are rigid, the reactions exerted by the grain to an imposed displacement will grow faster than proportionally, since the more compact the specimen the more difficult it is for the grains to change their position. A convenient way for relating volumetric strains to isotropic pressures is therefore (Butterfield [1])

$$\ln \frac{V}{V_0} = -B \ln \frac{p'}{p_0} \,, \tag{4.2}$$

where V is the volume of the specimen, p' the average effective isotropic pressure, B a material constant, while V_0 and p_0 are the reference volume and pressure, respectively. Note that in this case effective and total external pressures coincide, since the specimen is assumed to be dry. For a saturated specimen, we shall assume the validity of the effective stress principle (Terzaghi [62]), so that equation (4.2) holds true anyway.

For small volumetric strains, v, taken positive in compression, equation (4.2) can be also written in rate terms as

$$\dot{v} = -B \frac{\dot{p}'}{p'} \,. \tag{4.3}$$

By unloading the specimen, only part of the volumetric strain is recovered. Assume that the recoverable rate is

$$\dot{v}^e = -B^e \frac{\dot{p}'}{p'} \,. \tag{4.4}$$

The irreversible part will then be

$$\dot{v}^p = (B - B^e) \frac{\dot{p}'}{p'} = B^p \frac{\dot{p}'}{p'} \,; \tag{4.5}$$

B^e and B^p are material parameters and will be referred to as elastic and plastic logarithmic volumetric compressibility, respectively.

Consider now a similar test in which the displacement rates in the three directions are proportional to each other, but not equal. For the sake of simplicity, we shall assume that only the vertical one is different from the other two, so that the test conditions are axisymmetric and the vertical axis is the axis of symmetry.

The qualitative trend of the isotropic stress-strain relationship is similar, but the isotropic pressures generated by the same volumetric strains are smaller and smaller as the ratio between the vertical strain rate and the horizontal ones is larger and larger. As a counterpart, the vertical stress increases more than the horizontal ones. The force chains are in fact predominantly vertical, see again Figure 4.1 [2]. A deviator stress q can be defined as

$$q \equiv \sigma_v - \sigma_h \tag{4.6}$$

and the stress ratio, η, as

$$\eta \equiv \frac{q}{p'} \,. \tag{4.7}$$

Although the stress chains change continuously as the deformation proceeds, the pattern remains qualitatively similar. As a consequence, the stress ratio remains approximately constant too, as the stresses increase. It is possible therefore to define a relation between η and the dilatancy d defined as

$$d \equiv \frac{dv^p}{d\epsilon^p}\,, \tag{4.8}$$

where the plastic deviatoric strain $d\epsilon^p$ is given by

$$d\epsilon^p \equiv \frac{2}{3}(d\epsilon_v^p - d\epsilon_h^p)\,. \tag{4.9}$$

On the basis of experimental data, various expressions for such a relation, often called improperly stress-dilatancy relationship, have been proposed (see for instance Poorooshasb *et al.* [48, 49] and Stroud [59]). Recently Lagioia *et al.* [28] suggested the following one:

$$d = \frac{M - \eta}{\mu}\,(\frac{\alpha M}{\eta} + 1)\,, \tag{4.10}$$

which can be reduced as a special case ($\alpha = 0$) to the relationship proposed by Nova and Wood [46]. The expression used in the original Cam–Clay (Schofield and Wroth [58]) model can also be obtained putting $\alpha = 0$ and $\mu = 1$. A particular value of η is M, defined as that value for which no plastic volumetric strains take place. When a specimen of dense sand is sheared in such conditions, the deviatoric plastic strains vary in a way which is similar to that of the volumetric strains of equation (4.5) (Nova [36]), that is

$$\dot{\epsilon}^p = \frac{B^p}{D}\frac{\dot{p'}}{p'} \tag{4.11}$$

where D is another material constant. If D is positive, it is possible to shear sand at stress ratios even larger than M, at least for moderate p' values, so that plastic volumetric strain become negative (dilation) (El Sohby [17], Goldscheider [18]).

Note however that for very loose sand and normally consolidated clay, M is a stress ratio value that can be reached only asymptotically. For these materials D can be taken equal to 0.

4.4   A simple elasto-plastic strain-hardening model

The results discussed so far allow a simple elasto-plastic strain-hardening model to be defined. A micro-mechanical analysis can again guide our modelling (Calvetti [3]). Imagine, in fact, to load a specimen of a Schneebeli material from the stress free state to a given stress state. Imagine now to perturb such a state by means of small probes with the same intensity but

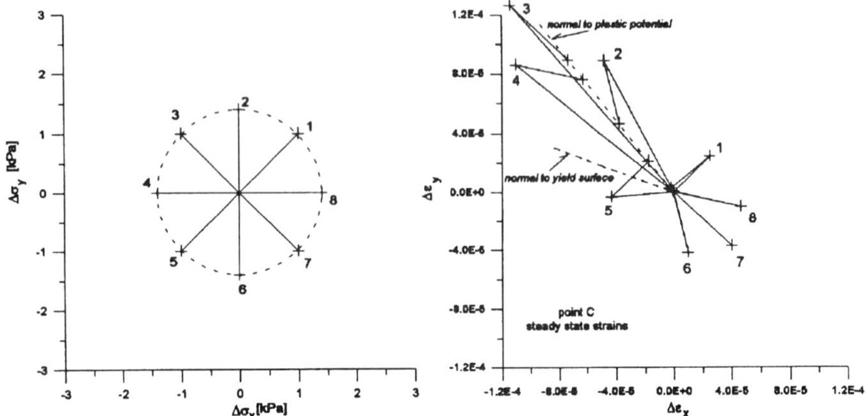

FIGURE 4.3. Strain responses of Schneebeli material to stress perturbations of a given stress state (after Calvetti [3]).

different directions and to remove them. The strain responses are illustrated in Figure 4.3. It is apparent that for a number of probes the strain upon loading-unloading is almost fully recoverable. For other probes, instead, a permanent strain remains at the end of the cycle. Although the modulus of such a strain depends on the direction of the probe, its direction is constant. The plastic strain increment vector can be therefore expressed as

$$d\epsilon^p = \Lambda\, m = \Lambda\, \mathrm{grad}\, g\,, \qquad (4.12)$$

where Λ is a non-negative scalar, m a unit vector and g is a function of the stress state which is called, by definition, plastic potential.

A vector n can also be defined as the unit vector orthogonal to the direction which delineates between probes causing plastic strains and probes causing only recoverable strains. We can define a loading function f such that

$$n = \mathrm{grad}\, f\,. \qquad (4.13)$$

Also f is a function of the stress state and of the previous loading history of the specimen. If in fact the specimen is loaded and then moderately unloaded prior to probing, only recoverable strains take place, independently from the probe direction.

Assume thus that there exists such a function $f(q, p, p_c)$ which can take only negative values or zero. The parameter p_c is a hidden variable, function of the plastic history of the material:

$$p_c = p_c(v^p, \epsilon^p)\,, \qquad (4.14)$$

whose physical meaning will be clarified later on. This relationship will be called the *hardening rule*.

The loading function f delineates between states for which plastic strains are possible and states for which only recoverable strains take place. Plastic strains can occur only when both the following conditions are fulfilled:

$$\begin{cases} f = 0 \,, \\ df = 0 \,. \end{cases} \tag{4.15}$$

From equation (4.14), the second of equations (4.15) and the definition of loading function, we get that, when plastic strain occurs:

$$\frac{\partial f}{\partial p'} dp' + \frac{\partial f}{\partial q} dq + \frac{\partial f}{\partial p_c} \left(\frac{\partial p_c}{\partial v^p} dv^p + \frac{\partial p_c}{\partial \epsilon^p} d\epsilon^p \right) = 0 \,. \tag{4.16}$$

Assume further that there exists a plastic potential such that

$$\left\{ \begin{array}{c} dv^p \\ d\epsilon^p \end{array} \right\} = \Lambda \operatorname{grad} g = \Lambda \left\{ \begin{array}{c} \dfrac{\partial g}{\partial p'} \\ \dfrac{\partial g}{\partial q} \end{array} \right\} \,. \tag{4.17}$$

By substituting equation (4.17) in equation (4.16) and solving for Λ, we get

$$\Lambda = -\frac{\dfrac{\partial f}{\partial p'} dp' + \dfrac{\partial f}{\partial q} dq}{\dfrac{\partial f}{\partial p_c} \left(\dfrac{\partial p_c}{\partial v^p} \dfrac{\partial g}{\partial p'} + \dfrac{\partial p_c}{\partial \epsilon^p} \dfrac{\partial g}{\partial q} \right)} \,. \tag{4.18}$$

Plastic strain rates can be determined therefore once loading function, plastic potential and hardening rule are specified.

It is readily apparent from equation (4.17) that the expression of the plastic potential in the p', q plane can be determined from the direct knowledge of the stress dilatancy relationship. It is in fact

$$dg = \frac{\partial g}{\partial p'} dp' + \frac{\partial g}{\partial q} dq = 0 \,, \tag{4.19}$$

thus

$$\frac{dq}{dp'} = -d = -\frac{Mp' - q}{\mu p'} \left[1 - \left(\frac{\alpha Mp'}{q} + 1 \right) \right] \tag{4.20}$$

that can be easily integrated. For instance, if $\alpha = 0$

$$g = q - \frac{Mp'}{1 - \mu} \left[1 - \left(\frac{p'}{p_g} \right)^{\frac{1-\mu}{\mu}} \right] = 0 \tag{4.21}$$

(Nova and Wood [46]), which coincides with the original expression of the Cam–Clay model in the limit for $\mu = 1$. As for any other potential, the parameter p_g is a dummy parameter, since only the derivatives of g are relevant.

The plastic potential expression could be used in principle as loading function, as well. Indeed, this was the original assumption of the earliest elasto-plastic models of soil behaviour (Schofield and Wroth [58], Roscoe and Burland [55]).

This assumption is in contrast with the micro-mechanical analysis and with the experimental data on actual granular media. For instance, Tatsuoka and Ishihara [61] conceived an experimental procedure for the determination of the yield locus in the p', q plane which is the locus for which the value of the loading function is nil. In such a procedure, a sand specimen is loaded up to a certain point, unloaded, $e.g.$, isotropically, and reloaded following a different path, $e.g.$, at constant horizontal stress. During such reloading phase a rather abrupt change of stiffness occurs. This is associated to yielding and the loading function can be determined by interpolating the yield points.

The size of the region delimited by the yield locus increases linearly with the level of the preloading isotropic pressure, while its shape remains approximately the same. This property is also shared by the expression of the plastic potential (equation 4.21), but the yield locus has a similar but not identical shape.

For instance, while the plastic potential has a horizontal tangent when $\eta = M$, the yield locus has a horizontal tangent when $\eta = M_f < M$. A convenient way for expressing the loading function is therefore to take the same formal expression of the plastic potential, but with different constitutive parameters. For instance

$$f = q - \frac{M_f p'}{1 - \mu_f} \left[1 - \left(\frac{p'}{p_c} \right)^{\frac{1 - \mu_f}{\mu_f}} \right] = 0 , \qquad (4.22)$$

where M_f and μ_f are material constants. The parameter p_c instead controls the size of the elastic domain delimited by the yield locus. It is now clear that its physical meaning is that of an isotropic preconsolidation pressure.

An expression of the plastic potential which is closer to experimental data and respects the symmetry condition of orthogonality to the p' axis (no deviatoric strains under isotropic loading) is obtained by taking $\alpha \neq 0$. In a similar way, an analogous parameter α_f can be defined. (Lagioia et $al.$ [28]).

In order to obtain Λ in equation (4.18) for a specified stress increment, it is only necessary at this point to specify the hardening rule and in particular the partial derivatives of p_c with respect to v^p and ϵ^p. It is easy to see, however, that such expressions are given by equations (4.5) and (4.11), respectively. In the former one in fact, p' coincides with p_c and varies with the plastic volumetric strain at constant (zero) plastic deviatoric strain. In the latter one, p' is proportional to p_c, since the shape of the yield locus is constant, and varies with the plastic deviatoric strains at constant plastic volumetric strain, since $d = 0$ (Nova [36]).

Finally, total strain rates can be determined by adding to plastic strain rates the elastic ones. Elastic volumetric strain rates are assumed to be given by equation (4.3) for any stress increment, while elastic deviatoric strain rates are taken as

$$\dot{\varepsilon} = \frac{\dot{q}}{3G} = \frac{\dot{q}}{3\lambda p'} \,, \tag{4.23}$$

where G is the shear modulus, assumed to be proportional to the isotropic pressure, through the material constant λ.

This hypothesis is in agreement with the experimental data obtained on normally consolidated clay and loose sand. An elastic potential cannot be defined, however, so that energy can be extracted from the specimen if convenient closed stress cycles are performed (Zytynski *et al.* [66]). The behaviour in unloading reloading is by no means elastic, in fact, and more complex descriptions are necessary if we want to accurately describe the material behaviour within the so-called "elastic" domain.

For the time being, however, we shall assume the validity of equation (4.3) (hypoelastic behaviour), limiting its use either to monotonic loading or to conditions when at most one unloading reloading cycle is performed. We shall discuss later the problem of the description of the behaviour in the small strain range.

Although very simple, a model like this can describe well monotonic tests on sand and normally consolidated clay. As an example Figures 4.4, 4.5, 4.6 show the comparison between calculated and observed results in various types of axisymmetric tests on different materials. (Nova [37], Nova [38], Nova and Wood [46]).

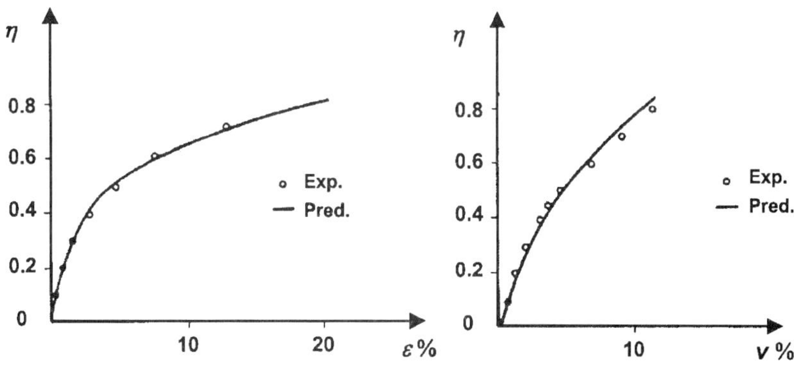

FIGURE 4.4. Simulation of a drained constant cell pressure test on a normally consolidated kaolin: data after Walker [63]; after Nova [46].

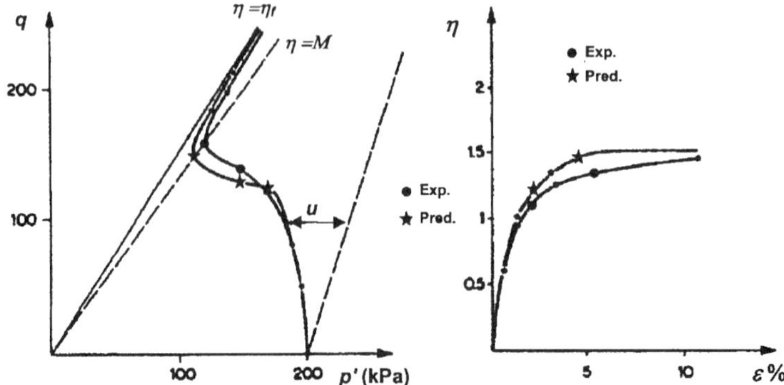

FIGURE 4.5. Simulation of an undrained test on Fuji River sand: data after Tatsuoka [60]; after Nova and Wood [46].

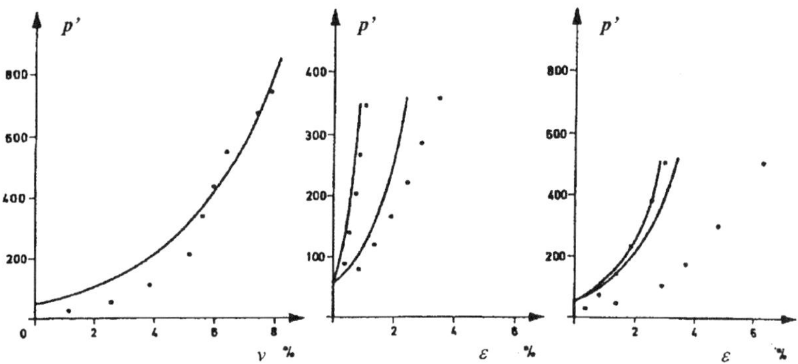

FIGURE 4.6. Simulation of consolidation tests on Karlsruhe sand a) oedometric, b) proportional deformation in compression, c) proportional deformation in extension: data after Goldsheider [18]; after Nova [38].

4.5 Derivation of the failure condition

In axisymmetric tests, the constitutive law in rate terms can be written as

$$\left\{ \begin{array}{c} \dot{v} \\ \dot{\varepsilon} \end{array} \right\} = \left[\begin{array}{cc} C_{pp} & C_{pq} \\ C_{qp} & C_{qq} \end{array} \right] \left\{ \begin{array}{c} \dot{p}' \\ \dot{q} \end{array} \right\}. \tag{4.24}$$

The parameters C_{ij} of the compliance matrix C of equation (4.24) depend on the state of stress, on the previous history of the specimen and on the stress increment direction. In what follows, we shall assume that the point representing the state of stress is always on the yield locus and that the stress increment is directed in such a way to produce plastic strains (virgin loading). From equations (4.3), (4.15), (4.16) and (4.23), equation (4.24) can be thus written as

$$\left\{ \begin{array}{c} \dot{v} \\ \dot{\varepsilon} \end{array} \right\} = \left[\begin{array}{cc} \dfrac{B^p}{p'} + \dfrac{1}{H}\dfrac{\partial g}{\partial p'}\dfrac{\partial f}{\partial p'} & \dfrac{1}{H}\dfrac{\partial g}{\partial p'}\dfrac{\partial f}{\partial q} \\[2mm] \dfrac{1}{H}\dfrac{\partial g}{\partial q}\dfrac{\partial f}{\partial p'} & \dfrac{1}{3\lambda p'} + \dfrac{1}{H}\dfrac{\partial g}{\partial q}\dfrac{\partial f}{\partial q} \end{array} \right] \left\{ \begin{array}{c} \dot{p}' \\ \dot{q} \end{array} \right\}, \qquad (4.25)$$

where the hardening modulus, H, is given by

$$H \equiv -\frac{\partial f}{\partial p_c}\left(\frac{\partial p_c}{\partial v^p}\frac{\partial g}{\partial p'} + \frac{\partial p_c}{\partial \epsilon^p}\frac{\partial g}{\partial q} \right) = h\,p'. \qquad (4.26)$$

By making use of the vectors m and n, equation (4.25) can be rearranged as

$$\left\{ \begin{array}{c} \dot{v} \\ \dot{\varepsilon} \end{array} \right\} = \frac{1}{H}\left[\begin{array}{cc} B^p h + m_p n_p & m_p n_q \\[2mm] m_q n_p & \dfrac{h}{3\lambda} + m_q n_q \end{array} \right] \left\{ \begin{array}{c} \dot{p}' \\ \dot{q} \end{array} \right\}. \qquad (4.27)$$

Consider first the determinant of the constitutive matrix, C:

$$\det C = \frac{1}{hp'^2}\left(\frac{B^p h}{3\lambda} + B^p m_q n_q + \frac{1}{3\lambda}m_p n_p \right). \qquad (4.28)$$

With ordinary values of the material constants, the quantity between parentheses is positive. The determinant therefore can be positive or negative, depending on the value of h, but not zero. In this case, in fact, for a fully displacement controlled test, either infinitely many stress increments would be associated to given strain increments or no solution of equation (4.25) would exist. This would mean that a total loss of control of the test would occur, which is not realistic, since the boundary displacements are under control, by hypothesis.

Consider now the stiffness matrix, S, *i.e.*, the inverse of the matrix of equation (4.27).

$$\left\{ \begin{array}{c} \dot{p}' \\ \dot{q} \end{array} \right\} = \qquad (4.29)$$

$$\frac{p'}{\dfrac{B^p h}{3\lambda} + B^p m_q n_q + \dfrac{1}{3\lambda}m_p n_p}\left[\begin{array}{cc} \dfrac{h}{3\lambda} + m_q n_q & -m_p n_q \\[2mm] -m_q n_p & B^p h + m_p n_p \end{array} \right] \left\{ \begin{array}{c} \dot{v} \\ \dot{\varepsilon} \end{array} \right\}.$$

The determinant of the stiffness matrix is given by

$$\det S = \frac{hp'^2}{\dfrac{B^p h}{3\lambda} + B^p m_q n_q + \dfrac{1}{3\lambda}m_p n_p} \qquad (4.30)$$

and becomes zero when the hardening modulus is zero. When this occurs therefore, in a load controlled test, which, when strains are small, practically coincides with a stress controlled test, arbitrarily large strains occur even under constant stresses. In particular

$$\dot{v} = \frac{m_p}{m_q}\dot{\varepsilon}\,, \tag{4.31}$$

as can be deduced from equation (4.29). This situation is usually associated to the concept of failure. The constitutive model adopted predicts therefore the failure condition

$$h = 0\,. \tag{4.32}$$

It is clear in fact from Section 4.4, that the existence of such a condition was not explicitly imposed. All the test data that were used for establishing the constitutive relationships were derived from proportional tests, which do not lead a specimen to failure.

We can further derive from equation (4.26), (4.5) and (4.11) that equation (4.32) is fulfilled when the dilatancy d is equal to

$$d = -D\,; \tag{4.33}$$

the material constant D is therefore associated to the value of the dilatancy at failure.

Furthermore, since from the stress-dilatancy relationship it is apparent that the dilatancy is uniquely associated to a value of the stress ratio, equation (4.33) implies that the failure condition in the p', q, plane is simply given by a particular value of the stress ratio, *i.e.*, it is a straight line passing through the origin.

We have then derived the Mohr–Coulomb failure condition, as a consequence of our hypothesis concerning the stress-dilatancy relationship and the hardening rule, not as an independent assumption.

In particular from equation (4.10) and α tending to zero, the limit condition is given by

$$q = (M + \mu D)p'\,. \tag{4.34}$$

Equation (4.34) can be compared with equation (4.1). Shear stresses at failure are proportional to the stresses that take the grains grouped together as the shear force in equation (4.1) is proportional to the normal force keeping two grains in contact. The proportionality constant of equation (4.32), which plays the role of the friction angle in equation (4.1), is the sum of two terms, a frictional constant M and a term which depends on the dilatancy. The extra strength of the grain assembly with respect to pure friction is therefore due to a geometric component. This depends on the interlocking of the grains and the consequent difficulty of moving each grain apart. When grains are unlocked, new voids are created and dilatancy occurs. Dilatancy at failure and strength of the assembly are then intimately connected. A dense specimen will dilate more than a looser one and is consequently stronger, although the grain to grain friction is the same.

4.6   Non-normality and material instabilities

The constitutive matrix of equation (4.27) is not symmetric. This is a direct consequence of the fact that $g \neq f$ and therefore $m \neq n$. At variance with classical plasticity, therefore, the plastic strain increment is not orthogonal to the yield locus.

This fact has profound consequences on the possible occurrence of material instabilities, even in the hardening regime ($h > 0$). If the stiffness matrix would be symmetric, in the hardening regime it would also be positive definite. Its determinant would be the least between the minors of the matrix for any positive value of h. This implies that when the determinant becomes zero (for $h = 0$) all the minors are either positive or become zero at the same moment.

If, however, the stiffness matrix is not symmetric, a minor can become zero when the determinant is still positive, *i.e.*, in the hardening regime. For instance, if $m \neq n$, the term $m_p n_p$ can be negative. When

$$h = -\frac{m_p n_p}{B_p} > 0 \qquad (4.35)$$

a minor of the stiffness matrix is zero. Consider now an isochoric test (*i.e.*, undrained test, if the specimen is fully saturated with an incompressible fluid, *e.g.*, water). From equation (4.27) we derive that when equation (4.35) is fulfilled, in a fully strain controlled test a peak in the deviator stress occurs. If therefore we perform are undrained test in which partly stresses and partly strains are controlled, as we do, for instance, by closing the drainage and monotonically increasing the axial load, when equation (4.35) is fulfilled, the control of the test is lost.

This can also be readily seen mathematically, by rewriting equation (4.27) in such a way that the control parameters are both put at the left-hand side. We have:

$$\left\{ \begin{array}{c} \dot{v} \\ \dot{q} \\ A \end{array} \right\} = \qquad (4.36)$$

$$\left[\begin{array}{cc} 1 & m_p n_q \\ -m_q n_p & \left(\dfrac{h}{3\lambda} + m_q n_q\right)(B^p h + m_p n_p) - m_p n_q m_q n_p \end{array} \right] \left\{ \begin{array}{c} \dot{p'} \\ A \\ \dot{\varepsilon} \end{array} \right\},$$

where

$$A \equiv \frac{\det S}{hp'}. \qquad (4.37)$$

It is readily apparent that, when equation (4.35)) is fulfilled, the determinant of the matrix of equation (4.36) is zero. Therefore, even under no change of the left-hand side quantities, the right-hand quantities can increase indefinitely. A loss of control of the loading procedure takes place.

In particular, since the isotropic pressure imposed externally (total isotropic pressure) does not change, an unlimited variation (decrease) of p' (the effective isotropic pressure) implies an opposite variation (increase) of the pore water pressure, leading to sample liquefaction.

The phenomenon of static liquefaction just described can be therefore seen as a material instability (Nova [40]). Small perturbations of the controlling parameters are in fact associated to large variations of the controlled quantities.

Since h is positive when this occurs, it means that static liquefaction occurs in the hardening regime, *i.e.*, before ordinary failure. Moreover, since the shape of the plastic potential is independent of the preconsolidation pressure, condition (4.35) in the plane p', q is a straight line passing through the origin. Its inclination is roughly one half that of the Coulomb failure line. Such a line, experimentally determined by Kramer and Seed [25], was called instability line by Lade [26].

It is finally worth mentioning that if normality holds the h value for which the minor is zero is negative. Therefore static liquefaction cannot occur in the hardening regime. In other words, non-normality is a necessary condition to describe this phenomenon, which occurs when loose sands are tested in axisymmetric undrained conditions (Castro [6]).

Of course, the undrained test is only a particular type of strain controlled test. We could conceive other tests, in which, for instance, the ratio between the volumetric and the deviatoric strain rates is constant

$$\frac{\dot{v}}{\dot{\varepsilon}} = \beta . \tag{4.38}$$

With varying β, different stress paths are obtained, some of them showing a peak in the deviator stress. From equation (4.27) this occurs for

$$h = (\beta m_q - m_p)\frac{n_p}{B^p} . \tag{4.39}$$

This is not an instability level, however. It is not possible in fact to control at the same time the variation of q (the axial force) and impose a kinematic condition such as that given by equation (4.35). The control variables can be either stresses or strains or a combination of stresses or a combination of strains, but it is necessary that controlling and response quantities correspond to each other in a work density equation. For instance, if we decide to control $\dot{\varepsilon}$ and $\dot{\chi}$ such that

$$\dot{\chi} \equiv \dot{v} - \beta\dot{\varepsilon} , \tag{4.40}$$

the corresponding stress variables are

$$\dot{\xi} \equiv \beta\dot{p'} + \dot{q} \tag{4.41}$$

and $\dot{p'}$. Under the kinematic condition (4.38) ($\dot{\chi} = 0$), an infinity of possible responses in terms of $\dot{\varepsilon}$ and $\dot{p'}$ will be possible if a peak in ξ is reached, this being the second controlling quantity.

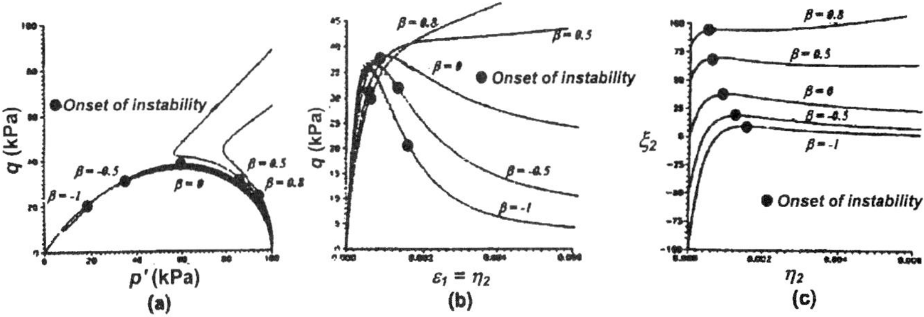

FIGURE 4.7. Rectilinear axisymmetric strain tests in which the volumetric strain rate is proportional to the axial one: a) stress path, b) axial stress strain law, c) stress strain law in generalised variables showing instability at peak.

Figure 4.7 (after Nova and Imposimato [44]) shows the results in a similar test in which the volumetric strain is proportional to the axial one. The critical value of h can be calculated as

$$h = \frac{-\beta^2 m_q m_q + \beta(m_q n_p + m_p n_q) - m_p n_p}{B^p + \dfrac{\beta^2}{3\lambda}}, \qquad (4.42)$$

which coincides with equation (4.35) for $\beta = 0$. The value of h for which such an instability occurs is clearly a function of β. It can be shown that its maximum value (corresponding to the lowest stress ratio) is given by

$$h_{max} = \frac{(m_p n_q + m_q n_p)^2}{4 B^p m_q n_q + \dfrac{(m_p n_q + m_q n_p)^2}{3\lambda m_q n_q}}. \qquad (4.43)$$

This coincides with the value for which the determinant of the symmetric part of the stiffness matrix is zero. Above this h value no type of instability is possible, whatever loading programme is followed. It was shown in fact by Nova [40, 43] that the positive definiteness of the symmetric part of the constitutive matrix is a necessary and sufficient condition for stability under any arbitrary loading programme, in which either partly stresses and partly strains are controlled or even linear combinations of them (for instance deviatoric stress and volumetric strain).

Since $\beta = 0$ is a particular case, in general, the lowest stress ratio associated to instability is less than that associated to equation (4.35), *i.e.*, to the so-called instability line of undrained tests.

A general instability line can be determined therefore in the p', q plane. Such a line corresponds to the locus of points in axisymmetric tests for which there exists a particular load increment for which the second-order work is zero (the eigenvector of the symmetric part of the stiffness matrix).

At variance with the "undrained" instability line existing for loose sands only, the general instability line is linked to the non-symmetry of the stiffness matrix and exists for any type of material (loose and dense sand, clay, silt, etc.) whose flow rule is non-associated. The Ostrowsky–Taussky [47] theorem guarantees in fact that when the determinant of the symmetric part of a matrix is zero, the determinant of the entire matrix cannot be negative. The loss of positive definiteness of the stiffness matrix, and the possibility of unstable responses, occurs therefore in the hardening regime.

Such macroscopic instabilities have a microscopic counterpart. We have seen that the force chains continuously change as the deformation proceeds. Each change of pattern is associated to a sudden collapse of one particular chain and the appearance of one, or more, new chains. The material "seeks" a more stable configuration.

If the boundary displacements are fully controlled, the test can be controlled from zero to very large displacements. The corresponding external forces change in a complex way, dictated by the deformation pattern. If on the contrary external displacements are only partly controlled and we try to enforce a loading programme, as for instance that given by equation (4.38), it may occur that at a certain loading level the force chains cannot rearrange themselves in such a way to satisfy at the same time equation (4.38) and equation (4.35).

In the attempt, force chains are continuously destroyed and created anew leading progressively to very large strains. A stable configuration is never achieved, therefore, under this loading programme.

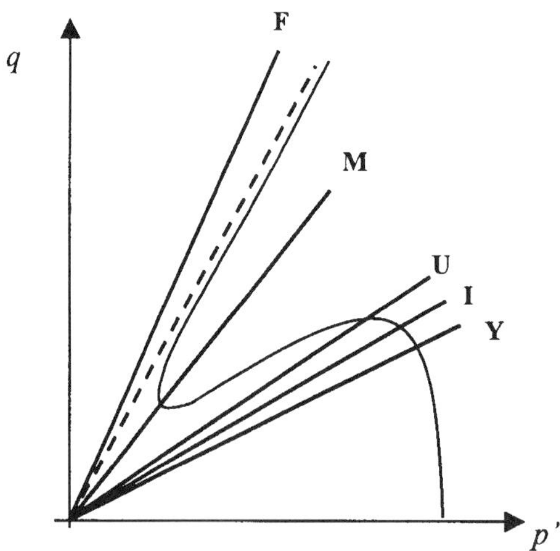

FIGURE 4.8. Calculated characteristic lines: F = failure line, M = phase transformation line, U = undrained instability line, I = general instability line, Y = locus of q maxima on yeld loci. The curve shows a typical undrained stress path for a medium dense sand.

Figure 4.8 shows the failure line (F), the undrained instability line (V), and the general instability line (I) for a medium loose sand.

For completeness, two other straight lines are plotted. The less inclined (line Y) is the locus of the maxima of q for the yield loci, whose size grows linearly with the preconsolidation pressure p_c. The other line (M) is the line at zero dilatancy, $i.e.$, from equation (4.10) the line for which $\eta = M$. Such a line corresponds to the so-called phase transformation line (Ishihara et $al.$ [22]) in undrained tests, as can be deduced from equations (4.4), (4.8), (4.10) and the conditions of no volume change. At this stress level the undrained effective stress path direction changes from right to left to left to right.

4.7   Three-dimensional loading conditions

The extension of the strain hardening model to 3D loading conditions can be easily performed by taking some material parameters to be functions of the Lode angle of stress, ($e.g.$, Nova [38], Lagioia et $al.$ [28]).

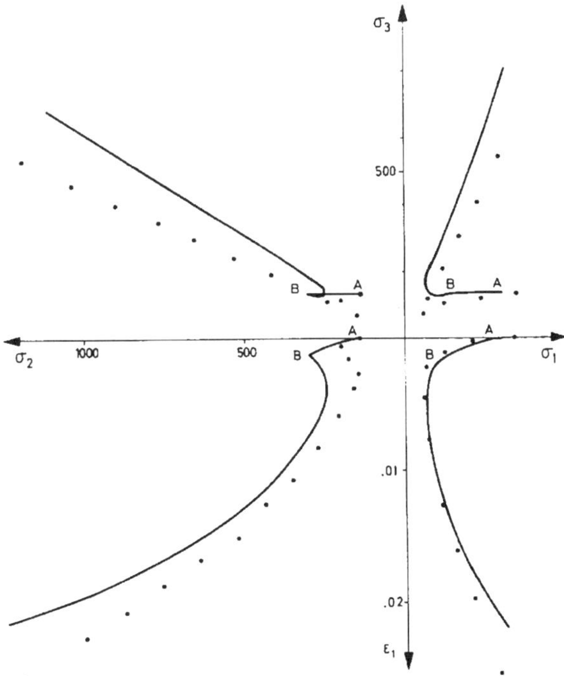

FIGURE 4.9. Prediction and experimental results for undrained plane strain test with rotation of axes of principal stress on Karlsruhe sand after Nova [38].

Alternatively, the loading function and the plastic potential are directly expressed in terms of the third invariant of the stress deviator, *e.g.*, in the model called "Sinfonietta classica" (Nova [39]). Plastic strain rates can be determined following the lines given in Section 4.4.

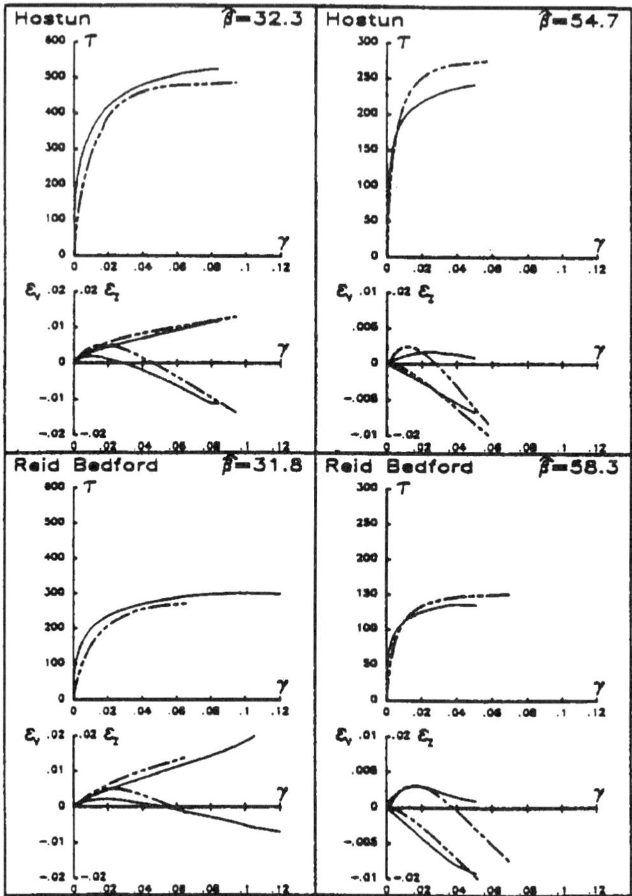

FIGURE 4.10. Comparison between predictions and actual test results on two silica sands-hollow cylinder after Nova [39].

Once the values of the constitutive parameters have been determined in elementary tests (isotropic loading-unloading and compression and extension tests at constant horizontal pressure, for instance) the behaviour in complex stress paths can be predicted.

Figure 4.9 shows for instance the comparison between predicted and observed strains in a plane strain constant volume test with principal stress inversion on a specimen of Karlsruhe sand (Nova [38]).

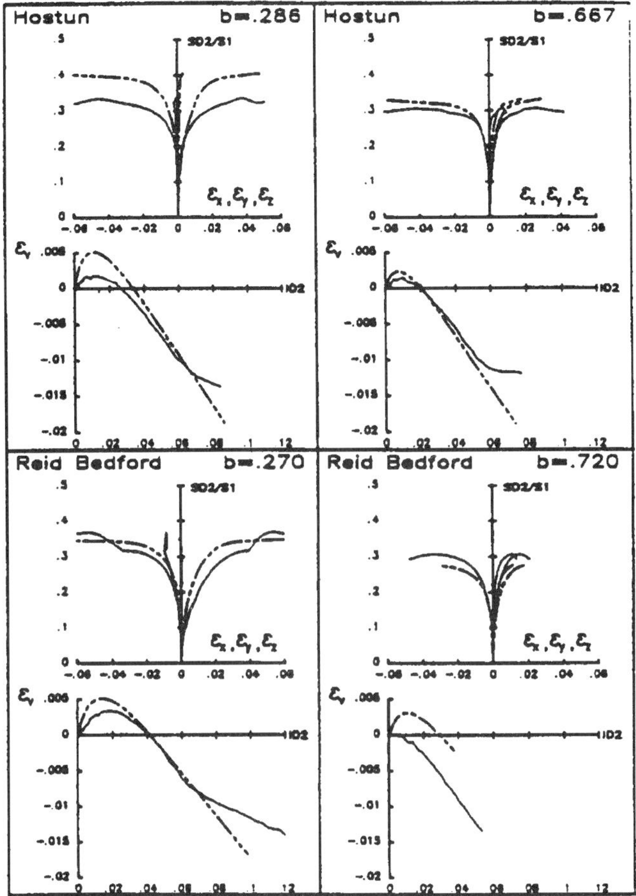

FIGURE 4.11. Comparison between prediction and actual test results on two silica sands-cubic triaxial cell after Nova [39].

Figures 4.10 and 4.11 show instead the comparison between observed and calculated results for two different sands, tested in two different apparatus: a hollow cylinder and a true (cubic) triaxial cell. In the former case the test was conducted at constant cell pressure and constant angle $\hat{\beta}$, defined as

$$\tan 2\hat{\beta} = \frac{2\tau_{\theta z}}{\Delta\sigma_z}.\tag{4.44}$$

The axial stress σ_z and the torsional shear stress $\tau_{\theta z}$ were increased or decreased starting from a spherical stress state.

In the latter test, the intermediate stress was kept constant as well as the parameter b, defined as

$$b \equiv \frac{\sigma_y - \sigma_x}{\sigma_z - \sigma_x},\tag{4.45}$$

where x, y, z are the axes of the cube and $\sigma_x, \sigma_y, \sigma_z$ are principal stresses. Tests started after isotropic compression.

In all mentioned cases, experimental results were disclosed to authors only after predictions were given. It is apparent that a model based on the premises discussed so far can predict soil behaviour in a monotonic loading test in a satisfactory way.

4.8   Unlimited pore pressure generation

An equally good agreement was obtained with different types of sand and normally consolidated clays in various, monotonic triaxial tests.

Among such tests, we can consider undrained axisymmetric compression tests on sand at different densities. For instance, Figure 4.12 shows a qualitative comparison between observed data (after Castro [7]) and calculated results (Nova [42]) for a specimen of Banding sand.

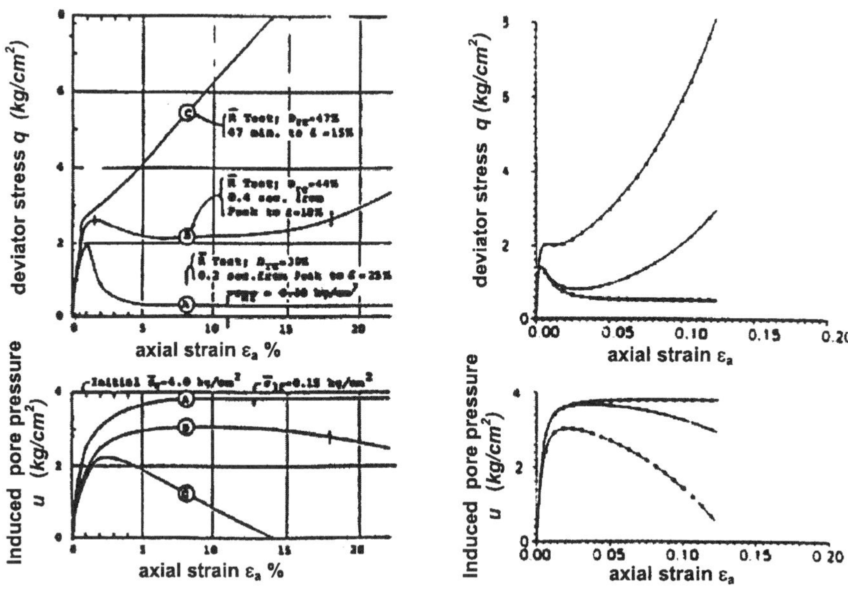

FIGURE 4.12. Qualitative comparison of experimental data after Castro [7] and back-calculated results after Nova [42].

The model can be used therefore as an adequate tool to investigate the occurrence of material instabilities prior to failure under displacement control. Figure 4.13 (Nova and Imposimato [44]) shows in fact a p' constant section in the space of principal stresses of three loci according to "Sinfonietta classica". The external one is the conventional displacement controlled failure locus, which is associated to the nullity of the hardening parameter

h and of the determinant of the stiffness matrix. As already noticed in Section 4.6 such a locus is associated to a loss of control when the test is load controlled.

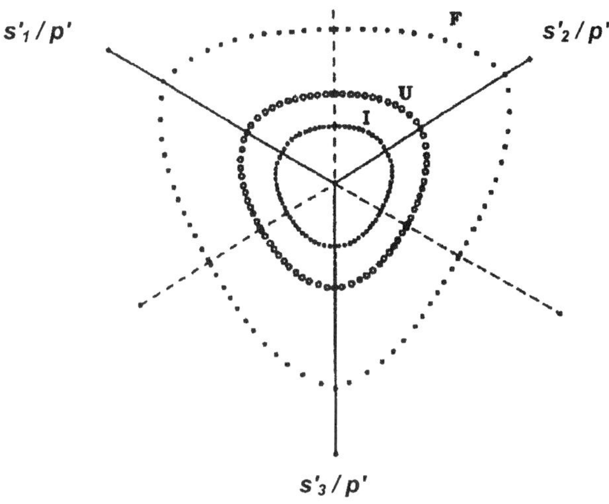

FIGURE 4.13. Instability loci in the deviatoric plane; $F\,(\det S = 0)$: failure $U\,(\Gamma_{33} = 0)$: unlimited pore pressure generation $I\,(\det S_s = 0)$: first homogeneous bifurcation surface.

The internal locus is associated to the nullity of the determinant of the symmetric part of the stiffness matrix, *i.e.*, to the loss of positive definiteness of S. When this condition is fulfilled, there exists one particular stress path for which the second-order work is zero. This implies that under a particular load control a homogeneous bifurcation occurs (infinite solutions of the response quantities for no variation of the controlling ones). This locus can be therefore also called first surface of homogeneous bifurcation.

The intermediate locus corresponds to a condition for which the volumetric compressibility is nil. This implies the possibility of unlimited increase of pore water pressure. Imagine in fact a test in a cubic triaxial cell in which volumetric strains are imposed to be nil. Furthermore imagine to give principal stress and strain increments such that

$$\begin{cases} \dot{\xi}_1 = \dot{\sigma}'_2 - \dot{\sigma}'_3 = 0 \\ \dot{\xi}_2 = \dot{\sigma}'_1 - \dot{\sigma}'_2 > 0 \quad ; \\ \dot{\eta}_3 = \dot{v} = 0 \end{cases} \tag{4.46}$$

the dual response variables are

$$\begin{cases} \dot{\eta}_1 = -\tfrac{1}{3}(\dot{\varepsilon}_1 + \dot{\varepsilon}_2) + \tfrac{2}{3}\dot{\varepsilon}_3 \\ \dot{\eta}_2 = \tfrac{2}{3}\dot{\varepsilon}_1 - \tfrac{1}{3}(\dot{\varepsilon}_2 + \dot{\varepsilon}_3) \\ \dot{\xi}_3 = \dot{p}' \end{cases} \qquad . \qquad (4.47)$$

Let C be the compliance matrix linking the stress rates $\dot{\sigma}'$ to the strain rates $\dot{\varepsilon}$, while Γ is the generalised compliance matrix linking the generalised stress rate variables $\dot{\xi}$ to the generalised strain rate variables $\dot{\eta}$. It can be shown (Imposimato and Nova [21]) that

$$\Gamma_{33} = \tilde{\delta} C \delta, \qquad (4.48)$$

where a tilde indicates transposition and the vector δ is analogous to the Kronecker symbol

$$\tilde{\delta} = \{\,1\,1\,1\,\}. \qquad (4.49)$$

When

$$\Gamma_{33} = 0, \qquad (4.50)$$

the determinant of the matrix linking the controlling quantities to the response variables (4.47) is nil. Therefore an infinity of solutions of (4.47) exists even under $\dot{\xi}_2 = 0$. In particular, if a specimen is saturated with an incompressible fluid (water), the unlimited variation (decrease) of p', at constant external isotropic pressure corresponds to an unlimited increase of the pore water pressure. This phenomenon was experimentally observed in an axisymmetric compression test by Imposimato and Nova [21].

The physical meaning of Γ_{33} is the soil volumetric compressibility under a purely isotropic stress decrease. This quantity can be zero since the elastic compressibility is positive, while the plastic one is negative. For that value of the stress obliquity, plastic volumetric strains are positive (compression), in fact, even under a decrease of the isotropic stress. From a micro-mechanical viewpoint this is because a decrease in the pressure packing the grains together allows contact sliding to occur with a consequent grain rearrangement and void reduction. This can be compensated however from the elastic strain relief generated by the pressure reduction, so that the overall volume can remain constant.

4.9   Drained shear banding

Another type of instability that can be described within the same framework is the shear band formation in plane strain tests. It is known (Rudnicki and Rice [56]) that when the determinant of the acoustic tensor is zero a shear band can be formed within the specimen. This type of bifurcation is different from those considered so far, since in this case non-homogeneous stress and strain states can coexist with the same boundary conditions, violating neither equilibrium nor compressibility.

The nullity of the determinant of the acoustic tensor coincides with the nullity of a particular minor of the constitutive matrix (Nova [40]). Take in fact a reference frame where x_n is the axis orthogonal to a plane and x_s, x_t are the other two axes. Let us choose as control variables $\tilde{\dot{\xi}}_1$ and $\tilde{\dot{\eta}}_2$ where

$$\left\{ \begin{array}{l} \tilde{\dot{\xi}}_1 = \left\{ \begin{array}{ccc} \dot{\sigma}_n & \dot{\tau}_{nt} & \dot{\tau}_{ns} \end{array} \right\} \\[2ex] \tilde{\dot{\eta}}_2 = \left\{ \begin{array}{ccc} \dot{\varepsilon}_t & \dot{\varepsilon}_s & \dot{\gamma}_{ts} \end{array} \right\} \end{array} \right. , \tag{4.51}$$

while $\dot{\xi}_2$ and $\dot{\eta}_1$ are vectors listing the complementary stress and strain variables. Let Δ be the generalised stiffness matrix connecting the generalised strain variables to the generalised stress variables

$$\left\{ \begin{array}{c} \dot{\xi}_1 \\ \dot{\xi}_2 \end{array} \right\} = \left[\begin{array}{cc} \Delta_{11} & \Delta_{12} \\ \Delta_{21} & \Delta_{22} \end{array} \right] \left\{ \begin{array}{c} \dot{\eta}_1 \\ \dot{\eta}_2 \end{array} \right\} . \tag{4.52}$$

Equation (4.52) can be rearranged so that the controlling quantities are put at the *l.h.s.*

$$\left\{ \begin{array}{c} \dot{\xi}_1 \\ \dot{\eta}_2 \end{array} \right\} = \left[\begin{array}{cc} \Delta_{11} - \Delta_{12}\Delta_{22}^{-1}\Delta_{21} & \Delta_{12}\Delta_{22}^{-1} \\ -\Delta_{22}^{-1}\Delta_{21} & \Delta_{22}^{-1} \end{array} \right] \left\{ \begin{array}{c} \dot{\eta}_1 \\ \dot{\xi}_2 \end{array} \right\} . \tag{4.53}$$

When

$$\det \Delta_{11} = 0 \tag{4.54}$$

the determinant of the matrix of equation (4.53) is also zero. Therefore, even under zero variation of the controlling variables the response quantities $\dot{\xi}_1$ and $\dot{\eta}_2$ can be different from zero. In particular

$$\dot{\xi}_2 = \Delta_{21}\dot{\eta}_1{}^* , \tag{4.55}$$

where $\dot{\eta}_1{}^*$ is the eigenvector of Δ_{11}. If stress and strain increments are seen as jumps of stress and strain rates between two adjacent regions divided by a plane, the condition of nullity of $\dot{\xi}_1$ and $\dot{\eta}_2$ is equivalent to the fulfilment of equilibrium and compatibility, as imposed by Rudnicki and Rice. The eigensolutions associated with equation (4.55) give the jumps in the mode of deformation across the plane (shear band). The submatrix Δ_{11} is the matrix form of the acoustic tensor.

We see therefore that the nullity of the determinant of the acoustic tensor corresponds not only to the occurrence of the discontinuity in the form of a shear band but also to the possibility of homogeneous deformations under a special loading programme, where stress normal to the band and strain along it are the controlling quantities.

The form of matrix Δ_{11} depends on the chosen orientation of the reference frame. In order to determine for which orientation equation (4.54) is fulfilled first all possible orientations should be examined. The deviatoric section of the locus for which the shear band condition is fulfilled first (equation (4.54)) is described in Figure 4.14 (Imposimato and Nova [21]).

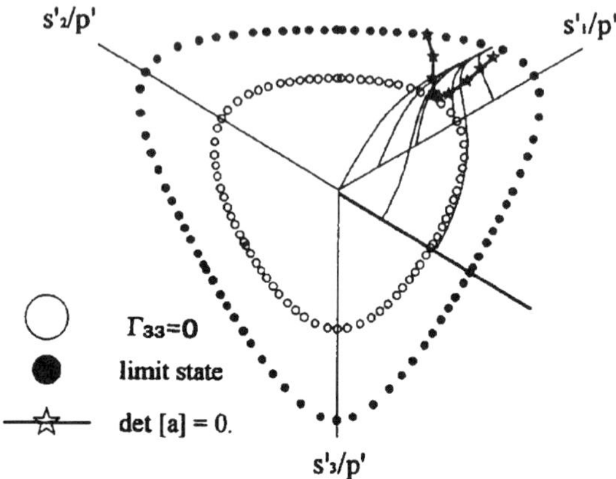

FIGURE 4.14. Loci for which the drained shear band condition is fulfilled first (after Imposimato and Nova [21]).

In the same picture the continuous thin lines give the stress paths followed in undrained plane strain tests, starting from different stress points, reached with different consolidation histories. Note that, although the tests considered are globally undrained, the shear band condition, equation (4.55), implies the possibility of local drainage, since $\dot{\eta}_1{}^*$ implies also $\dot{\varepsilon}_n \neq 0$.

It is apparent that, no matter which is the starting point, all stress paths converge when the limit locus is approached, so that the shear band locus is always crossed. In undrained plane strain tests drained shear banding may occur, if the boundary conditions allow such drainage to occur, when the stress state is still in the hardening regime. On the contrary, since the locus for shear banding does not cross the axes of axisymmetry, the model predicts that a drained shear band cannot occur in axisymmetric tests before the limit state is achieved. In other words, if a shear band occurs in such tests, that must be associated to a softening regime.

4.10   Locally undrained shear banding

In order to derive equation (4.54), no condition was imposed on volumetric strains, either within or outside the band. This implies that the local behaviour was implicitly assumed to be drained. If, on the contrary, we further impose that the material behaviour is locally undrained, a different shear band condition can be derived.

In this case, in fact, by imposing the conditions of plane strain ($\dot{\varepsilon}_s = \dot{\gamma}_{ts} = \dot{\gamma}_{ns} = 0$), strain continuity along the band ($\dot{\varepsilon}_t = 0$) and no volume change ($\dot{v} = 0$ and then $\dot{\varepsilon}_n = 0$) we realize that the only strain rate that can be different from zero is $\dot{\gamma}_{nt}$. Assume, without loss of generality that the stress and strain increments outside the band are zero. This implies that we are looking for the possibility of the spontaneous generation of stresses and strains in a band which violate neither equilibrium nor compatibility with the state of stress and strain, respectively, outside the band, keeping the volume constant.

The stress rates within the band (which coincide with the stress rate jumps for the assumed hypotheses) are therefore given by

$$
\left\{
\begin{array}{c}
\dot{\sigma'}_n \\
\dot{\sigma'}_t \\
\dot{\sigma'}_s \\
\dot{\tau}_{nt} \\
\dot{\tau}_{st} \\
\dot{\tau}_{ns}
\end{array}
\right\}
=
\left[
\begin{array}{c}
S_{14} \\
S_{24} \\
S_{34} \\
S_{44} \\
0 \\
0
\end{array}
\right]
\dot{\gamma}_{nt} \, .
\tag{4.56}
$$

The last two terms of the column vector are zero for the assumed coaxiality of stress and strain rates.

Equilibrium across the band imposes $\dot{\sigma}_n = \dot{\tau}_{nt} = 0$. In general, therefore, $\dot{\gamma}_{nt}$ should be zero and no shear band could occur. If however

$$
S_{44} = 0 \, ,
\tag{4.57}
$$

the equilibrium condition can be fulfilled even though $\dot{\gamma}_{nt} \neq 0$. The variation of the effective normal stress σ'_n can be compensated in fact by an equal and opposite variation of the pore water pressure, $\dot{u}$:

$$
\dot{u} = -S_{14}\dot{\gamma}_{nt} \, ,
\tag{4.58}
$$

so that total stresses are continuous across the band. The increment of the normal effective stresses in the directions x_s and x_t parallel to the band are self-equilibrated and overall equilibrium is guaranteed, as well as strain compatibility.

Equation (4.57) is therefore the shear band condition in locally undrained states. As for the drained case, the orientation of the band for which equation (4.57) is fulfilled first must be searched for by examining all possible orientations. The corresponding locus in the stress space is shown in Figure 4.15 (Imposimato and Nova [21]).

The locally undrained shear band condition always occurs slightly before the occurrence of the drained one. Of course, as in the drained case, by permuting the normal stress indices, the other five similar loci can be obtained.

It is finally interesting to note that the conditions imposed on the undrained shear band are the same imposed on a specimen in a constant

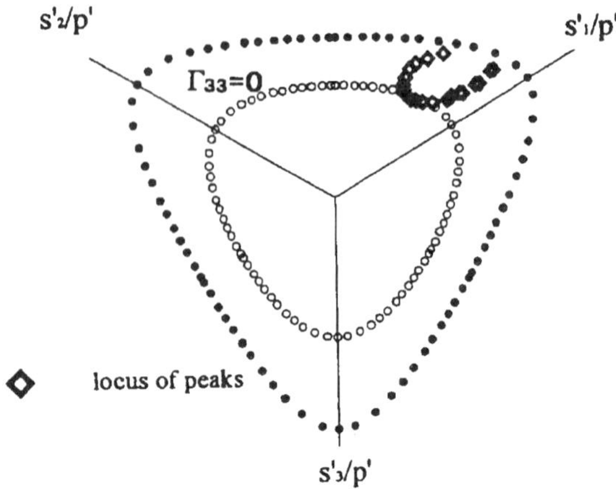

FIGURE 4.15. Locus for which a locally undrained shear band condition is fulfilled.

volume simple shear test. Equation (4.57) is also the condition for the occurrence of a peak in terms of shear stresses. We can therefore conclude that under shear force control an instability occurs in undrained plane strain well before the limit condition is achieved.

Undrained simple shear conditions occur in practice when a long slope is subjected to rapid external perturbations. If the slope is uniform and indefinitely long, all points at a certain depth from the free surface reach condition (4.57) at the same moment. A downwards planar movement can take place, therefore, and equation (4.57) gives what is usually called the failure condition of the slope. This is however an instability condition which can take place only if undrained conditions are locally enforced.

Di Prisco *et al.* [15] have shown that, by using equation (4.57) as an instability criterion, the failure of very flat subaqueous sandy slopes can be qualitatively explained. By extending this argument to rotational failures in clay slopes, Nova and Imposimato [45]) have shown that the use of equation (4.57) coincides with the traditional total stress analysis, in which the strength of the material is assumed to be given by a Tresca condition (purely cohesive material characterised by a given undrained cohesion).

4.11   Influence of induced anisotropy

A micro-mechanical analysis shows that the pattern of the force chains does not change much during a test in which either the external forces or the imposed displacement vary monotonically. If the 'direction' of loading is changed, however, the pattern of the force chains noticeably changes with associated irreversible strains. In particular, imagine consolidating

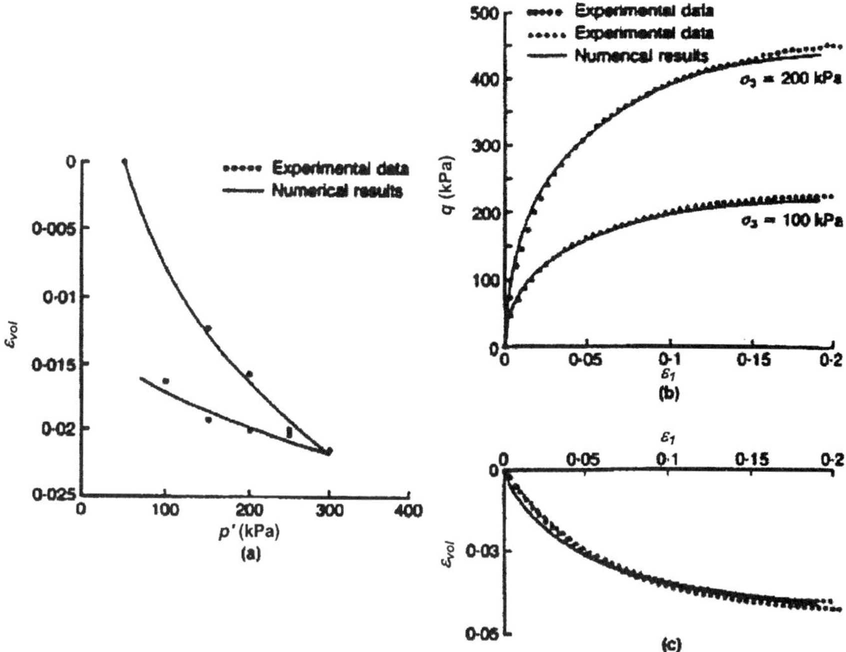

FIGURE 4.16. Comparison of experimental and numerical results on loose Hostun R.F. sand after di Prisco et al. [15]: a) isotropic loading unloading, b) drained triaxial compression deviatoric stress-strain relationship, c) drained triaxial compression volumetric strains.

isotropically a specimen of dense sand in a cubic triaxial cell. Then increase one stress component, say σ_z, and decrease the other two in order to keep constant the isotropic pressure.

The stress path in the deviatoric plane is given by a straight line stemming from the isotropic axis. Let θ be the angle between such line and the axis σ_z (*i.e.*, $\theta = 0$). Imagine further unloading the specimen to the isotropic consolidation state and reload it at constant isotropic pressure but with a different angle θ. The experimental evidence (Lanier *et al.* [29]) is that the larger is θ (between 0^0 and 180^0) the larger is the difference between the strains experienced by the specimen. This result cannot be predicted by the model described in the previous sections. The hardening rule assumed is isotropic, in fact, which implies that a homothetic expansion of the yield locus is the stress space. Cambou and Lanier [5] have shown instead that irreversible strains take place earlier for larger θ.

This difference is particularly relevant when complex cyclic loading is considered. The isotropic hardening model fails dramatically, for instance, in predicting the behaviour in circular tests in the deviatoric plane.

Di Prisco *et al.* [11] modified therefore the original model by taking induced anisotropy into account. The structure of the model is complex and cannot be presented in a few words. We shall mention only the fact that the structure of the model is elastic-plastic strain-hardening with a non-associated flow rule, but at variance with the isotropic model:

a) The yield function rotates and changes shape as long as deformation proceeds, taking therefore into account the stress path direction.

b) This can be achieved by introducing two more hidden variables (a scalar and a tensor). Each of these variables has an appropriate evolution rule, characterised by different constitutive parameters. Volumetric and deviatoric plastic strains govern such hidden variable evolutions.

c) The evolution rules are such that the deviatoric section of the yield locus changes size and moves in the deviatoric plane, as if hardening would be partly isotropic and partly kinematic.

d) The failure condition is determined a posteriori, as in the original model. The evolution rules are conceived in such a way that the failure condition is isotropic. It is therefore assumed that large strains erase the anisotropy induced by the previous phases of loading. This assumption is based on experimental evidence (Cambou and Lanier [5]).

e) The elastic behaviour is described by a non-linear law due to Lade and Nelson [27]. Altogether, 14 parameters characterise the material behaviour. They can be easily determined in three simple tests, however: isotropic compression, drained and undrained constant cell pressure standard triaxial tests.

The model is able to reproduce the observed behaviour in quite a satisfactory way.

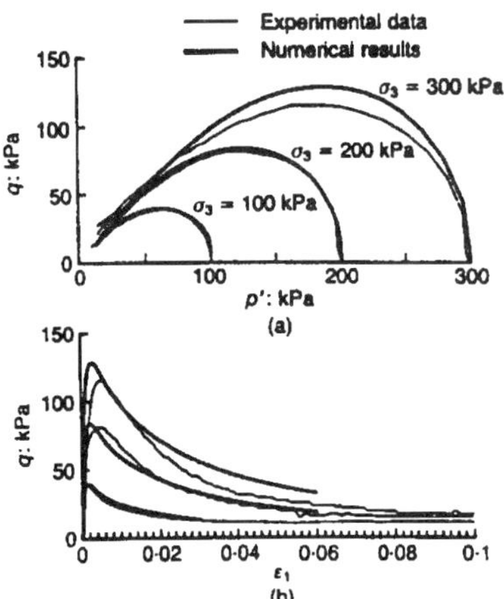

FIGURE 4.17. Comparison of experimental and numerical results on loose Hostun R.F. sand after di Prisco et al. [15] undrained triaxial compression: a) effective stress path, b) stress strain relationship.

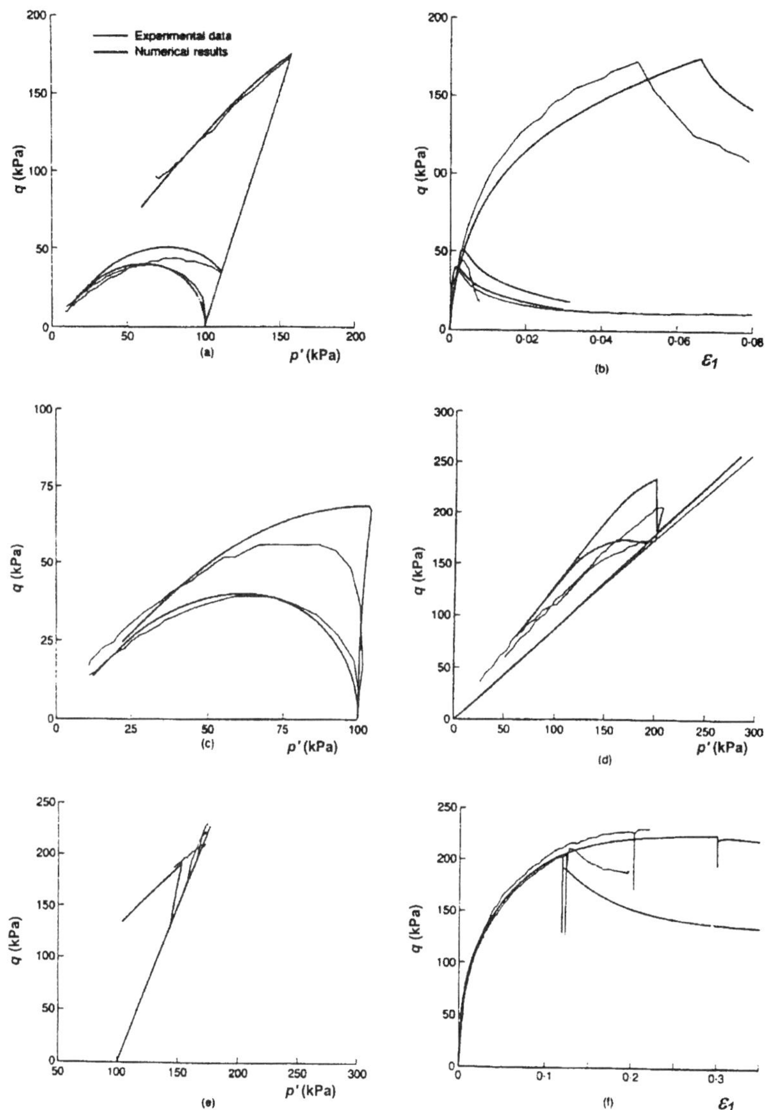

FIGURE 4.18. Comparison of numerical results and experimental data after di Prisco et al. [15]: a) stress path for a drained compression followed by an undrained compression on virgin loose Hostun R.F. sand, b) stress strain relationship for test (a), c) undrained stress path for a virgin and an isotropically preconsolidated specimen of loose Hostun R.F. sand, d) undrained stress paths for a virgin and preconsolidated specimen of loose Hostun R.F. sand after anisotropic consolidation, e) undrained stress paths as in (a) showing qualitative difference in behaviour for a small difference in starting point of the undrained phase, f) stress strain relationship for test (e).

116   R. Nova

Figures 4.16, 4.17 and 4.18 after di Prisco *et al.* [15] show comparisons
between drained and undrained axisymmetric tests with different preconsolidation histories. It is interesting to note that the model not only predicts
the differences in the behaviour of virgin and preloaded specimens, and the
influence of the type of preconsolidation (isotropic or anisotropic), but it
is also able to capture subtle differences. It is apparent that the behaviour
e) and *f*) depicted in Figure 4.18 is totally different for the two specimens
considered: one shows a peak and an unstable behaviour while for the other
one the deviator stress increases monotonically. This drastic difference in
behaviour is due only to an apparently minor difference in the starting
point of the undrained phase.

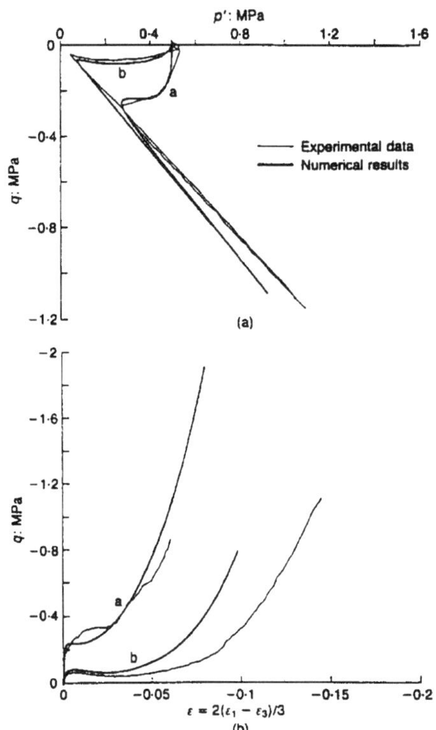

FIGURE 4.19. Comparison of numerical results and test data on dense
Hostun R.F. sand in undrained extension (after di Prisco et al. [15]):
a) stress path, b) stress strain-relationship.

Figure 4.19 after di Prisco *et al.* [15] shows a comparison for two specimens tested in undrained extension. Specimen *a*) is tested after isotropic compression, while specimen *b*) is first isotropically consolidated, then
loaded in compression, unloaded and reloaded in extension. It is quite clear
that induced anisotropy plays a relevant role: the specimen *b*) almost liq-

uefies although its relative density is close to 100%. The model is able to capture the difference between the two specimens in a convincing way.

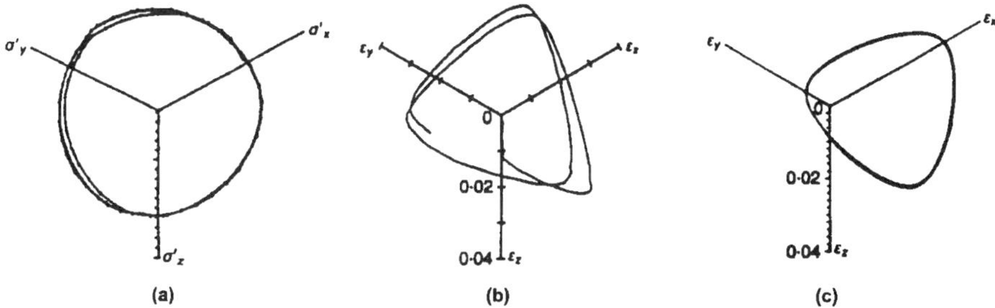

FIGURE 4.20. Comparison of numerical results and test data on dense Hostun R.F. sand in a deviatoric circular stress path (after di Prisco et al. [15]): a) stress path, b) experimental results, c) calculated strains.

Figure 4.20 after di Prisco [11] shows the comparison between experimental data and calculated results in a circular test in the deviatoric plane. Although not perfect, the simulation is again quantitatively close to the experimental evidence (data after Lanier and Zitouni [30]).

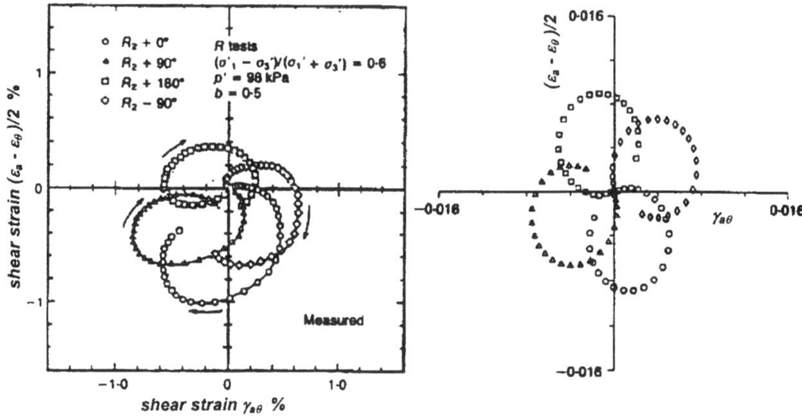

FIGURE 4.21. Comparison of numerical results and test data on Toyura sand, data from Miura et al. [33] after di Prisco [15].

Finally, Figure 4.21 shows a qualitative comparison between experimental data and calculated results in tests in which a continuous rotation of principal stresses is performed at constant principal stresses. The four curves are relative to different states at the beginning of the test that were reached following the same stress path in terms of principal stresses, but with different orientation of the principal axes. Due to induced material anisotropy different strain paths are obtained. It is apparent that also in this case the agreement is quite satisfactory.

4.12   Regularisation of the numerical response

The anisotropic model illustrated in the previous section predicts qualitatively the same type of instabilities of the isotropic hardening model (Imposimato and Nova [20]). When a model like that is introduced in finite element routines for the solution of boundary value problems, the inherent instabilities of the constitutive law make the computed solution mesh dependent. For instance, as long as deformation proceeds, the strains tend to localise in a narrow band which has the size of one element: the finer the mesh, the thinner the shear band and the more dramatic the loss of strength associated to shear band formation.

In order to avoid such a shortcoming several methods were proposed in the literature: use of fictitious soil viscosity, *e.g.*, Prévost and Loret [53], augmented continua (Cosserat media (Mühlhaus [35]) or second gradient theories (Zbib and Aifantis [65])) or non-local theories (Pijaudier-Cabot and Bazant [51]). The debate is still open on which of these procedures is the most convenient. Here an example is given which is based again on micro-mechanical considerations.

We have seen that when a soil specimen is perturbed at the boundaries either by imposing force or displacements, the grains tend to reach a different geometric configuration. Particularly, if external loads are controlled, the force chains continuously change under constant loading until a stable configuration is achieved. This takes time, shorter if the initial stress state is close to hydrostatic conditions and longer if the stress state is close to the limit condition. From a macroscopic viewpoint, therefore, strains develop in time at constant stress. The behaviour of soil is therefore physically time-dependent and the constitutive law must be expressed as a visco-plastic relationship.

Di Prisco and Imposimato [12] have modified the anisotropic model by assuming that strain rates are given by the sum of an elastic instantaneous component and a permanent, time-dependent, visco-plastic one

$$\dot{\varepsilon}_{ij} = \dot{\varepsilon}_{ij}{}^{e} + \dot{\varepsilon}_{ij}{}^{vp} \,. \tag{4.59}$$

The visco-plastic strain rates are assumed to be given by

$$\dot{\varepsilon}_{ij}{}^{vp} = \Lambda(f)\frac{\partial g}{\partial \sigma_{ij}} \,, \tag{4.60}$$

where, at variance with the ordinary plasticity theory illustrated in the previous sections, Λ is a non-negative function of the value of the loading function. The functions f and g are the same as those used in Section 4.7, while

$$\Lambda(f) = \Lambda_0 e^{kf} , \tag{4.61}$$

where Λ_0 and k are constitutive parameters.

It is apparent therefore that, according to this model, permanent strains can occur even when the value of the loading function is negative (in what it is usually considered an elastic regime).

Furthermore positive values of f are possible (a generalisation of the over-stress concept (Perzyna [50])). Di Prisco and Imposimato [12] have shown that in this way it is possible to reproduce accurately the experimental data both under displacement and force controlled tests.

Di Prisco and Imposimato [13] have shown further that the introduction of a time scale in the constitutive law largely reduces the original mesh dependency but does not fully eliminate it. Indeed the introduction of a microscopic length scale is also necessary. The definition of strain in a granular continuum is in fact somewhat arbitrary since a representative elementary volume of finite and a priori unknown size has to be defined. In a recent paper by Calvetti *et al.* [4] it is shown that by plotting the value of the calculated average strain in a specimen of a Schneebeli material with respect to the number of edges of the Voronoi tesselation, such a value tends to converge to a horizontal asymptote when the Voronoi polygon has more than ten edges. This means that, in order to define an average displacement in a point of a solid continuum equivalent to a two-dimensional granular medium, we have to consider the displacements of the centroids of about ten grains. For a three-dimensional granular assembly such a procedure has not been attempted yet, to the author's knowledge, at least. It is reasonable to expect that a similar result holds true.

This means that the representative elementary volume has a finite size; it is a sphere with a diameter equal to a few times (say 5) the average grain diameter. Such an internal length must be taken into account in formulating the stress-strain relationship. For instance di Prisco *et al.* [14] modify equation (4.57) by taking Λ as a function of an average value of f calculated as

$$\hat{f} = \int_{\bar{V}} f(x_i)\omega(x_i - x_{0i})dV , \tag{4.62}$$

where x_i are the spatial co-ordinates of a point in the volume $\bar{V}$ centred at the point x_{0i} where strain rates are calculated, $\bar{V}$ is a sphere whose diameter is proportional to the grain diameter and ω is a three dimensional bell shaped function centred at x_{0i}.

With such a non-local approach, mesh dependence is almost completely eliminated, at least in the cases considered. Figure 4.22 shows in fact the calculated results for a plane strain biaxial test in which either an elasto-plastic or a visco-plastic or finally a visco-plastic non-local model is employed (di Prisco and Imposimato [13]). It is apparent how big the differ-

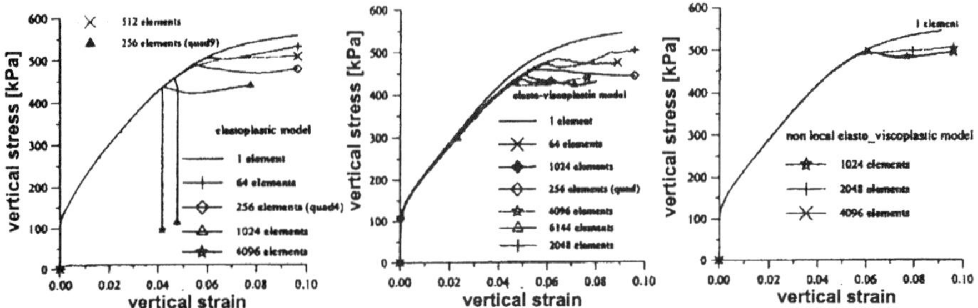

FIGURE 4.22. Mesh dependence in a plane strain biaxial test with different constitutive laws (after di Prisco and Imposimato [13]): a) elasto-plastic, b) elasto visco-plastic, c) non-local elasto visco-plastic.

ences are between the computed responses after the ontset of localisation.

Although very satisfactory, this is not, possibly, the end of the story. When localisation occurs, in fact, there is another physical fact that was not considered so far. We have seen in fact that together with sliding also rolling of the grain one over the other takes place, especially within the shear band. Desrues *et al.* [8] have clearly measured such rotations in biaxial tests on sand.

Such grain rolling implies non-coaxiality between stress and strain rates. A possible way of taking such deformation modes into account at a macroscopic level is that of using, together with the visco-plastic non-local approach, the possibility of couple stresses in the sense of the Cosserat theory.

Other tests are necessary to check whether the assumption of such an augmented continuum is mandatory or not. Presumably the tests where the influence of couple stress can be relevant are those in which principal stresses continuously rotate, like the aforementioned tests performed by Miura *et al.* [33].

4.13   Plasticity at very small strains

We have seen in the previous section that plastic strains take place even within what is considered in classical plasticity as an elastic domain. This result has been known for many years and many authors tried in different ways to take such strains into account in order to model soil behaviour under cyclic loading: *e.g.*, Prévost [52], Mroz *et al.* [34], Hueckel and Nova [19], Dafalias and Hermann [9].

Even under very small external displacements some local slippage of the contacts occurs so that a cycle of unloading-reloading is not perfectly elastic, but a hysteresis loop takes place. A schematic illustration of what is going on is given in Figure 4.23, modified after Iwan [23]. It is not surprising therefore that it is not easy to find an elastic potential, since truly elastic

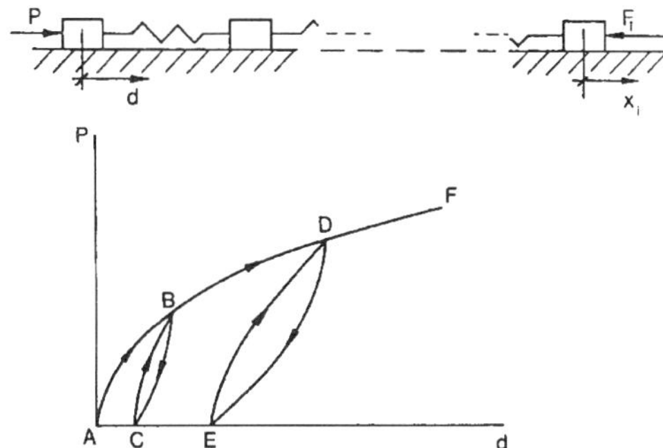

FIGURE 4.23. Simple model of soil behaviour and consequent hysteresis loop (modified after Iwan [23]).

strains are those smaller than 10^{-5}. The models discussed in the previous sections are instead conceived to model strain levels of the order of 10^{-3} or larger. The intermediate level needs a more refined description.

For instance, Jardine [24] and Puzrin and Kirschenboim [54] assume the existence of three yield surfaces. The first roughly corresponds to strains of the order of 10^{-5}, the third corresponds to the yield locus of the traditional elasto-plastic models, while the second is a surface defined within what is traditionally considered as an elastic domain. This latter surface changes size and moves around with the stress state as in a mixed isotropic-kinematic hardening model.

A correct description of the behaviour at such a strain level is very important, since this is the strain level at which most of the engineering structures 'work'. For engineering purposes, a further modification of the model described in the previous section for taking this fact into account is therefore necessary.

4.14 Conclusions

Soils can be considered, as a first approximation at least, as aggregates of spheroidal quasi-rigid particles. A micro-mechanical analysis of their intimate behaviour gives hints for the formulation of a comprehensive macroscopic constitutive model of soil behaviour. Starting from very simple assumptions, which take only friction and interlocking into account, it is possible to build up an elasto-plastic strain-hardening model with non-associate flow rule. Although very simple, such a model can reproduce quite well the behaviour of virgin soil in monotonic tests. It can further give a correct description of the instability phenomena which are microscopically linked to the continuous rearrangement of the pattern of force chains within a loaded specimen.

Micro-mechanics shows moreover the importance of other phenomena: the influence of induced anisotropy, the roles of time and of the particle size, the effect of microslips at small strain levels. A more complex model can be formulated which takes successively into account all the information retrieved from the micro-mechanical analysis. The more refined the model the more accurate the description of soil behaviour and the more objective the numerical solution of boundary value problems.

The endeavour of describing soil behaviour is a never-ending story, however. Most of the attention was focussed so far on saturated, virgin, uncemented materials, at medium large strains under static loading. Much remains to be done when soils are either unsaturated, over-consolidated, cemented or when very small strains are involved in static and dynamic conditions.

Acknowledgments: The results illustrated in this paper are a synthesis of the research activities on the modeling of soil behaviour conducted at Milan under my direction. Many people have given their contributions (in different proportions), namely C. di Prisco, R. Lagioia, R. Matiotti, F. Calvetti, S. Imposimato. To all of them my sincere thanks. This research was funded with several EU contracts and financial contribution of the Italian Ministry of University Research and Technology.

References

[1] R. BUTTERFIELD: A natural compression law for soils, *Géotechnique* **29**, pp. 469–480, 1979.

[2] F. CALVETTI: *Micromeccanica dei Materiali Granulari*, Ph.D. thesis, Dipartimento di Ingegneria Strutturale, Politecnico di Milano, 1998.

[3] F. CALVETTI: Micromechanical investigation of the visco-plastic behaviour of granular materials, in: G.N. PANDE, S. PIETRUSZCZAK, N.F. SCHWEIGER (eds.): *Numerical Models in Geomechanics - Proc. NUMOG VII, 1999, Graz, Austria*, Balkema, Rotterdam, 1999, pp. 59–64.

[4] F. CALVETTI, G. COMBE, J. LANIER: Experimental micromechanical analysis of a 2D granular material: relation between structure evolution and loading path, *Mech. Coh.-Frict. Mat.* **2**, pp. 121–164, 1998.

[5] B. CAMBOU, J. LANIER: Induced anisotropy in cohesionless soils: experiments and modelling, *Computers Geotech.* **6**, pp. 299–311, 1988.

[6] G. CASTRO: Liquefaction of sand, Ph.D. thesis, *Harvard Soil Mechanics Series* **81**, Harvard University, Cambridge, MA, 1969.

[7] G. CASTRO: Liquefaction and cyclic mobility of saturated sand, *J. Geotech. Eng. Div. ASCE* **101**, pp. 551–569, 1975.

[8] J. DESRUES, R. CHAMBON, M. MOKNI, F. MAZEROLLE: Void ratio evolution inside shear bands in triaxial sand specimens studied by computer tomography, *Géotechnique* **46**, pp. 529–546, 1996.

[9] Y.F. DAFALIAS, L.R. HERRMANN: Bounding surface formulation of soil plasticity, in: G.N. PANDE, O.C. ZIENKIEWICZ (eds.): *Soil Mechanics - Transient and Cyclic Loads*, Wiley, New York, 1982, pp. 253–282.

[10] F. DARVE: Liquefaction phenomenon: modelling, stability and uniqueness, in: ARULANANDAN, SCOTT (eds.): *Verification of Numerical Procedures for the Analysis of Soil Liquefaction Problems*, Balkema, Rotterdam, 1994, pp. 1305–1319.

[11] C. DI PRISCO: *Studio Sperimentale e Modellazione Matematica del Comportamento Sperimentale delle Sabbie*, Ph.D. thesis, Politecnico di Milano, 1993.

[12] C. DI PRISCO, S. IMPOSIMATO: Time dependent mechanical behaviour of loose sands, *Mech. Coh.-Frict. Mat.* **1**, pp. 45–73, 1996.

[13] C. DI PRISCO, S. IMPOSIMATO: Non-local numerical analyses of strain localisation in dense sand, in: G. CAPRIZ, V.N. GHIONNA, P. GIOVINE, N. MORACI (eds.): Mathematical Models in Soil Mechanics, *Mathematical and Computer Modelling*, Elsevier Science Publishers, Amsterdam, 2002, in press.

[14] C. DI PRISCO, S. IMPOSIMATO, E.C. AIFANTIS: A visco-plastic constitutive model for granular soils modified according to non-local and gradient approaches, *Int. J. Num. Anal. Meth. Geomech.*, **26**, pp. 121–138, 2002.

[15] C. DI PRISCO, R. MATIOTTI, R. NOVA: Theoretical investigation of the undrained stability of shallow submerged slopes, *Géotechnique* **45**, pp. 479–496, 1995.

[16] C. DI PRISCO, R. NOVA, J. LANIER: A mixed isotropic-kinematic hardening constitutive law for sand, in: D. KOLYMBAS (ed.): *Modern Approaches to Plasticity*, Elsevier Science Publishers, Amsterdam, 1993, pp. 83–124.

[17] M.A. EL SOHBY: *The Behaviour of Particulate Materials under Stress*, Ph.D. thesis, University of Manchester, 1964.

[18] M. GOLDSCHEIDER: True triaxial tests on dense sand, in: F. DARVE, G. GUDEHUS, I. VARDOULAKIS (eds.): *Constitutive Relations for Soils*, Balkema, Rotterdam, 1984, pp. 11–54.

[19] T. HUECKEL, R. NOVA: Some hysteresis effects of the behaviour of geological media, *Int. J. Solids Struct.* **15**, pp. 625–642, 1979.

[20] S. IMPOSIMATO, R. NOVA: An investigation on the uniqueness of the incremental response of elastoplastic models for virgin sand, *Mech. Coh.-Frict. Mat.* **3**, pp. 65–873, 1998.

[21] S. IMPOSIMATO, R. NOVA: Instabilities of loose sand specimens in undrained tests, in: T. ADACHI, F. OKA, A. YASHIMA (eds.): *Localisation and Bifurcation Theory for Soils and Rocks*, Balkema, Rotterdam, 1998, pp. 313–322.

[22] K. ISHIHARA, F. TATSUOKA, S. YASUDA: Undrained deformation and liquefaction of sand under cyclic stresses, *Soils and Foundations*, **15**, pp. 29–44, 1975.

[23] K. IWAN: On a class of models for the yielding behaviour of continuous and composite systems, *J. App. Mech. Ser.E* **3**, pp. 612–617, 1967.

[24] R.J. JARDINE: Some observations of the kinematic nature of soil stiffness, *Soils and Foundations* **32**, pp. 111–124, 1992.

[25] S.L. KRAMER, H.B. SEED: Initiation of soil liquefaction under static loading conditions, *J. Geo. Engng. ASCE* **114**, pp. 412–430, 1988.

[26] P.V. LADE: Static instability and liquefaction on loose fine sandy slopes, *J. Geo. Engng. ASCE* **118**, pp. 51–71, 1992.

[27] P.V. LADE, R.B. NELSON: Modelling the elastic behaviour of granular materials, *Int. J. Num. Anal. Meth. Geom.* **11**, pp. 521–542, 1987.

[28] R. LAGIOIA, A.M. PUZRIN, D.M. POTTS: A new versatile expression for yield and plastic potential surfaces, *Computers and Geotechnics* **19**, pp. 171–193, 1996.

[29] J. LANIER, C. DI PRISCO, R. NOVA: Etude experimentale et analyse théorique de l'anisotropie induite du sable d'Hostun, *Revue Française de Géotechnique* **57**, pp. 59–74, 1991.

[30] J. LANIER, Z. ZITOUNI: Development of a data base using the Grenoble true triaxial apparatus, in: A.S. SAADA, G.F. BIANCHINI (eds.): *Constitutive Equations for Granular Non-Cohesive Soils*, Balkema, Rotterdam, 1988, pp. 47–58.

[31] R. MATIOTTI: *Analisi Sperimentale del Fenomeno della Liquefazione Statica delle Sabbie Sciolte e sue Applicazioni Numeriche*, Ph.D. thesis, Politecnico di Milano, 1996.

[32] M. MEGACHOU: *Stabilité des Sables Laches. Essais et Modélisations*, Ph.D. thesis, INPG, Grenoble, 1993.

[33] K. MIURA, S. TOKI, S. MIURA: Deformation prediction for anisotropic sand during the rotation of principal stress axes, *Soils and Foundations* **26**, pp. 42–56, 1986.

[34] Z. MROZ, V.A. NORRIS, O.C. ZIENKIEWICZ: Application of an anisotropic hardening model in the elasto-plastic deformation of soils, *Géotechnique* **29**, pp. 1–34, 1979.

[35] H.-B. MÜHLHAUS: Scherfugenanalyse bei granularem material im rahmen der Cosserat-theorie, *Ingenieur Archiv* **56**, pp. 389–399, 1986.

[36] R. NOVA: On the hardening of soils, *Archiw. Mech. Stos.* **29**, pp. 445–458, 1977.

[37] R. NOVA: A constitutive model for soil under monotonic and cyclic loading, in: G. N. PANDE, O. C. ZIENKIEWICZ (eds.): *Soil Mechanics: Transient and Cyclic Loads*, John Wiley and Sons, New York, 1982, pp. 343–373.

[38] R. NOVA: A model of soil behaviour in plastic and hysteretic ranges. Part I - Monotonic loading, in: G.GUDEHUS, F. DARVE, I. VARDOULAKIS (eds.): *Constitutive Relations for Soils*, Balkema, Rotterdam, 1984, pp. 289–309.

[39] R. NOVA: Sinfonietta classica: an exercise on classical soil modelling, in: A.S. SAADA, G.F. BIANCHINI (eds.): *Constitutive Equations for Granular Non-Cohesive Soils*, Balkema, Rotterdam, 1988, pp. 510–519.

[40] R. NOVA: Liquefaction, stability, bifurcations of soil via strain-hardening plasticity, *Proc. Int. Works. Numer. Meth. Localis. and Bifurcat. of Granular Bodies*, Gdansk, 1989, pp. 117–132.

[41] R. NOVA: A note on sand liquefaction and soil stability, 3rd *Int. Conf. Constitutive Laws of Eng. Mat.*, Tucson, 1991, pp. 53–156.

[42] R. NOVA: Mathematical modelling of natural and engineered geomaterials, General lecture of 1st E.C.S.M. Munchen, *Eur. J. Mech. A/Solids*, Special issue **11**, pp. 135–154, 1992.

[43] R. NOVA: Controllability of the incremental response of soil specimens subjected to arbitrary loading programmes, *J. Mech. Behav. Mater.* **5**, pp. 193–201, 1994.

[44] R. NOVA, S. IMPOSIMATO: Non-uniqueness of the incremental response of soil specimens under true-triaxial stress paths, in: G.N. PANDE, S. PIETRUSZCZAK (eds.): *Numerical Models in Geomechanics - Proc. NUMOG VI, 1997, Montreal, Canada*, Balkema, Rotterdam, 1997, pp. 193–197.

[45] R. NOVA, S. IMPOSIMATO: Analysis of instability conditions for normally consolidated soils, in: W. EHLERS (ed.): *Proc. IUTAM Symposium on "Theoretical and Numerical Methods in Continuum Mechanics of Porous Materials"*, Kluwer Academic Publishers, Dordrecht, 2001, pp. 265–272.

[46] R. NOVA, D.M. WOOD: A constitutive model for sand in triaxial compression, *Int. J. Num. Anal. Meth. Geomech.* **3**, pp. 255–278, 1979.

[47] A. OSTROWSKI, O. TAUSSKY: On the variation of the determinant of a positive definite matrix, *Neder. Akad. Wet. Proc.* **A54**, pp. 333–351, 1951.

[48] H.B. POOROOSHASB, I. HOLUBEC, A.N. SHERBOURNE: Yielding and flow of sand in triaxial compression (part I), *Can. Geot. J.* **4**, pp. 179–190, 1966.

[49] H.B. POOROOSHASB, I. HOLUBEC, A.N. SHERBOURNE: Yielding and flow of sand in triaxial compression (part II and part III), *Can. Geot. J.* **4**, pp. 376–397, 1966.

[50] P. PERZYNA: The constitutive equations for rate sensitive plastic materials, *Quart. Appl. Math.* **20**, pp. 321–332, 1963.

[51] G. PIJAUDIER-CABOT, Z.P. BAZANT: Nonlocal damage theory, *J. Eng. Mech. ASCE* **113**, pp. 1512–1533, 1987.

[52] J.-H. PRÉVOST: Mathematical modelling of monotonic and cyclic undrained clay behaviour, *Int. J. Num. Anal. Meth. Geomech.* **1**, pp. 195–216, 1977.

[53] J.-H. PRÉVOST, B. LORET: Dynamic strain localisation in elasto (visco) plastic solids: part 2: plane strain examples, *Computer Methods App. Mech. Eng.* **83**, pp. 275–294, 1990.

[54] A.M. PUZRIN, E. KIRSCHENBOIM: Kinematic hardening model for overconsolidated clays, *Computers and Geotechnics* **28**, pp. 1–36, 2001.

[55] K.H. ROSCOE, J.B. BURLAND: Stress strain behaviour of wet clay, in: C.R. CALLADINE (ed.): *Engineering Plasticity*, Cambridge University Press, Cambridge, 1968, pp. 535–609.

[56] J.W. RUDNICKI, J.R. RICE: Conditions for the localisation of deformation in pressure sensitive dilatant materials, *J. Mech. Phys. Solids* **23**, pp. 371–394, 1975.

[57] G. SCHNEEBELI: Une analogie mécanique pour les terres sans cohésion, *Comptes Rendus Acad. Sc.* **243**, p. 125, 1956.

[58] A.N. SCHOFIELD, C.P. WROTH: *Critical State Soil Mechanics*, McGraw-Hill, Chichester, 1968.

[59] M.A. STROUD: *The Behaviour of Sand at Low Stress Levels in the Simple Shear Apparatus*, Ph.D. thesis, University of Cambridge, 1971.

[60] F. TATSUOKA: *Shear Tests in a Triaxial Apparatus. A Fundamental Research on the Deformation of Sand*, Ph.D. thesis, University of Tokyo (in Japanese), 1972.

[61] F. TATSUOKA, K. ISHIHARA: Yielding of sand in triaxial compression, *Soils and Foundations* **12**, pp. 63–76, 1974.

[62] K. TERZAGHI: Die berechnung der durchlässigkeitsziffer des tones aus dem verlauf der hydrodynamischen spannungserscheinungen, *Sitz. Akad. Wissen. Wien. Math.-Naturw.-Schaf. Kl.*, Abt. IIa **132**, pp. 125–138, 1923.

[63] A.F. WALKER: *Stress Strain Relationships for Clay*, Ph.D. thesis, University of Cambridge, 1965.

[64] R.K.S. WONG, J.R.F. ARTHUR: Induced and inherent anisotropy in sand, *Géotechnique* **35**, pp. 471–481, 1985.

[65] H.M. ZBIB, E.C. AIFANTIS: A gradient dependent flow theory of plasticity: application to metal and soil instabilities, *Appl. Mech. Rev.* **42**, pp. 295–304, 1989.

[66] M. ZYTYNSKY, M.F. RANDOLPH, R. NOVA, C.P. WROTH: On modelling the unloading-reloading behaviour of soils, *Int. J. Num. Anal. Meth. Geomech.* **2**, pp. 87–94, 1978.

ROBERTO NOVA
Dipartimento di Ingegneria Strutturale
Politecnico di Milano
Piazza Leonardo da Vinci, 32
I-20133 Milano, ITALY
E-mail: nova@stru.polimi.it

Chapter 5

Dynamic Thermo-Poro-Mechanical Stability Analysis of Simple Shear on Frictional Materials

Ioannis Vardoulakis

ABSTRACT In this paper, the basic mathematical structure of a thermo-poro-mechanical model for faults under rapid shear is discussed. The analysis is 1D in space and concerns the infinitely extended fault. The gauge material is considered as a two-phase material consisting of a thermo-elastic fluid and of a thermo-poro-elasto-viscoplastic skeleton. The governing equations are derived from first principles expressing mass, energy and momentum balance inside the fault. They are a set of coupled diffusion-generation equations that contain three unknown functions, the pore-pressure, the temperature and the velocity field inside the fault. The original mathematically ill-posed problem is regularized using a viscous-type and a second-gradient regularization. Numerical results are presented and discussed.

5.1 Introduction

According to Habib [6], mechanical energy dissipated in heat inside a fault zone may lead to vaporization of pore-water, thus creating a cushion of zero friction. The idea that a heat generating mechanism might account for the loss of strength of large earth slides due to vaporization has been discussed in the past by Romero and Molina [14], Habib [7] and Goguel [5]. Within 1D-analyses of sliding block-mechanisms (Figure 5.1), Anderson [1] first and later Voight and Faust [20] showed that, even if vaporization does not take place, heat generation may give rise to high pore-water pressures inside the shear-band. Finally, it should be mentioned that at the same time a series of papers appeared in the geophysics community, which dealt with the problem of pore-fluid pressures and frictional heating, as well as frictional melting in relation to seismic fault rupture [9, 11]. Recently Vardoulakis [18] formulated the set of governing equations that account for heat-generated pore-pressures inside rapidly deforming shear-bands, starting from first principles. The soil was considered as a two-phase mixture of solids and fluid, and the governing equations were derived from the corresponding balance laws for mass, momentum and energy.

In general, clays, when heated slowly under fully drained conditions, suffer a net volume reduction. This is a form of micro-"structural collapse", which in turn is attributed to "failure of some inter-particle ties". This internal collapse mechanism is due to changes in water absorption by the clay particles and changes in the equilibrium between "attractive and repulsive electrostatic forces", which act between ions forming the "double layer" at a particle scale [2, 12]. This phenomenon is documented in both isotropic and oedometric compression tests. In particular, it is observed that slow heating of normally consolidated clay material, tested in fully drained constant-load oedometric or isotropic conditions, results usually in volume reduction. Over-consolidated clays on the other hand, at relatively low temperatures, behave elastically and expand. At elevated temperatures, however, over-consolidated clays contract ('collapse') as well. This internal thermoplastic collapse mechanism of clayey gauges is considered to be the main mechanism responsible for the thermo-poro-mechanical instabilities considered here.

In this paper we first summarize and discuss the equations that govern the phenomenon of heat-generated pore pressures inside a rapidly deforming fault (shear-band). They are a set of coupled diffusion-generation equations that contain three unknown functions, the heat-generated excess pore-fluid pressure, the temperature and the velocity. We remark that these equations constitute, together with a set of initial and boundary conditions at the shear-band boundaries, a mathematically ill-posed problem. This means in turn that, in order to solve the corresponding boundary-value problem, some kind of mathematical regularization is needed. In our previous paper [18] a "crude" method of regularization was used, namely that of replacing the momentum equation for the velocity field with an ad-hoc solution for the velocity that is correct to the leading, linear-in-z term and is compatible with the boundary conditions.

A more rigorous way to approach this problem, which after all justified the previous approximation, is to reconsider the friction law and introduce friction rate sensitivity. In the last section we discuss critically the viscous regularization and point to its limitations in the light of some experimental results reported recently by Tika and Hutchinson [16], which suggest that for some clay fault materials, frictional rate softening is taking place. In that case a second-gradient viscous regularization is proposed and discussed [19].

5.2   Mass balance

5.2.1   Formulation

Water-saturated soil is modeled as a two-phase mixture consisting: (1) of solids, and (2) of water. Let $\rho^{(\alpha)}$, $v_i^{(\alpha)}$ be the partial densities and partial velocities of the two constituents ($\alpha = 1, 2$). The partial densities are expressed in terms of the porosity n of the soil, the density ρ_s of the solids and the density ρ_w of water:

$$\rho^{(1)} = (1 - n)\rho_s \, ; \qquad \rho^{(2)} = n\rho_w \, . \tag{5.1}$$

The total density of the mixture is

$$\rho = n\rho_w + (1 - n)\rho_s \, . \tag{5.2}$$

When there is internal fluid flow, the partial velocities differ ($v_i^{(1)} \neq v_i^{(2)}$) and mass balance is expressed for each constituent separately. For non-diffusing, non-reacting species the local mass balance equations are [17]

$$\frac{D^{(\alpha)}}{Dt}\rho^{(\alpha)} + \rho^{(\alpha)}\frac{\partial}{\partial x_k}v_k^{(\alpha)} = 0 \, . \tag{5.3}$$

In these expressions, $\frac{D^{(\alpha)}(\cdot)}{Dt} = \frac{\partial(\cdot)}{\partial t} + v_k^{(\alpha)}\frac{\partial(\cdot)}{\partial x_k}$ denotes the material time-derivative operator with respect to the species (α). By combining equations (5.1) and (5.3), we obtain a) the evolution equation for the porosity

$$\frac{D^{(1)}n}{Dt} = (1 - n)\left(\frac{1}{\rho_s}\frac{D^{(1)}\rho_s}{Dt} + D_{kk}^{(1)}\right) \tag{5.4}$$

and b) the continuity equation for the interstitial fluid flow

$$-\frac{\partial q_k}{\partial x_k} = (1 - n)\frac{1}{\rho_s}\frac{D^{(1)}\rho_s}{Dt} + n\frac{1}{\rho_w}\frac{D^{(2)}\rho_w}{Dt} + \dot{\varepsilon}, \tag{5.5}$$

where

$$q_k = n(v_k^{(2)} - v_k^{(1)}) \tag{5.6}$$

is the relative specific discharge, and

$$\dot{\varepsilon} = D_{kk}^{(1)} = \frac{\partial v_k^{(1)}}{\partial x_k} \tag{5.7}$$

is the total volumetric deformation rate of the solid skeleton, where

$$D_{ij}^{(1)} = \frac{1}{2}\left(\frac{\partial v_i^{(1)}}{\partial x_j} + \frac{\partial v_j^{(1)}}{\partial x_i}\right) \tag{5.8}$$

is the rate of deformation of the solid skeleton.

In order to evaluate the mass balance equations (5.4) and (5.5), constitutive assumptions must be formulated concerning the variations of the densities of the constituents and of the flow of the pore-fluid. In order to do this, we first postulate a unique temperature field $\theta = \theta(x_k, t)$ by assuming that, locally, grains and fluid are always in thermal equilibrium.

The grains are assumed to be incompressible, thus only grain thermal expansion is taken into account, and the following constitutive equation is postulated for the density of the solids:

$$\frac{1}{\rho_s}\frac{D^{(1)}\rho_s}{Dt} \approx -\alpha_s\frac{D^{(1)}\theta}{Dt}. \tag{5.9}$$

As a typical value for the coefficient of thermal expansion for clay particles, $\alpha_s = 3 \cdot 10^{-5}[1/^\circ\mathrm{C}]$ is taken here.

On the other hand, we assume that density variations in the aqueous phase are due both to changes in pore-water pressure and to temperature changes:

$$\frac{1}{\rho_w}\frac{D^{(2)}\rho_w}{Dt} \approx c_w\frac{D^{(2)}p_w}{Dt} - \alpha_w\frac{D^{(2)}\theta}{Dt} \tag{5.10}$$

with

$$c_w = \left.\frac{\partial}{\partial p_w}\ln\rho_w\right|_{\theta=\mathrm{const.}} \qquad \text{and} \tag{5.11}$$

$$\alpha_w = -\left.\frac{\partial}{\partial\theta}\ln\rho_w\right|_{p_w=\mathrm{const.}} \tag{5.12}$$

If we neglect convective terms, then no distinction is made between the various material time derivatives and the local time derivative, $\frac{D^{(1)}(\cdot)}{Dt} \approx \frac{D^{(2)}(\cdot)}{Dt} \approx \frac{\partial(\cdot)}{\partial t}$. Accordingly, we have

$$\frac{\partial n}{\partial t} = (1 - n)\left(-\alpha_s\frac{\partial\theta}{\partial t} + \dot{\varepsilon}\right), \tag{5.13}$$

$$-\frac{\partial q_k}{\partial x_k} = -\alpha_m\frac{\partial\theta}{\partial t} + nc_w\frac{\partial p_w}{\partial t} + \dot{\varepsilon}, \tag{5.14}$$

where

$$\alpha_m = (1 - n)\alpha_s + n\alpha_w \tag{5.15}$$

is the coefficient of thermal expansion of the soil-water mixture.

In order to eliminate the relative specific discharge vector from the continuity equation (5.14) we assume the validity of Darcy's law for the pore-water flow:

$$q_i = \frac{1}{f}\left(-\frac{\partial p_w}{\partial x_i} - \rho_w g_i\right), \tag{5.16}$$

where

$$f = \frac{\gamma_w}{k_w} \tag{5.17}$$

and $\gamma_w = \rho_w g$ is the unit weight of water, g the acceleration of gravity and k_w is the coefficient of permeability of the soil with respect to water. In equation (5.16), g_i is the i-th component of the acceleration of gravity.

Introducing Darcy's law into the continuity equation (5.14) results in

$$\frac{1}{f}\left(\nabla^2 p_w + g_i \frac{\partial \rho_w}{\partial x_i}\right) = -\alpha_m \frac{\partial \theta}{\partial t} + n c_w \frac{\partial p_w}{\partial t} + \dot{\varepsilon}. \tag{5.18}$$

The second term on the left-hand side of equation (5.18) refers to the contribution of the elevation head. We notice that, from the assumption that the fluid density varies with pressure and temperature, equations (5.11) and (5.12), we have that

$$\begin{aligned}
\frac{1}{\rho_w}\frac{\partial \rho_w}{\partial x_i} &= \frac{1}{\rho_w}\frac{\partial \rho_w}{\partial p_w}\bigg|_\theta \frac{\partial p_w}{\partial x_i} + \frac{1}{\rho_w}\frac{\partial \rho_w}{\partial \theta}\bigg|_{p_w} \frac{\partial \theta}{\partial x_i} \\
&= c_w \frac{\partial p_w}{\partial x_i} - \alpha_w \frac{\partial \theta}{\partial x_i}.
\end{aligned} \tag{5.19}$$

5.2.2 Skeleton volumetric response

Volume changes of the soil skeleton obey an incremental thermo-elasto-plastic constitutive law in terms of the effective mean stress and temperature. In terms of rates this law reads[1]

$$\dot{\varepsilon} = c_c \dot{p}' + \alpha_c \dot{\theta}, \tag{5.20}$$

where p' is the mean effective stress, $p' = \frac{1}{3}\sigma'_{kk}$, which in turn is introduced through Terzaghi's operational definition, that decomposes the total stress in effective stress and pore-water pressure:

$$\sigma_{ij} = \sigma'_{ij} - p_w \delta_{ij}. \tag{5.21}$$

Isothermal swelling

Under conditions of effective-stress reduction due to pore-pressure increase the skeleton is deforming almost elastically. In that case, the compressibility coefficient c_c is identified here with the swelling constant

$$c_{sw} = \frac{\partial \varepsilon}{\partial p'}\bigg|_{\substack{\theta=\text{const.}\\ \text{OCR}>1}} \tag{5.22}$$

This swelling constant in turn is identified with the elastic (unloading) compressibility in isotropic compression and can be estimated from the

---

[1] Superimposed dots in equation (5.20) denote material time derivatives with respect to the velocity of the solid phase.

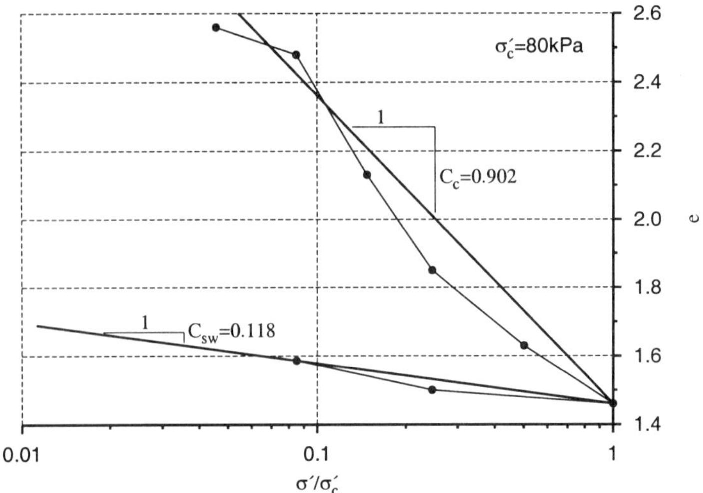

FIGURE 5.1. Example of oedometric compression-swelling line for San Francisco Bay Mud, after Holtz and Kovacs [8].

corresponding value of the swelling index,

$$c_{sw} = \frac{1}{1 + e_c} \frac{\bar{C}_{sw}}{|p'_c|}, \quad \bar{C}_{sw} = \frac{C_{sw}}{\ln 10}, \quad C_{sw} = -\frac{\Delta e}{\log\left(\frac{p'_c + \Delta p'}{p'_c}\right)}, \quad (5.23)$$

with $\Delta e > 0$ and where $e = \frac{n}{1-n}$ is the void ratio.

Isobaric expansion and collapse

In equation (5.20), α_c is the *thermal 'expansion'* coefficient of the soil skeleton,

$$\alpha_c = \left.\frac{\partial \varepsilon}{\partial \theta}\right|_{\sigma' = \text{const.}} \quad (5.24)$$

The thermal volumetric strains of an over-consolidated clay may be approximated by a bilinear law, as shown in Figure 5.2. Sultan [15] gives valuable information concerning the thermo-elastic and the thermo-elasto-plastic 'expansion' coefficients of 'Boom' clay. Accordingly, thermo-plastic collapse takes place as soon as the temperature is above a critical value:

$$\alpha_c = \begin{cases} \alpha_c^e, & \text{if } \theta \leq \theta_{cr} \quad \text{or if } \theta \leq \theta_Y, \\ \alpha_c^{ep}, & \text{if } \theta = \theta_Y > \theta_{cr}. \end{cases} \quad (5.25)$$

These tests suggest also that the elastic thermal expansion coefficient is well reproduced by the above mixtures-theory formula (5.15)

$$\alpha_c^e \approx \alpha_m. \quad (5.26)$$

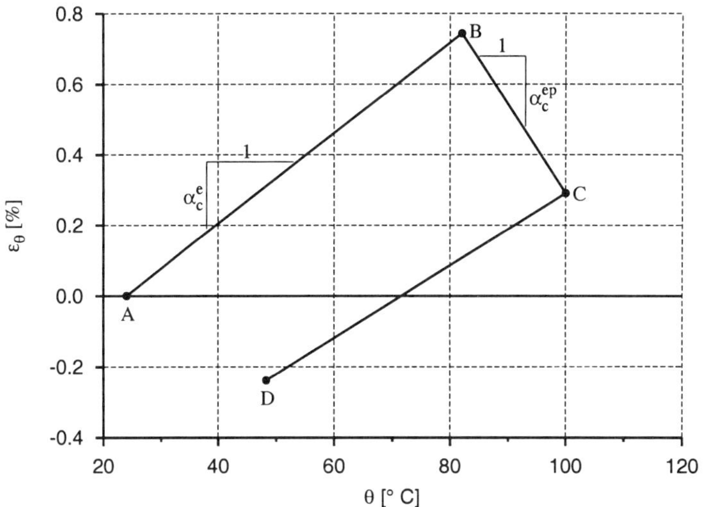

FIGURE 5.2. Isotropic thermal volumetric deformation of an over-consolidated Boom clay specimen ($p'_c = 4.2\text{MPa}$, $p' = 0.35\text{MPa}$, (OCR) $= 12$) after Sultan [15]; a) Heating phase: (AB) thermo-elastic expansion; (BC) thermoplastic collapse ($\theta_{cr} = 82°\text{C}$). b) Cooling phase: (CD) thermo-elastic contraction.

Notice that in our previous paper [18] the limiting case was considered of ideal thermo-plasticity, with $\alpha_c^{ep} = 0$.

According to the experiment results shown in Figures 5.3 and 5.4, both the critical thermo-plastic collapse temperature and the thermo-plastic contraction coefficient are functions of the over-consolidation ratio (OCR):

$$\alpha_c^{ep} = \alpha_c^e + \alpha_c^p \Rightarrow$$

$$\alpha = \alpha_m - \alpha_c = \begin{cases} 0, & \text{if} \quad \theta \leq \theta_{cr} \quad \text{or if} \quad \theta \leq \theta_Y, \\ -\alpha_c^p, & \text{if} \quad \theta = \theta_Y > \theta_{cr}. \end{cases} \qquad (5.27)$$

5.2.3 *Pore-pressure diffusion-generation equation*

With the above constitutive assumptions the mass balance equation (5.18) yields, within a good approximation, the pore-pressure, diffusion-generation, partial differential equation

$$\frac{nc_w}{c_c}\frac{\partial p_w}{\partial t} + \frac{\partial p'}{\partial t} = c_v \frac{\partial^2 p_w}{\partial z^2} + \lambda_m \frac{\partial \theta}{\partial t}, \qquad (5.28)$$

where c_v and λ_m are the *consolidation* and the *pore-pressure-temperature* coefficients, respectively.

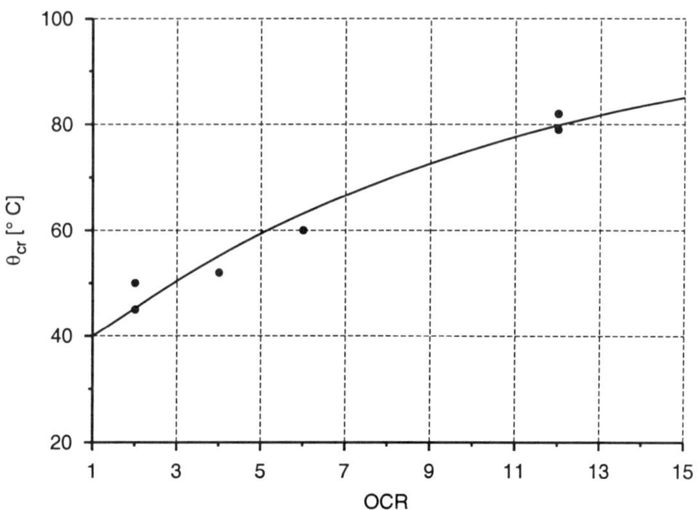

FIGURE 5.3. Critical thermo-plastic collapse temperature as a function of OCR (data taken from [15]).

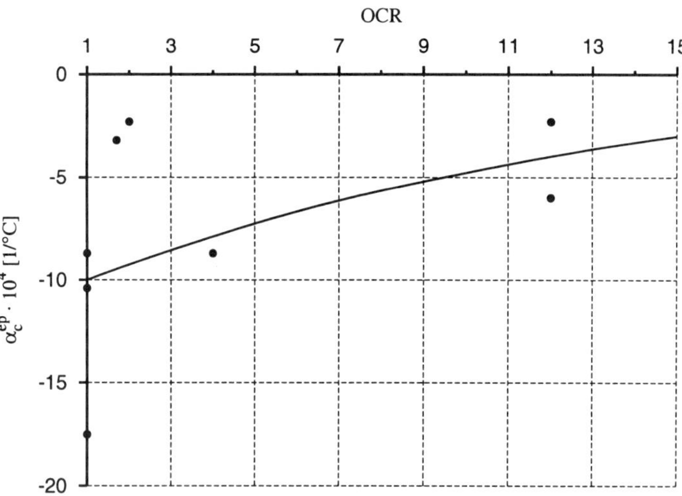

FIGURE 5.4. Thermoplastic contraction coefficient as a function of OCR (data taken from [15]).

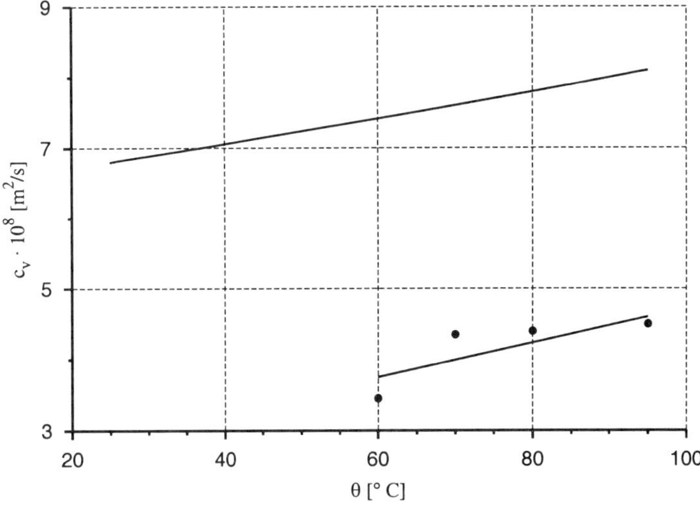

FIGURE 5.5. Example of variation of the 'consolidation' coefficient as function of temperature.

The consolidation coefficient

The consolidation coefficient is given as a function of the soil compressibility and the soil permeability, as

$$c_v = \frac{1}{f c_c}. \tag{5.29}$$

Following the experimental results by Delage *et al.* [4], the consolidation coefficient of a clay is expected to be a weakly-increasing function of temperature, as indicated in Figure 5.5. The assumed relatively 'large' values of c_v are justified because we consider here the case where the gauge material becomes over-consolidated, since it undergoes an 'elastic' unloading $(c_c \equiv c_{sw})$, as far as the mean effective stress is concerned. This results in turn in a relatively small value for the soil compressibility and to 'large' values for the consolidation coefficient. The corresponding permeability coefficient is computed from the equation

$$k_w = c_v \gamma_w c_{sw}. \tag{5.30}$$

Pore-pressure-temperature coefficient

The pore-pressure-temperature coefficient is given in terms of thermal expansion and compressibility coefficients as

$$\lambda_m = \left. \frac{\partial p}{\partial \theta} \right|_{\substack{V=\text{const},\\ \text{undrained}}} = \frac{\alpha}{c}, \tag{5.31}$$

with

$$\alpha = \alpha_m - \alpha_c;$$
(5.32)

we get

$$\lambda_m = \begin{cases} 0, & \text{if} \quad \theta \leq \theta_{cr} \quad \text{or if} \quad \theta \leq \theta_Y, \\ -\dfrac{\alpha_c^p}{c} < 0, & \text{if} \quad \theta = \theta_Y > \theta_{cr}. \end{cases}$$
(5.33)

Notice that the coefficient λ_m should be directly calibrated in isochoric, undrained-heating tests on water-saturated specimens with specially designed equipment that will allow for compensated pore-pressure measurements. There is no such experiment in the pertinent soil mechanics literature today.

5.3   Energy balance in porous soils

We consider a two-phase, porous, soil-like material, consisting of solids and fluid (water). Energy balance for the solid skeleton in local form is expressed in terms of the rate of specific internal energy $e(x_i, t)$, the effective-stress power $P(x_i, t)$ and the heat flux vector $Q_i(x_k, t)$ [17]:

$$\rho \dot{e} = P - \frac{\partial Q_k}{\partial x_k}.$$
(5.34)

The effective-stress power is defined as the inner product of the effective stress (defined in the sense of Terzaghi) and of the rate of deformation:

$$P = \sigma'_{ij} D_{ij}.$$
(5.35)

In order to evaluate the energy balance equation we need to introduce a set of additional constitutive assumptions. First we assume that the rate of deformation and with that the stress power are decomposed additively into two parts

$$D_{ij} = D^e_{ij} + D^p_{ij} \Rightarrow P = \dot{w}^{(e)} + D.$$
(5.36)

We identify D^e_{ij} as the *elastic deformation*, which is responsible for the mechanical energy stored in the material in a recoverable way[2]:

$$\dot{w}^{(e)} = \sigma'_{ij} D^e_{ij}.$$
(5.37)

The part D^p_{ij} is identified as the *visco-plastic deformation*, which corresponds to that part of the mechanical energy that is dissipated in heat:

$$D = \sigma'_{ij} D^p_{ij}.$$
(5.38)

---

[2]This energy may be recovered by performing isothermal *loading-unloading* cycles. This means that the "elastic" deformation is reversible.

Secondly we assume that, in a first approximation, the rate of the specific internal energy depends on the changes in temperature $\theta(x_i, t)$ and on the rate of 'elastic' deformation

$$\rho\dot{e} = \rho j C\dot{\theta} + \dot{w}^{(e)}. \qquad (5.39)$$

In this expression, C is the specific heat of the soil. Notice that in the internal energy constitutive equation (5.39) the factor $j = 4.2\,\mathrm{J/cal}$ is the mechanical equivalent of heat. Accordingly, the energy balance law, equation (5.34), becomes

$$\rho j C\dot{\theta} = -\frac{\partial Q_k}{\partial x_k} + D. \qquad (5.40)$$

The heat flux vector is related to the temperature gradient, according to Fourier's constitutive law of heat conduction,

$$Q_i = -j k_F \frac{\partial\theta}{\partial x_i}. \qquad (5.41)$$

In equation (5.41), k_F is Fourier's coefficient of thermal conductivity of the soil; for a water-saturated clay a typical value is $k_F = 0.1\,\mathrm{cal}/(^\circ\mathrm{C\,m\,s})$. With Fourier's law, the energy balance equation (5.40) results in the heat conduction/heat generation equation

$$\rho j C\dot{\theta} = j k_F \nabla^2\theta + D. \qquad (5.42)$$

The heat generation term is given by the dissipation function D, defined above through equation (5.38). By neglecting again convective terms, equation (5.42) leads to the heat conduction equation

$$\frac{\partial\theta}{\partial t} = \kappa\nabla^2\theta + \frac{1}{\rho j C}D, \qquad (5.43)$$

where $\kappa = k_F/(\rho C)$ is Kelvin's coefficient of thermal diffusivity of the soil-water mixture with dimensions $[\kappa_m] = \mathrm{L}^2\mathrm{T}^{-1}$.

5.4   The infinite slide

Landslides move due to the action of gravity and seepage forces. Here we consider an extended landslide at constant base slope angle β and constant height h (Figure 5.6). The x-coordinate points in the long direction of the slide and the z-coordinate is taken normal to it. The analysis is 1D and accordingly all variations in x-direction are neglected, $\partial/\partial x \equiv 0$. The deformation is assumed to be localized at the bottom of the landslide, within a shear-band. This means that the shear-band is primarily considered as a low-friction material layer. Since soils are frictional materials, obeying

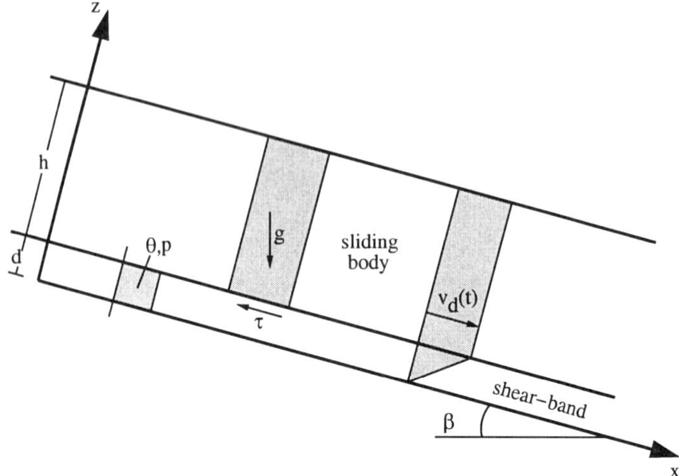

FIGURE 5.6. Geometric layout of a long landslide and the dynamics of landslide motion.

Terzaghi's effective stress principle, a slice is now cut free inside the shear-band at a level where maximum pore pressures develop. We notice that the pore-pressure, which controls the dynamics of the problem, is not necessarily hydrostatic. Excess pore-pressures produced by frictional heating and thermo-plastic collapse of the shear-band material may not have the time to dissipate rapidly and thus they will affect essentially the stability of the system. This is true as soon as the shear-band material has relatively low permeability and thermal conductivity in the z-direction, which is indeed the case in most clayey gauges. Indeed conditions of low friction parallel to the band axis and low permeability as well as thermal conductivity normal to the band axis will develop inside as soon as the gauge material is rich in clay, whose particles are elongated and align themselves parallel to the direction of motion. In addition to that, clays suffer thermo-plastic skeleton collapse. Unless the base of the band material is impermeable, the maximum excess pore-water pressures will develop in the middle of the shear-band. Thus we envision cutting the slice free at the mid-level of the shear-band and there we consider the shear stress $\tau = \tau_h(t)$, which resists to the landslide motion.

The considered slice of the slide is infinitesimal in the x-direction. Since the shear-band thickness d is assumed to be small as compared to the total height h of the slide ($d \ll h$), we conclude that the thickness in the z-direction of the slice is also constant, $h' = h + d/2 \approx h$. It is assumed that the bulk of the slide moves as a rigid body with a speed $v_x = \bar{v}(t)$. Velocity and acceleration in the z-direction are negligible (this constitutes the so-called "shallow-water" approximation of the slide dynamics). Equilibrium in the z-direction yields that the normal total stress at the bottom of the slice is constant, $\sigma_{zz} \approx -\sigma_h = \gamma h \cos\beta$, where $\gamma = \rho g$ is the total unit weight of the soil.

In the x-direction, dynamic equilibrium of the considered slice yields the following equation of motion[3]:

$$\frac{d\bar{v}}{dt} = g\left(\sin\beta - \frac{\tau_h(t)}{\gamma h}\right). \tag{5.44}$$

We notice that, due to frictional heating, pore-pressures $\tilde{p} = \tilde{p}_w(z,t)$ will be generated inside the shear-band, which will be in excess to the hydrostatic ones. Accordingly, for $0 \leq z \leq d$ and $t > 0$, the hydrostatic pore-water pressure is perturbed:

$$p_w = \gamma_w h \cos\beta + \tilde{p}_w(z,t). \tag{5.45}$$

Normal effective and shear stresses inside the shear band are equally perturbed. Let

$$\sigma'_{zz} = \sigma'(z,t); \quad \sigma_{zx} = \tau(z,t) = \tau_h(t) + \tilde{\tau}(z,t). \tag{5.46}$$

For simplicity, we assume conditions of simple shear inside the shear-band, with $\sigma'_{zz} = \sigma'_{xx}$ (Figure 5.7). The out-of-plane effective stress is set proportional to the mean in-plane effective normal stress, and thus the mean effective normal stress becomes

$$p' = \frac{1}{3}(\sigma'_{xx} + \sigma'_{yy} + \sigma'_{zz}) = \frac{2}{3}(1+\nu)\sigma', \tag{5.47}$$

where ν is the drained Poisson's ratio of the gauge.

We notice that, for low values of the frictional shear stress due to either small values of the friction angle and/or large values of the excessive thermo-plastic pore-water pressures, the above equation (5.44) yields asymptotically to the solution

$$\frac{d\bar{v}}{dt} \approx g\sin\beta. \tag{5.48}$$

Inside the shear-band, the velocity field varies in the z-direction:

$$v_x^{(1)} = v(z,t); \quad v_z^{(1)} = w(z,t); \quad v_y^{(1)} = 0. \tag{5.49}$$

In order to make this field compatible with the slide motion, we impose the boundary conditions

$$v(0,t) = 0; \quad v(d,t) = \bar{v}(t). \tag{5.50}$$

Similarly we assume that pore-water can only flow in and out of the shear-band in the z-direction, normal to the shear-band axis:

$$q_x = q_y = 0; \quad q_z = q(z,t). \tag{5.51}$$

---

[3] The differential operator d/dt, acting on a function of time only, is not to be confused with the material time derivative D/Dt, acting on a field that is a function of x_i and t.

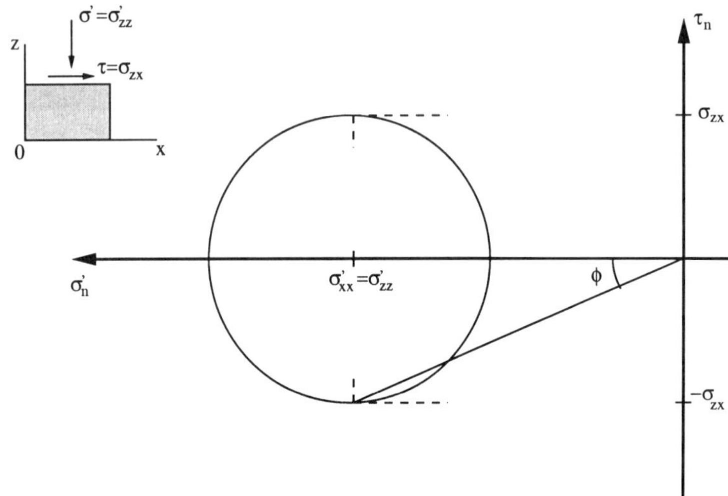

FIGURE 5.7. State of stress inside the shear-band.

5.5 Drained soil behavior

The soil inside the shear-band is assumed to have undergone already large amounts of shear, reaching the so-called *critical state* as far as its plastic volumetric strains is concerned; *i.e.*, we will assume here zero plastic dilatancy. As we will show in this paper, during rapid shear the heat production due to plastic work dissipation leads to a net increase of the pore-water pressure. Since the total vertical stress is constant, this means a decrease in effective stresses. Assuming full plastification of the soil under shear, the corresponding effective stress path is similar to the one that loose soils follow in undrained "liquefaction" tests. However, the material behavior here is different. In order to illustrate the considered behavior, we assume here the existence of two distinct yield surfaces in stress space: a) The so called "cup" describing the volumetric behavior:

$$F_1 = \sigma' - \sigma'_{v0} = 0, \tag{5.52}$$

where $\sigma_{v0} = \sigma'_h$ is the so-called *pre-consolidation stress*, and b) the so-called "Coulomb" yield surface, describing the deviatoric, frictional response:

$$F_2 = \tau - \mu_C |\sigma'_{zz}| = 0, \tag{5.53}$$

where

$$\mu_C = \tan \phi \tag{5.54}$$

is Coulomb's friction coefficient for the shear-band material.

In summary, we assume here that the soil inside the shear band is undergoing unloading with respect to the cup $F_1 = 0$ and that at any instant

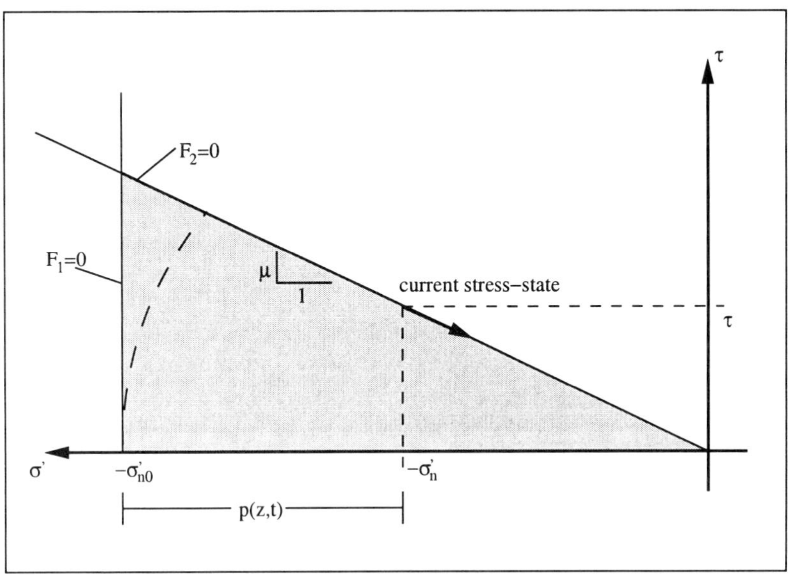

FIGURE 5.8. Assumed effective stress-path.

the effective stress state S lies on the Coulomb yield surface $F_2 = 0$. The corresponding effective stress path points towards the origin (the stress-free or "liquefied" state).

5.6   Governing equations

In the considered case, the pore-pressure diffusion equation (5.28) becomes

$$\frac{\partial p_w}{\partial t} \approx c_v \frac{\partial^2 p_w}{\partial z^2} + \lambda_m \frac{\partial \theta}{\partial t}, \tag{5.55}$$

where c_v and λ_m are the *"consolidation"* and *"pore-pressure-temperature"* coefficients respectively.

For evaluation of the heat equation we set

$$\dot{\gamma} = D_{13} = \frac{\partial v}{\partial z} \quad \text{and} \quad \dot{\varepsilon} = D_{33} = \frac{\partial w}{\partial z} \tag{5.56}$$

for the total shear and volumetric strain-rate inside the shear-band, respectively. As already mentioned, volumetric strains are purely elastic since a) unloading takes place with respect to the cup, and b) with respect to the Coulomb yield surface the material is at critical state,

$$\dot{\varepsilon}^p = 0 \Rightarrow \dot{\varepsilon} = \dot{\varepsilon}^e. \tag{5.57}$$

Similarly, we have the decomposition of the shear strain-rate into an elastic and a plastic part. Assuming that the elastic shear strain-rate is small as compared to the plastic one, we get

$$|\dot{\gamma}^e| \ll \dot{\gamma}^p \Rightarrow \dot{\gamma}^p \approx \dot{\gamma}. \qquad (5.58)$$

From these equations, the dissipation function is computed as

$$D = \sigma'\dot{\varepsilon}^p + \tau\dot{\gamma}^p = \tau\dot{\gamma}^p \approx \tau\frac{\partial v}{\partial z}. \qquad (5.59)$$

In the considered problem, the shear stress is given by the expression

$$\tau = -(\sigma_{zz} + p_w)\mu \Rightarrow \tau = \mu\gamma h\left(\cos\beta - \frac{p_w}{\gamma h}\right). \qquad (5.60)$$

Combining the above equations with the energy balance equation (5.43), we obtain the heat equation

$$\frac{\partial\theta}{\partial t} = \kappa_m\frac{\partial^2\theta}{\partial z^2} + \mu_C\frac{\gamma h}{\rho j C_m}\left(\cos\beta - \frac{p_w}{\gamma h}\right)\frac{\partial v}{\partial z}. \qquad (5.61)$$

Finally, we remark that momentum balance inside the shear-band together with the assumed friction law for the shear stress gives

$$\frac{\partial v}{\partial t} = -\frac{1}{\rho}\mu\frac{\partial p}{\partial z}. \qquad (5.62)$$

The boundary conditions for the velocity field are given above through equations (5.50).

5.7 Viscous regularization

5.7.1 Formulation of the problem

First, we observe an asymmetry as far as the mathematical structure of the momentum balance equation (5.62) is concerned in comparison to the other two balance laws, equations (5.55) and (5.61). Due to the Darcy and Fourier gradient-type laws, mass and energy balance are expressed by diffusion-type, second-order partial differential equations (p.d.e.), whereas momentum balance for an ideally plastic frictional material results in a first-order wave-type p.d.e. These equations, together with a set of initial and boundary conditions for the fields $p(z,t)$, $\theta(z,t)$ and $v(z,t)$ at the shear-band boundaries, constitute a mathematically ill-posed problem. This is clear from the fact that the velocity appears in the above system of p.d.e. with its first spatial derivative only, and accordingly there is no way to absorb two distinct boundary conditions for it. This, in turn, means that, in order to solve the corresponding boundary-value problem, some kind of mathematical regularization is needed. In our previous paper [18] we used a "crude" method of regularization, namely that of replacing the momentum equation for the velocity field with an ad-hoc solution for the velocity that is linear-in-z and is compatible with the boundary conditions. Here we justify this assumption by resorting to a more "rigorous" approach. Accordingly, we first consider a viscous-type regularization by assuming that the friction coefficient μ is rate sensitive, $i.e.$, also a function of the shearing velocity gradient

$$\mu = \hat{\mu}(\dot{\gamma}) \,; \quad \dot{\gamma} = \frac{\partial v}{\partial z}. \tag{5.63}$$

If we introduce such a shearing velocity-gradient dependency of the friction coefficient into the friction law for the shear stress, the momentum equation (5.62) is drastically modified, resulting in a diffusion-generation type p.d.e. for the velocity:

$$\frac{\partial v}{\partial t} = \nu_m \frac{\partial^2 v}{\partial z^2} - \mu \frac{\partial p}{\partial z}. \tag{5.64}$$

The coefficient ν_m in front of the second spatial derivative of the velocity plays the role of a kinematic viscosity, $\nu_m = (\sigma'_{n0} - p)\frac{H\nu}{\rho_m}$, and is given in terms of the friction-rate sensitivity modulus $H = \frac{d\hat{\mu}}{d\dot{\gamma}}$.

5.7.2   Discussion of the model

The problem is now re-formulated in terms of a set of dimensionless variables. For this purpose we first select a set of reference quantities. The geometric scale in the vicinity of the shear-band is set by its thickness d. Thus, the shear-band thickness is selected as a reference length $d_{\mathrm{ref}} = d$. The mean initial effective normal stress σ'_{n0} also serves as a reference pressure or stress $p_{\mathrm{ref}} = \sigma'_{n0}$. We introduce a 'geostatic' depth $h_{\mathrm{ref}} = \frac{\sigma'_{n0}}{\gamma'}$, where γ' is the effective or buoyant unit weight. For gravity driven phenomena, we may use this length scale parameter to define an appropriate reference velocity $v_{\mathrm{ref}} = \sqrt{gh_{\mathrm{ref}}}$ and from that we may compute a reference time $t_{\mathrm{ref}} = \frac{d_{\mathrm{ref}}}{v_{\mathrm{ref}}}$. Finally, we use the initial ambient temperature θ_0 as reference, $\theta_{\mathrm{ref}} = \theta_0$. With these reference values we define the following set of non-dimensional quantities:

$$z^* = \frac{z}{d_{\mathrm{ref}}} \,, \quad t^* = \frac{t}{t_{\mathrm{ref}}} \geq 0 \,,$$

$$v^* = \frac{v}{v_{\mathrm{ref}}} \,, \quad p^* = \frac{p}{p_{\mathrm{ref}}} \,, \quad \theta^* = \frac{\theta}{\theta_{\mathrm{ref}}} \,. \tag{5.65}$$

In addition we define the following dimensionless numbers:

$$\kappa_p = \frac{c_v t_{\text{ref}}}{d_{\text{ref}}^2}\,, \qquad \lambda = \frac{\lambda_m \theta_{\text{ref}}}{p_{\text{ref}}}\,,$$
$$\kappa_\theta = \frac{\kappa_m t_{\text{ref}}}{d^2}\,, \qquad \eta_F = \frac{p_{\text{ref}}}{\rho_m j C_m \theta_{\text{ref}}}\mu\,, \qquad (5.66)$$
$$\kappa_v = \frac{p_{\text{ref}} t_{\text{ref}}}{\rho_m d_{\text{ref}}^2}H\,, \quad \eta_p = \frac{p_{\text{ref}}}{\rho_m v_{\text{ref}}^2}\mu\,.$$

For simplicity in notation, in the resulting set of governing equations we drop the superimposed asterisk and assume that all quantities are dimensionless. The resulting set of the three coupled partial differential equations reads as follows:

$$\frac{\partial p}{\partial t} = \frac{\partial}{\partial z}\left(\kappa_p \frac{\partial p}{\partial z}\right) + \lambda \frac{\partial \theta}{\partial t}\,, \qquad (5.67)$$

$$\frac{\partial \theta}{\partial t} = \kappa_\theta \frac{\partial^2 \theta}{\partial z^2} + \eta_F(1-p)\frac{\partial v}{\partial z}\,, \qquad (5.68)$$

$$\frac{\partial v}{\partial t} = \kappa_v(1-p)\frac{\partial^2 v}{\partial z^2} - \eta_p \frac{\partial p}{\partial z}\,. \qquad (5.69)$$

The initial conditions for the concerned fields, in the domain of definition $0 \le z \le 1$, are

$$v(z,0) = 0\,, \quad p(z,0) = 0\,, \quad \theta(z,0) = 1\,. \qquad (5.70)$$

The boundary conditions are

$$\begin{aligned} p(0,t) &= p(1,t) = 0\,, \\ \theta(0,t) &= \theta(1,t) = 1\,, \\ v(0,t) &= 0\,, \quad v(1,t) = v_1(t)\,. \end{aligned} \qquad (5.71)$$

As far as the boundary velocity is concerned, we may assume for simplicity the example of constant acceleration, equal to some fraction α $(0 < \alpha < 1)$ of the gravity acceleration $v_d = \alpha g t$, $\alpha = \sin\beta$ (*cf.* [18]). In Figures 5.9 (a) to (c) we see the computed isochrons for the temperature and heat-generated pore-pressure and velocity fields. The corresponding material and system parameters are listed in Table 5.1.

One important feature we observe from this typical computation is that, in order to achieve stability, the diffusivity κ_v has to assume relatively high values. If this is not the case, then the velocity profile becomes, after some finite time, non-linear and results in mathematical instability. The value of the diffusivity κ_v is mainly controlled by the hardening modulus H.

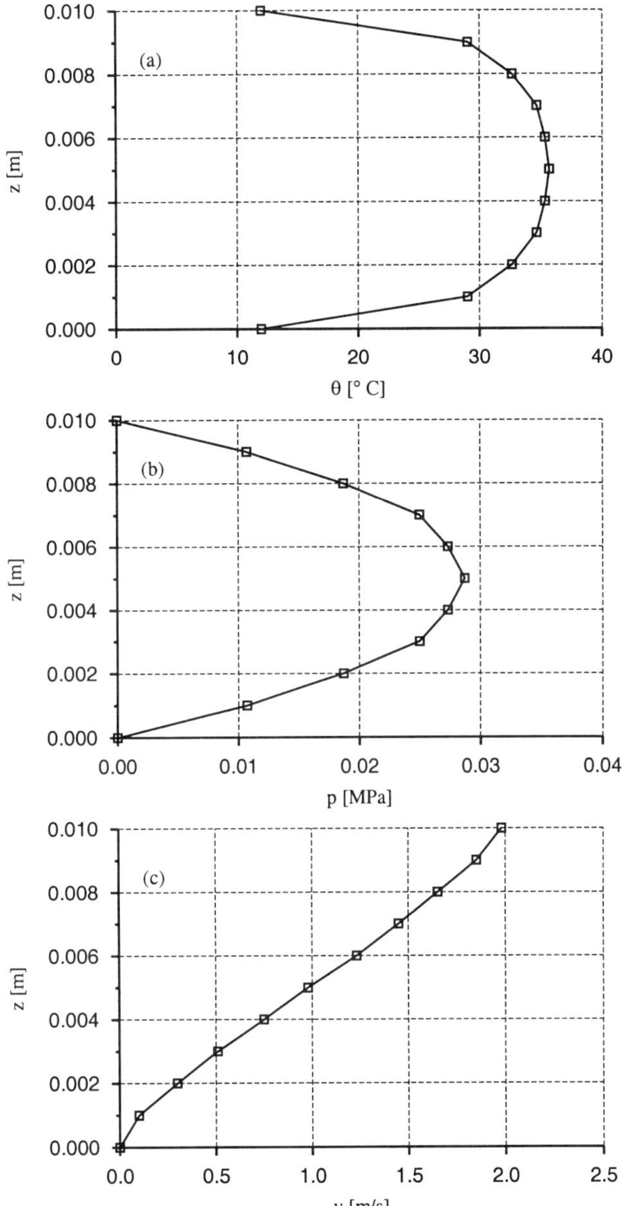

FIGURE 5.9. Viscous regularization: (a) isochron for the temperature ($t = 1\,s$); (b) isochron for the excess pore-pressure ($t = 1\,s$); (c) early-time mathematical instability resulting in non-linear velocity profile across the shear-band ($t = 1\,s$).

ρ_m	2.435 gr/cm^3	d_{ref}	0.01 m
γ_m	23.89 kN/m^3	h_{ref}	169.07 m
n	0.18	v_{ref}	40.725 m/s
ϕ_{sta}	10 °	t_{ref}	$2.46 \cdot 10^{-4}$ m/s
ϕ_{dyn}	15 °	t_{ref}	12 °C
χ	0.103 s	σ'_{n0}	2.38 MPa
c	$1.5 \cdot 10^{-3}$ 1/MPa	α	0.2
$k_{w,\text{ref}}$	$3.0 \cdot 10^{-7}$ cm/s		
λ_m	$2.0 \cdot 10^{-2}$ MPa/°C	κ_p	$4.9 \cdot 10^{-5}$
$(\rho jC)_m$	2.8 MPa/°C	κ_θ	$7.4 \cdot 10^{-7}$
c_w	$4.9 \cdot 10^{-4}$ 1/MPa	κ_{v0}	$2.2 \cdot 10^1$
$c_{w,\text{ref}}$	$2.0 \cdot 10^{-5}$ 1/MPa	λ	$1.0 \cdot 10^{-1}$
κ_m	$3.0 \cdot 10^{-7}$ m^2/s	η_{F0}	$1.3 \cdot 10^{-2}$
ν_m	9.0 m^2/s	η_{p0}	$1.0 \cdot 10^{-1}$

TABLE 5.1. System and material parameters for rapid shearing of a shear band with shearing velocity hardening.

5.8 Gradient regularization

In a recent paper, Tika and Hutchinson [16] reported the strain- and strain-rate softening of clay soil that was sampled from the exposed failure surface of the Vaiònt slide and was tested in the ring shear apparatus of Imperial College, Figure 5.10. We remark that in another recent paper, Di Prisco *et al.* [3] reported and modeled a similar frictional rate-softening behavior in loose sands, tested in triaxial creep tests.

Friction strain-rate softening results formally in negative 'kinematic' viscosity. The corresponding momentum balance equation, in terms of dimensionless quantities, has a negative diffusivity coefficient ($\kappa_v \propto H < 0$) and becomes accordingly an 'uphill' diffusion equation

$$\frac{\partial v}{\partial t} = -|\kappa_v|(1-p)\frac{\partial^2 v}{\partial z^2} - \eta_p \frac{\partial p}{\partial z}, \tag{5.72}$$

since, for the time running backwards ($t' = -t < 0$), this is an ordinary diffusion-generation equation. Uphill diffusion is known to be mathematically ill posed [10], and in the considered setting has no physical meaning.

Following Vardoulakis and Sulem [17], we may regularize such a problem by introducing a second-gradient modification of the visco-plastic friction law, by assuming that the friction coefficient is not only a function of the shear strain rate, but also of its second-spatial derivative,

$$\mu = \hat{\mu}(\dot{\gamma}) - r\nabla^2\dot{\gamma}, \quad r > 0. \tag{5.73}$$

The parameter r has the dimensions $[r] = \text{TL}^2$. If we normalize r by the

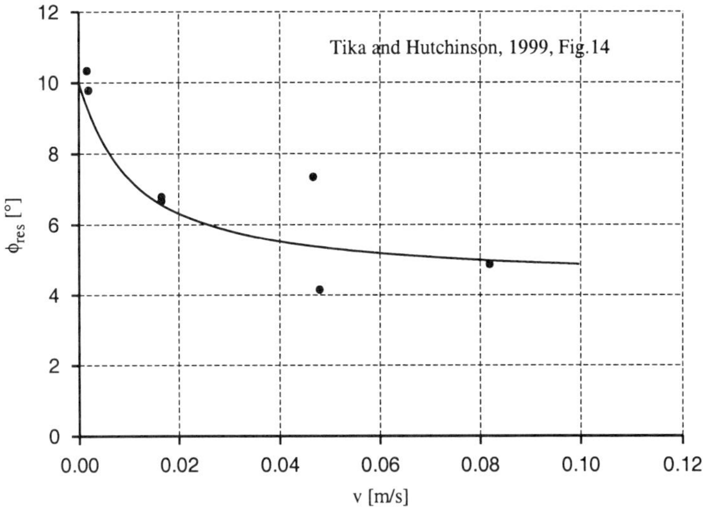

FIGURE 5.10. Friction-velocity softening in ring-shear tests after Tika and Hutchinson [16].

rate sensitivity parameter

$$\chi = \frac{\partial \mu}{\partial \dot{\gamma}} \tag{5.74}$$

with $[\chi] = \mathrm{T}$, we obtain the corresponding 'material length':

$$\ell_c = \sqrt{\frac{r}{\chi}} \;\Rightarrow\; r = \chi \ell_c^2. \tag{5.75}$$

This second-gradient extension of the viscous friction law yields to the following modification of the momentum equation:

$$\frac{\partial v}{\partial t} = -\varepsilon \frac{\partial^4 v}{\partial z^4} - |\kappa_v|(1-p)\frac{\partial^2 v}{\partial z^2} - \eta_p \frac{\partial p}{\partial z} \tag{5.76}$$

with

$$\varepsilon = r \frac{p_{\mathrm{ref}}\, t_{\mathrm{ref}}}{\rho_m\, d_{\mathrm{ref}}^4}. \tag{5.77}$$

According to Lattes and Lions [10], the set of additional boundary conditions that complies with this regularization scheme is a statement for the second spatial derivative of the velocity, *e.g.*,

$$\left. \frac{\partial^2 v}{\partial z^2}\right|_{z=0,1} = 0. \tag{5.78}$$

Equation (5.76) was solved together with equations (5.67) and (5.68), with initial conditions (5.70) and boundary conditions (5.50) and (5.71), using a Crank–Nicolson integration scheme. In Figures 5.11, 5.12 and 5.13 we show the computational results for the parameter values listed in Table 5.1. Notice that critical for the integration of equation (5.76) is the appropriate selection of the singular perturbation parameter ε. The selection of ε depends in turn on the assumed value of the softening modulus

$$H_s = \left|\frac{d\hat{\mu}}{d\dot{\gamma}}\right|. \tag{5.79}$$

This analysis suggests that reasonable results were obtained for a value of the perturbation parameter $\varepsilon = 0.1$. For $\chi = 0.1 \cdot 10^{-3}$ s, this value corresponds to $\ell_c \approx 0.65$mm. With a mean particle size for clay material of $d_{50\%} \approx 7\mu$m, this analysis suggests

$$\ell_c \approx 100 \cdot d_{50\%}. \tag{5.80}$$

This means that the most appropriate material length seems to be again a large multiple of the clay-particle characteristic dimension.

5.9   Summary of main results

In rapid clayey-fault-gauge shearing, the effects of pore-water heating and consequent pore-pressure rise to the thermo-plastic gauge-skeleton collapse cannot be disregarded. It has been demonstrated above that mass and energy balance together with Darcy's and Fourier's laws are mathematically incompatible with the simple Coulomb–Terzaghi soil model. We also notice that a simple viscous regularization due to friction rate sensitivity[4] does not rectify the mathematical ill-posedeness and one has to resort on top of that to second-gradient regularization schemes, which are justified on physical grounds by resorting to micro-structural continuum models for soil behavior as proposed earlier by Vardoulakis and Sulem [17].

Acknowledgments: The Author wants to acknowledge the European Union project: *Fault, Fractures and Fluids: Gulf of Corinth,* in the framework of program Energy (ENK6-2000-0056).

References

[1] D. L. ANDERSON: *An Earthquake Induced Heat Mechanism to Explain the Loss of Strength of Large Rock and Earth Slides,* Int. Conf. on Engineering for Protection from Natural Disasters, Bangkok, 1980.

---

[4]Rice and Tse [13] suggested, for example, frictional strain-rate softening for the mechanical description of joint rock sliding.

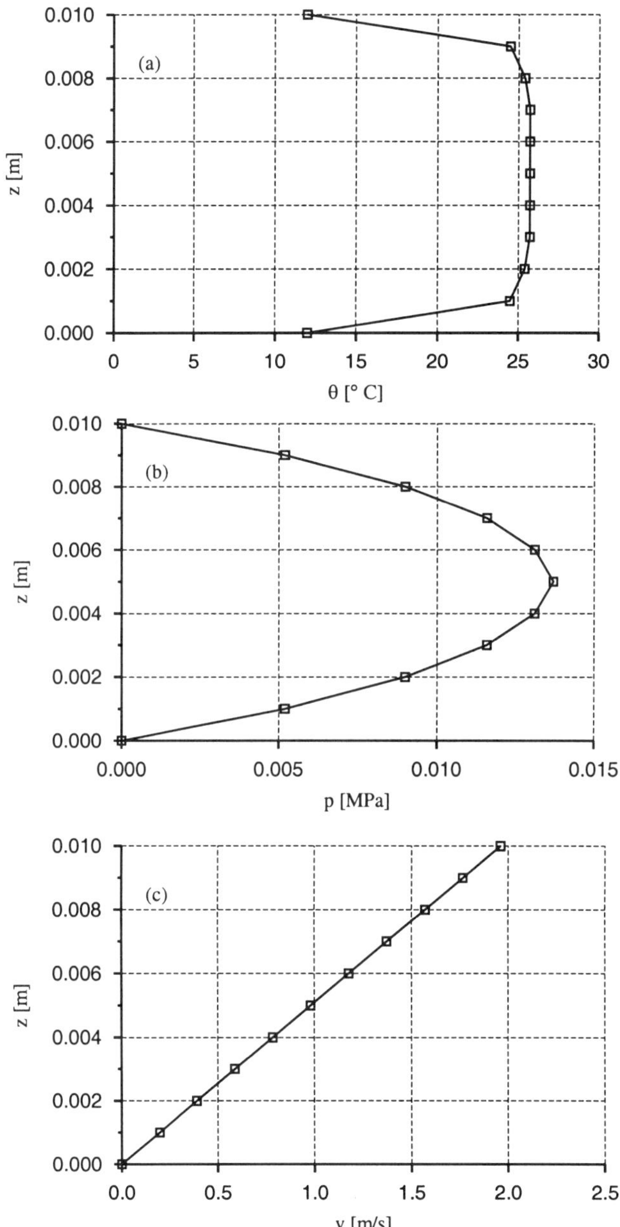

FIGURE 5.11. Second-gradient regularization: (a) isochron for the temperature; (b) isochron for the excess pore-pressure; (c) stable solution resulting in linear velocity profile across the shear-band.

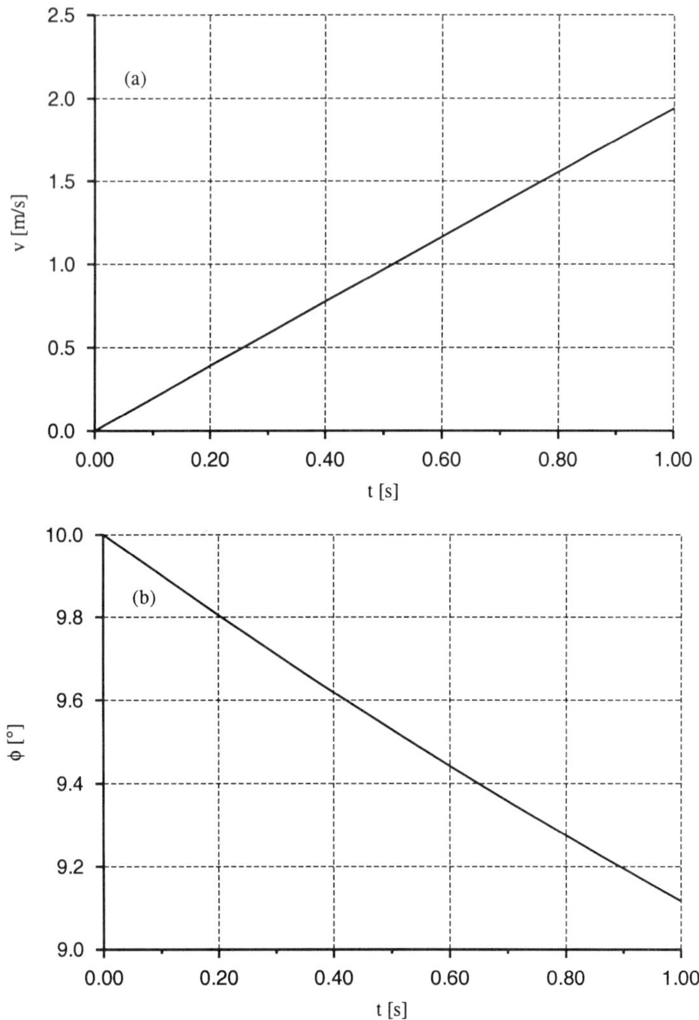

FIGURE 5.12. Second-gradient regularization: (a) boundary velocity (input); (b) friction softening inside the shear-band (input);

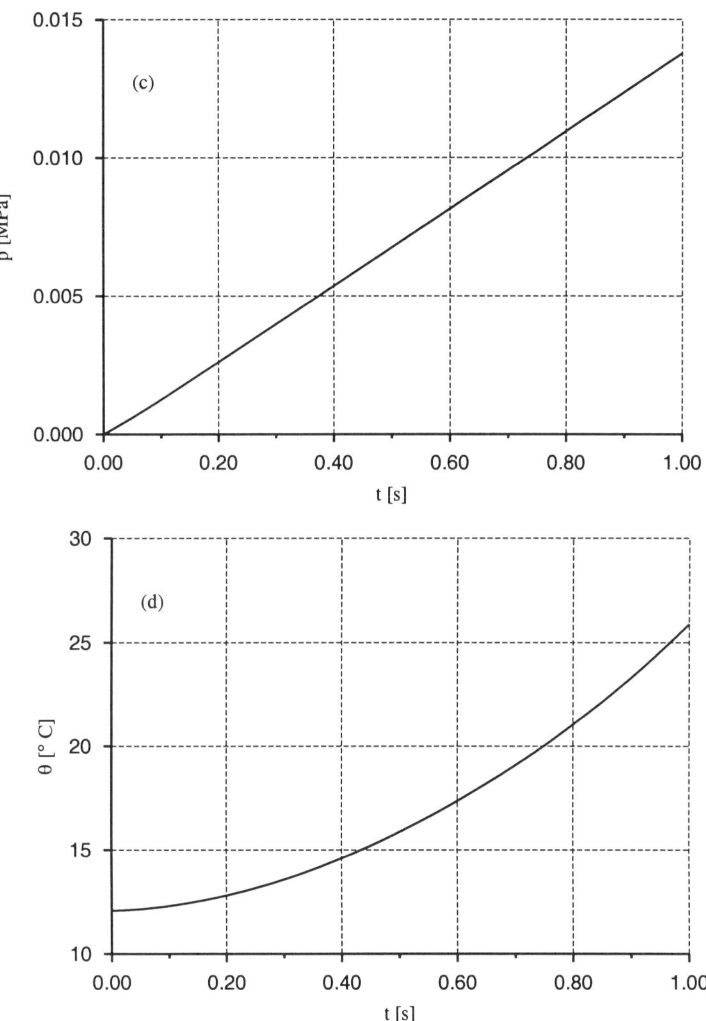

FIGURE 5.13. Second-gradient regularization: (c) excess-pore-pressure inside the shear-band (output); (d) temperature evolution inside the shear-band (output).

[2] R.G. CAMPANELLA, J.K. MICHELL: Influence of temperature variations on soil behavior, *J. Soil Mech. Found. Div. ASCE* **94**, pp. 709–734, 1968.

[3] C. DI PRISCO, S. IMPOSIMATO, I. VARDOULAKIS: Mechanical modeling of drained creep triaxial tests on loose sand, *Geotechnique* **50**, pp. 73–82, 2000.

[4] P. DELAGE, S. NABIL, Y.J. CUI: On the thermal consolidation of Boom clay, *Can. Geotech. J.* **37**, pp. 343–354, 2000.

[5] J. GOGUEL: Scale-dependent rockslide mechanisms, with emphasis on the role of pore fluid vaporization, in: B. VOIGHT (ed.): Rockslides and avalances; natural phenomena, *Developments in Geotechnical Engineering* **14**, Elsevier Science Publishers, Amsterdam, 1978, pp. 693–705.

[6] P. HABIB: Sur un mode de glissement des massifs rocheaux, *C. R. Hebd. Seanc. Acad. Sci.* **264**, pp. 151–153, 1967.

[7] P. HABIB: Production of gaseous pore pressure during rock slides, *Rock Mech.* **7**, pp. 193–197, 1975.

[8] R.D. HOLTZ, W.D. KOVACS: *An Introduction to Geotechnical Engineering*, Prentice-Hall, Englewood Cliffs, New Jersey, 1981.

[9] A.H. LACHENBRUCH: Frictional heating, fluid pressure and the resistance to fault motion, *J. Geophys. Res.* **85**, pp. 6097–6112, 1980.

[10] R. LATTES, J.-L. LIONS: *Methode de Quasi-Reversibilite et Applications*, Dunod, Paris, 1967.

[11] C.W. MASE, L. SMITH: Pore-fluid pressures and frictional heating on a fault surface, *Pure Appl. Geoph.* **122**, pp. 583–607, 1985.

[12] H. MODARESSI, L. LALOUI: A thermo-viscoplastic constitutive model for clays, *Int. J. Num. Anal. Meth. Geomech.* **21**, pp. 313–335, 1997.

[13] J.R. RICE, S.T. TSE: Dynamic motion of a single degree of freedom system following a rate and state dependent friction law, *J. Geophys. Res.* **91**, pp. 521–530, 1986.

[14] S.U. ROMERO, R. MOLINA: Kinematic aspects of the Vaiònt slide, *Proc. 3rd Congress Int. Soc. Rock Mechanics, Denver* **IIB**, 1974, pp. 865-870.

[15] N. SULTAN: *Etude du Comportement Thermo-Mechanique de l'Argile de Boom: Experiences et Modelisation*, Ph.D. thesis, E.N.P.C., C.E.R.M.E.S, Paris, 1997.

[16] T.E. TIKA, J.N. HUTCHINSON: Ring shear tests on soil from the Vaiònt landslide slip surface, *Geotechnique* **49**, pp. 59–74, 1999.

[17] I. VARDOULAKIS, J. SULEM: *Bifurcation Analysis in Geomechanics*, Chapman&Hall, London, 1995.

[18] I. VARDOULAKIS: Catastrophic landslides due to frictional heating of the failure plane, *Mech. Coh.-Frict. Mat.* **5**, pp. 443–467, 2000.

[19] I. VARDOULAKIS: Dynamic thermo-poro-mechanical analysis of catastrophic landslides, *Geotechnique* **52**, 2002, to appear.

[20] B. VOIGHT, C. FAUST: Frictional heat and strength loss in some rapid landslides, *Geotechnique* **32**, pp. 43–54, 1982.

IOANNIS VARDOULAKIS
Department of Mechanics
National Technical University of Athens
GR-15773, Zographou, GREECE

E-mail: I.Vardoulakis@mechan.ntua.gr

Part II

Flow and Transport Phenomena in Particulate Materials

Chapter 6

Mathematical Models for Soil Consolidation Problems: a State of the Art Report

---

Davide Ambrosi, Renato Lancellotta and Luigi
Preziosi

ABSTRACT The mathematical modeling of the consolidation theory is
outlined moving from the fundamentals of the mechanics of porous media.
Starting from averaging approaches or from the postulates of the theory of
mixtures, we introduce the volume fraction concept, the balance equations
for the components of the mixture (written in Eulerian and Lagrangean
frames of reference) and the concept of effective stress. Initial boundary
value problems are considered for typical geotechnical applications, and
in this context Biot's theory is illustrated. A final discussion concerns the
mathematical priciples that are to be accomplished when formulating constitutive relationships.

6.1 Introduction

The object of the consolidation theory is, in a broad sense, any movement of
the soil that yields a non-negligible effect at the macroscale. This definition
includes prediction of settlement rate of structures interacting with the soil,
subsidence, oil production, diffusion of pollutants, transient phenomena
occurring in earthquakes and wave propagation.

Since the pioneering work of Terzaghi [37, 38], there has been a growing
interest in understanding the physical basis and the related mathematical
modeling of the consolidation theory. As outlined in the recent book by de
Boer [19], where the reader can find an introduction to the subject in a
historical perspective, this interest arises from both a theoretical basis, the
mechanics of porous media, and from engineering applications. By the way,
we note that the consolidation is the most ancient field of application of a
more general science, the porous media theory, that is currently applied in a
surprisingly large number of different fields. As an example, in biomechanics
[24] bone can be represented as a porous medium infiltrated by blood and
in composite material manufacturing the injection moulding process is in

essence the infiltration of a non-Newtonian fluid into a solid porous matrix (see [1]).

The class of problems we are interested in is characterized by the motion of solids and fluids at different velocities. The global mechanical behavior of the fluid-solid system is then dictated by both the mechanical properties of the single constituents and by the interaction occurring between them, due to their relative displacement. The aim is to describe the property of the solid-fluid system at a spatial scale much larger than the one characterizing the velocity fields. Therefore, in its essence, the problem is relevant in all scientific areas where a "coupling between the evolution of a porous deformable medium and the motion of fluids in the connected porosity" occurs.

The aim of this chapter is to provide a summary of the main results of the porous media theory that can be obtained rigorously moving from the principles of continuum mechanics. As a by-product, we make an effort to put into a unique framework equations and notations that are used in the mathematical as well engineering literature, stressing the assumptions that lead to the different models. The presentation is mainly intended for an engineering audience, so that the mathematical formalism is kept at a minimum.

This chapter is divided into five sections. Section 1 outlines the general framework of the modeling of porous materials, *i.e.*, the "averaging approaches", emphasizing the related basic assumptions. Section 2 discusses the "effective stress" concept. Section 3 deals with kinematics and balance equations at a macroscopic level according to the *mixture theory*, and derives a consistent Lagrangian model for both the soil skeleton and the fluid. Section 4 formulates the initial boundary value problem of typical geotechnical applications, such as consolidation of a stratum between draining and impervious boundaries with stress or velocity boundary conditions: in this context Biot's theory is illustrated. Finally, Section 5 discusses aspects related to the constitutive equations, which is of crucial importance when dealing with reliable quantitative predictions.

6.2   Modeling porous media

In the current literature the mechanics of a porous medium is typically described by two different approaches. The *averaging approach* starts from equations describing the motion of all phases constituting the porous medium separately: each of them is considered in its own domain as a single body. The link between this description and the macroscopic behavior is obtained by the averaging procedure. Conversely, the *macroscopic approach* starts from an *a priori* assumption of homogenized phases and is based on the axioms of the theory of mixtures supplemented by the concept of volume fraction.

It is to be noted that the final goal of both these approaches is the description of predictable and repeatable macroscopic features of the dynamics of the porous medium. In fact, even though it might be at some extent desirable to achieve a detailed knowledge of the flow of the fluid in the pores and the displacement of the solid particles at each point, this appears to be an impossible task: we can not know the detailed geometry at the microscopic scale and, more important, this geometry will be different from specimen to specimen, *i.e.*, there is a lack of repeatability. For these reasons it is relevant to focus attention on some macroscopic features independent of the specific configuration of the pores which can be verified by repeatable macroscopic experiments. Despite the aim to provide a large scale description, it is not to be understood that the smaller scale can be completely disregarded: a knowledge of the microscopic mechanisms that drive the macroscopic repeatable processes is of great interest in stating an effective macroscopic model.

The continuum mechanics approach starts from these basic considerations and any experimental determination of the constitutive response of a given medium yields relationships between "gross properties". Therefore, in their essence, the methods illustrated in this chapter are aimed at "the replacement of a micro-heterogeneous medium with a homogeneous one, which macroscopically behaves in the same manner". Such a process poses a fundamental question: how can one describe the macroscopic behavior of a medium which exhibits microscopic heterogeneity, on the basis of microstructural information?

In the following section we first discuss the conceptual passage from a microscopic to a macroscopic viewpoint through the averaging procedure, while in the next one we present a description based on porous media theory. A comparison of the results obtained by these approaches enables one to clarify the meaning and the role of several terms appearing in the field equations.

6.2.1   Volume and ensemble averaging approach

We start considering the real non-homogeneous structure of the porous medium, in which the scale of heterogeneity is of the same order of magnitude of the pore or grain dimension. Then we focus on fields describing the status of a single phase, defined only at the points occupied by that phase. Because, as already mentioned, this level of detail is not needed for our purposes, we use an averaging technique in order to get a macroscopic description (see [36], [5] and [28]). In this respect the crucial concept is the Representative Elementary Volume (REV). This idea was introduced for the first time by Lorentz, in his book on the theory of electrons [29]. Later Bear [4] discussed the same concept in relation to porous media and further rigorous contributions were given by Nemat-Nasser *et al.* [32] and Drugan

162   D. Ambrosi, R. Lancellotta and L. Preziosi

and Willis [21]. We introduce the definition of REV in a rather heuristic manner: "It is a volume that, at a macroscopic scale, is small enough to be treated as a point of the heterogeneous medium, and, on the reverse, at a microscale, is large enough to contain a significant number of single heterogeneities". The interested reader can find a comprehensive discussion on this subject in the paper of Markov [30].

From the above definition it follows that the position of a REV is identified, within the domain which mathematically defines the medium of interest, by the vector position $\mathbf{x}$. Any field q of interest at the micro-level, *i.e.*, within the REV, depends on the local coordinate $\mathbf{r}$ spanning the REV:

$$q = q(\mathbf{x}, \mathbf{r}, t). \tag{6.1}$$

The dependence expressed by (6.1) of any internal field from both the macro-coordinate $\mathbf{x}$ and the micro-coordinate $\mathbf{r}$ implies that the internal field can behave in different ways within different REVs of the medium. The relation between macro and micro quantities is obtained through volume averaging with respect to the micro-coordinate $\mathbf{r}$, which allows us to define the so-called phase average and intrinsic phase average of any field q:

$$\overline{q_\alpha(\mathbf{x}, t)} =: \frac{1}{V} \int_v H_\alpha(\mathbf{r}, t) q(\mathbf{r}, t) d\mathbf{r}, \tag{6.2}$$

$$\overline{q(\mathbf{x}, t)}^\alpha =: \frac{1}{V_\alpha} \int_v H_\alpha(\mathbf{r}, t) q(\mathbf{r}, t) d\mathbf{r}, \tag{6.3}$$

where v is the REV of volume V centered on $\mathbf{x}$ and H_α is an indicator function for the α-th constituent, its value being 1 if, at time t, the constituent is in $\mathbf{r}$, and 0 otherwise. According to this definition, the volume of the constituent within a REV is given by

$$V_\alpha(\mathbf{x}, t) =: \int_v H_\alpha(\mathbf{r}, t) d\mathbf{r}, \tag{6.4}$$

and the corresponding volume fraction n_α is

$$n_\alpha(\mathbf{x}, t) =: \frac{V_\alpha}{V} = \frac{1}{V} \int_v H_\alpha(\mathbf{r}, t) \, d\mathbf{r}. \tag{6.5}$$

As a consequence, we define $\rho_\alpha^R =: \overline{\rho}^\alpha$ material density and $\rho_\alpha = \overline{\rho} = n_\alpha \rho_\alpha^R$ is the partial density. From the above definitions, it follows that the material incompressibility is not equivalent to the bulk incompressibility of each component: as $\rho_\alpha = n_\alpha \rho_\alpha^R$ the partial density of the α-th component in the mixture can change due to changes in the volume fraction even though ρ_α^R is constant.

Once the REV average property has been obtained, in order to define the properties of the whole porous medium we should in principle perform experiments with N different REVs and obtain the so-called *ensemble average*

$$\overline{q_i^*} = \frac{1}{N}\left(\overline{q_{i,v_1}} + \overline{q_{i,v_2}} + \cdots + \overline{q_{i,v_N}}\right).$$ (6.6)

Provided that the microscale length of a typical inhomogeneity is small enough when compared with the volume of the medium, it can be expected that the ensemble and the volume averages coincide. This implies the introduction of the hypothesis of ergodicity, *i.e.*, a macroscopic or statistical homogeneity of the porous medium. Roughly speaking, this hypothesis means that the macroscopic properties of all the REV are the same and the REV has the same properties of the porous medium as a whole. For this reason in the following we make use of the volume averaging approach. Moreover, in some cases we also introduce the hypothesis that the medium is statistically isotropic, *i.e.*, the macroscopic properties are independent of the orientation of the medium in the space.

6.2.2 Balance equations

In order to derive macroscopic field equations, a general strategy can be pursued starting from the microscopic balance equations of any extensive quantity in a continuum and obtaining the corresponding macroscopic balance equation by averaging rules. In this section we follow the approach of Bear and Bachmat [5].

Let q be the volume density of any extensive phase quantity Q which is assumed to be a differentiable function of time and space within the phase domain. The general balance microscopic equation takes the form

$$\frac{\partial q}{\partial t} + \nabla\cdot(q\mathbf{V}^Q) = \rho\Gamma^Q,$$ (6.7)

where ρ is the mass density of the phase, Γ^Q is the rate of internal production of Q per unit mass and $\mathbf{V}^Q$ is the velocity attached to Q. If we introduce the volume averaged velocity $\mathbf{V}$ we can define the diffusive flux

$$\mathbf{j}^Q := q(\mathbf{V}^Q - \mathbf{V}),$$ (6.8)

and the balance equation (6.7) can be rewritten in the form

$$\frac{\partial q}{\partial t} + \nabla\cdot(q\mathbf{V} + \mathbf{j}^Q) = \rho\Gamma^Q.$$ (6.9)

Now in equation (6.9) one can distinguish the contribution of the advective flux from the diffusive one. Taking the volume average of (6.9), and by using the definitions (6.3) and (6.5), one gets

$$n_\alpha\overline{\frac{\partial q}{\partial t}}^\alpha + n_\alpha\overline{\nabla\cdot(q\mathbf{V} + \mathbf{j}^Q)}^\alpha = n_\alpha\overline{\rho\Gamma^Q}^\alpha.$$ (6.10)

By applying standard averaging rules to both sides of (6.10) (for more details, see [5]), the general macroscopic differential balance equation is obtained:

$$\frac{\partial}{\partial t}(n_\alpha \overline{q}^\alpha) + \nabla \cdot \left(n_\alpha(\overline{q\mathbf{V}}^\alpha + \overline{\mathbf{j}^Q}^\alpha)\right)$$

$$+ \frac{1}{V}\int_{S_\alpha}\left[q(\mathbf{V} - \mathbf{v}^{S_\alpha}) + \mathbf{j}^Q\right] \cdot \mathbf{n}\, dS = n_\alpha \overline{\rho \Gamma^Q}^\alpha.$$

$$(6.11)$$

The surface S_α, moving with velocity $\mathbf{v}^{S_\alpha}$, separates the phase under consideration from the other phases. The term containing the surface integral represents the flux of the extensive quantity under consideration through S_α. It is worth noticing that a boundary condition on the interface becomes, thanks to the averaging process, a source term in the macroscopic equation.

6.2.3 Mass balance equation of a phase

Now we specialize the definitions above to the mass of the α-th component (*i.e.*, $Q = m$). The corresponding density is ρ and we assume that $\Gamma^m = 0$. We chose the surface S_α as a material surface for the phase under consideration, so that there is no mass flux across it. In addition we decompose the average advective flux $\overline{\rho \mathbf{V}}^\alpha$ into a macroscopic advective flux $\overline{\rho}^\alpha \overline{\mathbf{V}}^\alpha$ and a dispersive flux which is the difference between the former and the latter. If the diffusive ($\mathbf{j}^Q$) and dispersive fluxes can be neglected when compared to the advective one, the general macroscopic balance equation (6.11) specifies to

$$\frac{\partial}{\partial t}(n_\alpha \overline{\rho}^\alpha) + \nabla \cdot \left(n_\alpha \overline{\rho}^\alpha \overline{\mathbf{V}}^\alpha\right) = 0.$$

$$(6.12)$$

6.2.4 Momentum balance equation of a phase

Following the interpretation of Bear and Bachmat, by similar arguments the extensive quantity *linear momentum* (*i.e.*, $\mathbf{Q} = m\mathbf{V}^m$) has density $q = \rho \mathbf{V}^m$, the rate of momentum production is the body force per unit mass $\mathbf{b}$ and the diffusive flux of linear momentum is the stress tensor $\mathbb{T}$, where $\mathbf{V}^m$ is the mass weighted velocity defined by

$$\rho \mathbf{V}^m(\mathbf{x}, t) =: \overline{\rho^R \mathbf{v}(\mathbf{r}, t)}.$$

$$(6.13)$$

Under the assumption that the momentum dispersive flux can be neglected when compared with the advective one and that the interface surface is material for the component at hand, equation (6.11) gives the macroscopic momentum balance

$$n_\alpha \overline{\rho}^\alpha \frac{D\overline{\mathbf{V}^m}^\alpha}{Dt} - \nabla \cdot \left(n_\alpha \overline{\mathbb{T}}^\alpha\right) = \frac{1}{V}\int_\alpha \mathbb{T} \cdot \mathbf{n}\, dS + n_\alpha \overline{\rho \mathbf{b}}^\alpha,$$

$$(6.14)$$

where

$$\frac{D}{Dt} =: \frac{\partial}{\partial t} + (\mathbf{V}^m \cdot \nabla). \qquad (6.15)$$

The surface integral term at the r.h.s. of equation (6.14) represents the interaction force between the phases across the microscopic interface that separates them. In general, such an interaction occurs between wetting and non-wetting fluid phases, as well as between solid and fluid. Usually the assumption of continuity of traction at the interface between the solid and the fluids $S_{s,f}$ is made, so that no jump exists across this microscopic interface, $i.e.$,

$$[\![\mathbb{T}]\!]_{s,f} \cdot \mathbf{n} = 0. \qquad (6.16)$$

On the contrary it is known that the jump at the interface between two immiscible fluids $S_{f,f}$ is balanced by the surface tension

$$\sigma = [\![\mathbb{T}]\!]_{f,f} \cdot \mathbf{n}. \qquad (6.17)$$

Remark. When considering the specific problem of two phases (solid and fluid), it is convenient to assume as a basic kinematic variable the solid displacement $\mathbf{u}^s$ and to introduce the motion of the fluid relative to the solid. With such an assumption, balance equations for the two phases are

$$\rho_s \ddot{\mathbf{u}}^s - \nabla \cdot \mathbb{T}^s = \rho_s \mathbf{b} - \hat{\mathbf{p}}, \qquad (6.18)$$

$$\rho_f \frac{\partial}{\partial t} \left(\dot{\mathbf{u}}^s + \mathbf{w}^f \right) + \rho_f \mathbf{w}^f \cdot \nabla \left(\dot{\mathbf{u}}^s + \mathbf{w}^f \right) - \nabla \cdot \mathbb{T}^f = \rho_f \mathbf{b} + \hat{\mathbf{p}}, \qquad (6.19)$$

where

$$\mathbf{w}(\mathbf{x}, t) =: \mathbf{v}^f(\mathbf{x}, t) - \mathbf{v}^s(\mathbf{x}, t). \qquad (6.20)$$

The term $\hat{\mathbf{p}}$ represents the local interaction per unit volume between the phases.

6.3   The effective stress

The interaction between the soil skeleton and the pore water was first described by Terzaghi [37] with the introduction of the *principle of effective stress*. Such a principle states that the total stress acting in a solid-liquid mixture can be decomposed into two additive contributions. The former acts both in the water and in the solid in every direction and is called *neutral stress* (or pore-water pressure), the latter represents an excess over the neutral one, it has its seat exclusively in the solid phase of the soil and is called *effective stress*. In a more formal presentation of this principle, Terzaghi [38] also stated that "porous materials (such as sand, clay and concrete) react to a change of pore pressure as if they were incompressible and as if their internal friction were equal to zero. All the measurable effects of a change of stress, as a compression, distortion and a change of shearing resistance are exclusively due to changes in the effective stress".

Note that Terzaghi's definition of effective stress does not follow from any theoretical investigation of basic porous medium balance equations, but it is introduced in a rather heuristic manner in terms of cause and measurable effects.

For sake of simplicity, in the following we refer to a two-phase medium, *i.e.*, a solid skeleton saturated by water. This is the case of major interest in soil mechanics, but obviously the methodological approach can be extended to more than two phases. Moreover, we consider a saturated medium

$$n_s + n_f = 1. \tag{6.21}$$

In particular, summing up equations (6.12) for two incompressible constituents it turns out that the following kinematic constraint has to be satisfied:

$$\nabla \cdot \left(n_f \overline{\mathbf{V}}^w + (1 - n_f) \overline{\mathbf{V}}^s \right) = 0. \tag{6.22}$$

In order to clarify the meaning of effective stress in the present framework, we consider first the case of a single fluid phase that saturates the void space. In this case, the principle of effective stress reads

$$\overline{\mathbb{T}} = (1 - n_f)\overline{\mathbb{T}}^s - n_f \overline{p}^f \mathbb{I}. \tag{6.23}$$

According to the definition of effective stress, this equation is usually written in soil mechanics in the form

$$\overline{\mathbb{T}} = \overline{\mathbb{T}}' - \overline{p}^f \mathbb{I}, \tag{6.24}$$

where $\overline{\mathbb{T}}'$ is then defined by

$$\overline{\mathbb{T}}' = (1 - n_f) \left(\overline{\mathbb{T}}^s + \overline{p}^f \mathbb{I} \right). \tag{6.25}$$

At this stage it is relevant to note that the partial stress tensor assumes in the saturated mixtures theory the more general expression (see Section 4)

$$\mathbb{T}_\alpha = -n_\alpha p \mathbb{I} + \mathbb{T}_\alpha^E \tag{6.26}$$

where the partial stress for each phase is split into two contributions: the pressure p and an extra term linked to the deformations of the α-th phase corresponding to the effective stress, or to the fluid dissipation. The interest in comparing (6.25) and (6.26) arises from the fact that the latter expression is derived on a thermodynamical basis, requiring that the entropy inequality of the overall medium be satisfied in any admissible process (see [22], [45], [46] and [47]). Note that, in this framework, p is the pressure of the *mixture* as a whole, that is the internal reaction force necessary for the accomplishment of the kinematic constraint (6.22) and there is no apparent reason for identifying it as the pressure of the water in the pores.

Wilmański [47] has proved that such a constraint is thermodynamically admissible only if the constitutive equations depend on ∇n_f. The following form of the constitutive relations is often used:

$$\mathbb{T}_s = -(1 - n_f)p\,\mathbb{I} + \mathbb{T}_s^E(n_f, \mathbb{F}_s), \tag{6.27}$$

$$\mathbb{T}^f = -n_f p\,\mathbb{I}, \tag{6.28}$$

where $\mathbb{F}_s$ is the tensor gradient of deformation of the soil skeleton and the material coefficients (that is the parameters characterizing the mechanical behavior of the material) depend on the porosity n_f.

When considering the three phasic flow (solid with two immiscible fluids), equation (6.23) can be rewritten in the form

$$\overline{\overline{\mathbb{T}}} = (1 - n_f)\overline{\mathbb{T}}^s - S_w\overline{p}^w\mathbb{I} - S_n\overline{p}^n\mathbb{I}, \tag{6.29}$$

where S_w and S_n are the volume fractions of the wetting and non-wetting phase referred to the porosity n_f (i.e., $S_w + S_n = n_f$). Accordingly, an average fluid pressure can be defined as

$$\overline{p}^f =: \frac{S_w\overline{p}^w + S_n\overline{p}^n}{n_f} \tag{6.30}$$

and the effective stress tensor again formally satisfies the relationship (6.24). A relevant consequence of (6.30) is that when considering unsaturated soils, the air is the non-wetting fluid and for pressure values in excess to the atmospheric one the expression is obtained for the effective stress

$$\overline{\overline{\mathbb{T}'}} = \overline{\overline{\mathbb{T}}} - \frac{S_w}{n_f}\overline{p}^w\mathbb{I}. \tag{6.31}$$

The agreement of equation (6.31) with experimental results is still a matter of investigation, also considering that in the literature different expressions have been suggested for the coefficient that multiplies the pore pressure.

The interaction force $\hat{\mathbf{p}}$ typically involves a Fickian contribution, accounting for diffusion, plus a Darcian one, so that the interaction force is usually modeled as

$$\hat{\mathbf{p}} = p\,\nabla n_f - (n_f)^2\,\mu\mathbb{K}^{-1}\mathbf{w}^f \tag{6.32}$$

where μ is the dynamic viscosity of the fluid and $\mathbb{K}(\mathbf{x})$ is the permeability tensor (possibly dependent on t). When neglecting the inertial terms the momentum equation for the fluid phase simplifies to

$$\rho_f n_f\left(\mathbf{v}^f - \mathbf{v}^s\right) = -\frac{\mathbb{K}}{\mu}\nabla p + \rho_f\mathbf{g}. \tag{6.33}$$

Equation (6.33) is commonly referred to as Darcy's law for deformable porous media.

6.4   Lagrangian description of porous media

As already mentioned in the introduction, the macroscopic approach starts from an *a priori* assumption of homogenized phases and is based on the axioms of the theory of mixtures, supplemented with the concept of volume fraction. Fundamentals of the theory of mixture date back to the works of Truesdell [39], Truesdell and Toupin [43], Atkin and Craine [3], Bedford and Drumheller [6], Bowen [9, 10, 11], de Boer and Ehlers [20], Ehlers [22], Coussy [14], Müller [31], Rajagopal and Tao [35], Wilmański [45]. The paper by de Boer [18] is particularly relevant for its review character.

The basic assumption of the theory of mixture is that the individual components of the porous medium are statistically distributed over the control space, *i.e.*, each spatial point $\mathbf{x}$ of the control space is simultaneously co-occupied by particles of all components. As a consequence, mathematical functions describing both geometrical and physical properties of each constituent are field functions defined over the entire control space. In addition, principles of mixtures theory to be used in balance equations are clearly stated by Truesdell in "Rational Thermodynamics" [41]:

- All properties of the mixture must be mathematical consequences of properties of the constituents.

- So as to describe the motion of a constituent, we may in imagination isolate it from the rest of the mixture, provided we allow properly for the actions of the other constituents upon it.

- The motion of the mixture is governed by the same equations as for a single body.

6.4.1   Kinematics

Usually, the kinematics of the fluid phase is described by an Eulerian approach, whereas for the kinematics of the solid structure reference is made to a Lagrangian frame of reference. In order to overcome the shortcoming deriving from this mixed description, recently Coussy [14, 15], Bourgeois and Dormieux [12] and Wilmański [45] have suggested to introduce a unified Lagrangian description, by assuming the soil skeleton as a material reference volume.

For each phase we define the motion function

$$\mathbf{x} = \chi_\alpha(\mathbf{X}_\alpha, t), \tag{6.34}$$

and a corresponding velocity and acceleration field

$$\dot{\mathbf{x}}_\alpha = \frac{\partial}{\partial t}\chi_\alpha(\mathbf{X}_\alpha, t), \tag{6.35}$$

$$\ddot{\mathbf{x}}_\alpha = \frac{\partial^2}{\partial t^2}\chi_\alpha(\mathbf{X}_\alpha, t). \tag{6.36}$$

Referring to the solid matrix and dropping the index $\alpha = s$ for simplicity, the deformation gradient tensor, defined by

$$\mathbb{F}_s =: \frac{\partial \chi_i}{\partial X_j} \mathbf{e}_i \otimes \mathbf{e}_j, \tag{6.37}$$

maps the material vector $d\mathbf{X}$ onto its current configuration

$$d\mathbf{x} = \mathbb{F}_s d\mathbf{X}. \tag{6.38}$$

Similarly, the infinitesimal initial volume dV_0 is transformed into the corresponding area in the current configuration thanks to the relation

$$dV = J dV_0, \tag{6.39}$$

where

$$J =: \det \mathbb{F}_s > 0. \tag{6.40}$$

The determinant of $\mathbb{F}_s$ is restricted to take positive values, so that the motion function of each phase is invertible. A material area $d\mathbf{A}$, oriented by its normal $\mathbf{N}$ (*i.e.*, $d\mathbf{A} = \mathbf{N}dA$), is transformed into the current configuration according to the rule

$$\mathbf{n}\, da = J \mathbb{F}_s^{-T} \mathbf{N}\, dA \tag{6.41}$$

where $\mathbb{F}_s^{-T}$ indicates the transpose of the inverse of $\mathbb{F}_s$.

6.4.2 *Mass balance*

In the Lagrangian description, the conservation of mass of the solid component is identically satisfied,

$$\frac{\partial}{\partial t}\left[J\rho_s(1 - n_f) \right] = 0, \tag{6.42}$$

where $\rho_s =: \rho_s^R$. The balance of mass for the fluid phase over a material control volume V fixed on the solid matrix and bounded by the surface S is written, in integral form,

$$\frac{\partial}{\partial t} \int_V \rho_f \frac{e}{1 + e} dV + \int_S \rho_f \frac{e}{1 + e} \left(\mathbf{v}^w - \mathbf{v}^s \right) \cdot \mathbf{n}\, da = 0, \tag{6.43}$$

where $\rho_f =: \rho_f^R$, $\mathbf{v}^w$ and $\mathbf{v}^s$ are the velocity of the liquid and solid constituent, respectively, and e is the void ratio defined as $e =: n_f/(1 - n_f)$.

By recalling (6.41), transforming the volume integral at the left-hand side of (6.43) into an integral over the reference volume and by using the Gauss theorem in order to transform the surface integral into an integral over the reference volume, one gets (if the void ratio in the reference configuration is constant)

$$\frac{\partial e}{\partial t} + \mathrm{div}\left[e\mathbb{F}_s^{-1} \left(\mathbf{v}^w - \mathbf{v}^s \right) \right] = 0, \tag{6.44}$$

where the symbol "div" indicates the divergence operator in Lagrangian coordinates.

According to Biot [8] and Coussy [14], it is sometimes convenient to introduce the change of fluid mass referred to the initial volume m. Accordingly, the quantity $m\,dV_0$ indicates the difference of the fluid mass passing from the initial to the current configuration,

$$m =: J\rho_f n_f - \rho_{w0} n_f^0. \tag{6.45}$$

By defining the Eulerian vector

$$\mathbf{w} =: \rho_f n_f \left(\mathbf{v}^w - \mathbf{v}^s\right), \tag{6.46}$$

representing the mass flux of water relative to the solid skeleton, the mass balance can also be expressed as

$$\frac{\partial m}{\partial t} + \operatorname{div}\left(J\mathbb{F}_s^{-1}\mathbf{w}\right) = 0. \tag{6.47}$$

6.4.3 Momentum balance equation

When denoting by ρ_s, $\mathbf{a}^s$ and $\mathbf{a}^w$ the density of the porous medium, the acceleration of the soil particles and the acceleration of the fluid, respectively, the Eulerian form of Cauchy's first law of motion for the porous medium as a whole is written

$$\int_V \left[\rho_s \mathbf{a}^s + n\rho_f \left(\mathbf{a}^w - \mathbf{a}^s\right)\right] dV = \int_V \rho \mathbf{b}\, dV + \int_S \mathbb{T}\, dS, \tag{6.48}$$

where V is a control volume fixed on the solid skeleton and S its surface. As the relationship (6.48) holds for any control volume, for regular enough fields the local form of the momentum equation is obtained thanks to the Gauss theorem

$$\nabla\cdot\mathbb{T} + \rho_s\left(\mathbf{b} - \mathbf{a}^s\right) - n\rho_f\left(\mathbf{a}^w - \mathbf{a}^s\right) = 0. \tag{6.49}$$

In order to obtain the corresponding Lagrangian formulation, we introduce the second Piola–Kirchhoff stress tensor

$$\tilde{\mathbb{T}} =: J\mathbb{F}^{-1}\mathbb{T}\mathbb{F}^{-T}. \tag{6.50}$$

The relationship (6.41) enables us to write the transport formula

$$\mathbb{F}\tilde{\mathbb{T}}\cdot\mathbf{N}dA = \mathbb{T}\cdot\mathbf{n}\,da, \tag{6.51}$$

so that (6.48) reads

$$\int_{A_0} \mathbb{F}\tilde{\mathbb{T}}\cdot\mathbf{N}dA + \int_{V_0} \left[\left(\rho_0 + m\right)\left(\mathbf{b} - \mathbf{a}^s\right) - \left(\rho_{w0} n_0 + m\right)\left(\mathbf{a}^w - \mathbf{a}^s\right)\right] dV = 0,$$
$$\tag{6.52}$$

where ρ_0, A_0, V_0 are the density, surface and volume in the reference configuration. The local Lagrangian form is

$$\operatorname{div}(\mathbb{F}\tilde{\mathbb{T}}) + (\rho_0 + m)\,(\mathbf{b} - \mathbf{a}^s) - (n_0\rho_{w0} + m)\,(\mathbf{a}^w - \mathbf{a}^s) = 0. \qquad (6.53)$$

In the Lagrangian framework, Darcy's law (6.33) takes the generalized expression

$$\rho_f n_f(\mathbf{v}^w - \mathbf{v}^s) = -\frac{\mathbb{K}}{\mu}\left(\mathbb{F}^{-T}\operatorname{grad}p\right) + \rho_f\mathbf{g} \qquad (6.54)$$

where $\mathbb{K}$ is the permeability tensor, g is the gravity acceleration, μ is the dynamic viscosity of the fluid and "grad" represents the gradient operator in Lagrangian coordinates.

6.5   Consolidation theories

The first attempt to describe the consolidation of a deformable porous medium with pores completely filled by water is due to Terzaghi [37]. He introduced the concept of effective stress in a strictly one-dimensional framework. Later Biot [7] generalized the theory to the three-dimensional case in a framework consistent with the basic principles of continuum mechanics. His work deals with small strains and elastic behavior of the soil skeleton, and it is briefly recalled in this section as a basis to outline fields of further developments.

6.5.1   Biot's theory

Within the general framework illustrated in the previous sections, Biot's formulation reads as a macroscopic one. In particular, the behavior of the soil skeleton is described by global deformation characteristics, which include all local deformations (*i.e.*, rolling and sliding of particles).

The momentum balance equation of the fluid phase is expressed by Darcy's law (6.33). By combining it with the saturation condition (6.22) one gets

$$\frac{\partial}{\partial t}(\nabla\cdot\mathbf{u}) = \frac{K}{\rho_f}\nabla p, \qquad (6.55)$$

where
   $\mathbf{u}$ is the displacement vector of the soil skeleton;
   p is the pore pressure in excess to the initial equilibrium value;
   K is the (constant) hydraulic conductivity of the porous medium.
The total stress tensor $\mathbb{T}$ can be decomposed into the effective stress tensor and the pore pressure according to Terzaghi's equation

$$\mathbb{T} = \mathbb{T}' - p\mathbb{I}. \qquad (6.56)$$

Assuming an elastic behavior for the soil skeleton for small deformations, the equilibrium equations can be expressed in terms of displacement components (Navier equations). Introducing the tensor of the linear elasticity

$$\epsilon =: \mu(\nabla \mathbf{u} + (\nabla \mathbf{u})^T) + \lambda \nabla \cdot \mathbf{u}, \tag{6.57}$$

for the effective stress, one obtains

$$(\lambda + \mu)\nabla \nabla \cdot \mathbf{u} + \mu \nabla^2 \mathbf{u} - \nabla p = \rho \mathbf{b}. \tag{6.58}$$

Equations (6.55) and (6.58) provide a system of four differential equations for the unknowns $(\mathbf{u}, p)$. The boundary conditions can be specified as

-prescribed pore pressure: $p = h$,

-prescribed flux: $\dfrac{K}{\rho_f g} \nabla p \cdot \mathbf{n} = f$,

-prescribed traction: $\mathbb{T} \cdot \mathbf{n} = \mathbf{t}$,

-prescribed displacement: $\mathbf{u} = \mathbf{a}$,

while, without loss of generality, $\mathbf{u}$ and p can be taken to be null at $t = 0$.

Remark. Known exact solutions of the consolidation problem are limited to simple boundary conditions under the assumption of elastic behavior of the soil skeleton. A general treatment of the problem, involving a more realistic soil behavior, coupled with the diffusion process of the fluid phase calls for numerical simulation.

6.5.2 One-dimensional finite deformation theory

There are two major areas of developments in which departures from Terzaghi and Biot's theories occur: finite deformations and non-linear soil response. In this paragraph we deal with the former aspect, with special reference to the one-dimensional case.

Consider a soil stratum of infinite extent in the 0XY plane and transversely isotropic about the 0Z axis. The upper surface is supposed to be a free draining boundary, so that

$$p(t, Z = h) = 0. \tag{6.59}$$

The lower boundary is supposed to be impermeable and the corresponding condition is

$$\frac{\partial p}{\partial Z}(t, Z = 0) = 0. \tag{6.60}$$

At time $t = 0$ the upper boundary is uniformly loaded:

$$\mathbf{t}(t, Z = h) = -\Delta q \, \mathbf{e}_z. \tag{6.61}$$

Because of the one-dimensional nature of the problem, the deformation gradient reduces to

$$\mathbb{F}_s = \mathbb{I} + \frac{\partial u_z}{\partial Z} \mathbf{e}_z \otimes \mathbf{e}_z, \tag{6.62}$$

where u_z is the displacement in the Z direction, so that

$$\det \mathbb{F}_s = 1 + \frac{\partial u_z}{\partial Z}. \tag{6.63}$$

The unique non-trivial component of the second Piola–Kirchhoff tensor is related to the Cauchy tensor according to equation (6.50)

$$\tilde{T}_{ZZ} = T_{ZZ} \left(1 + \frac{\partial u_z}{\partial Z} \right)^{-1}. \tag{6.64}$$

After integration in time, the Lagrangian mass balance equation for the solid phase (6.42) reduces to

$$\det \mathbb{F} = \left(1 + \frac{\partial u_z}{\partial Z} \right) = \frac{1+e}{1+e_0}, \tag{6.65}$$

where e_0 is the void ratio in the reference configuration. The mass balance equation for the fluid phase in Lagrangian coordinates (6.44) reads

$$\frac{\partial e}{\partial t} + (1 + e_0) \frac{\partial}{\partial Z} \left[\frac{e}{1+e} (v_z^w - v_z^s) \right] = 0. \tag{6.66}$$

By recalling the definition of m (6.45) and neglecting the inertial terms, the balance of linear momentum of the porous medium (6.53) reads

$$\frac{\partial \tilde{T}_{ZZ}}{\partial Z} + g \frac{e\rho_f + \rho_s}{1+e_0} = 0. \tag{6.67}$$

The Lagrangian formulation of Darcy's law is given by

$$\rho_f \frac{e}{1+e} (v_z^w - v_z^s) = -\frac{K}{\mu} \left[\frac{1+e_0}{1+e} \frac{\partial p}{\partial Z} \right] - \rho_f g. \tag{6.68}$$

Plugging equation (6.68) into (6.66) and by using the definition of effective stress into (6.67), one obtains the Lagrangian equation of one-dimensional consolidation:

$$\frac{\partial e}{\partial t} + (1 + e_0) \frac{\partial}{\partial Z} \left[\frac{K}{\mu} \left(\frac{\rho_s - \rho_f}{1+e} g + \frac{\partial T'_{ZZ}}{\partial Z} \frac{1+e_0}{1+e} \right) \right] = 0. \tag{6.69}$$

The equation above represents the most general formulation of the problem (see Gibson $et\ al.$ [23], Lancellotta and Preziosi [27], Ambrosi [2] and Pane [34]). The one-dimensional models currently used in the literature can be interpreted as a special case of equation (6.69).

As shown by Carter *et al.* [13], the solution of equation (6.69) under the assumption of an elastic soil behavior characterized by the Young modulus E' and Poisson's ratio ν' depends on the following dimensionless parameters only:

$$\frac{\Delta q}{E'}, \qquad \frac{\rho_f g H}{E'}, \qquad \nu', \qquad e_0, \qquad \frac{\rho_s}{\rho_f}. \tag{6.70}$$

Terzaghi's solution is obtained when both the parameters $\dfrac{\delta q}{E'}$ and $\dfrac{\rho_f g H}{E'}$ approach zero, thus indicating that the behavior of deep soft layers can be significantly different from the one predicted by Terzaghi's theory.

6.6   Constitutive equations

According to Truesdell and Noll [42] three fundamental postulates are assumed to be valid for any constitutive theory:

-Principle of determinism for stress: the stress in a body is determined by its motion history.

-Principle of local action: in determining the stress at a given particle, the motion outside an arbitrary neighborhood of the particle can be disregarded.

-Principle of material frame-indifference or principle of material objectivity: constitutive equations must be invariant under changes of frame of reference.

In the following we concentrate on the latter principle, because, despite its effectiveness in limiting the generality of the constitutive equations of elasticity as well as the functionals appearing in the constitutive equations of simple materials with memory, its use is not yet common in the engineering literature.

A mathematical statement of this principle is the following: if a constitutive equation is satisfied for a process with a motion and a symmetric stress tensor given by

$$\mathbf{x} = \chi(\mathbf{X}, t), \qquad \mathbb{T} = \mathbb{T}(\mathbf{X}, t), \tag{6.71}$$

it must also be satisfied for any equivalent process, given by the motion and the stress tensor

$$\mathbf{x}^* = \mathbf{c}(t) + \mathbb{Q}(t)\mathbf{x}, \tag{6.72}$$

$$\mathbb{T}^* = \mathbb{Q}(t)\mathbb{T}\mathbb{Q}^T(t), \tag{6.73}$$

$$t^* = t - a, \tag{6.74}$$

for any arbitrary position vector $\mathbf{c}(t)$, any arbitrary real number a and any arbitrary orthogonal tensor $\mathbb{Q}(t)$. In this respect, functions and fields of scalar, vector and tensor are objective if both the dependent and the independent vectors and tensors transform according to equation (6.73), while the scalar variables are unchanged. Note that the coordinate transformation equations between various coordinate systems in the same frame of reference are time-independent, so that the invariance under this kind of coordinate change is ensured by the requirement that the constitutive equations have to be tensor equations. But this does not ensure invariance under the time dependent change of reference frame, unless the relationships above are satisfied.

A constitutive equation allows us to obtain the stress field from the knowledge of the history of the deformation of a body. In elasticity the stress is a function of deformation, so that the deformation history is immaterial and the stress depends only on the current value of deformation. Soil behavior is not elastic and, in order to account for its mechanical history, a general formulation should provide the stress increment in terms of some non-integrable function of the deformation increment

$$d\mathbb{T} = f(d\epsilon).$$
(6.75)

If we consider a material without internal time scale (*i.e.*, a rate independent behavior), equation (6.75) can also be represented as a rate equation

$$\dot{\mathbb{T}} = f(\dot{\epsilon}).$$
(6.76)

This is the usual approach in plasticity and hypoplasticity. Truesdell [40] introduced for the constitutive relationships the general form

$$\dot{\mathbb{T}} = f(\mathbb{T}, \mathbb{D}).$$
(6.77)

If the function f is linear in $\mathbb{T}$ and $\mathbb{D}$, the behavior is called hypoelastic: equation (6.77) can produce a non-linear response, but it is inappropriate to describe plastic behavior. In order to overcome these limits (see Kolymbas [25]), the function f is required to be non-linear in $\mathbb{D}$. In particular, when f is first order homogeneous in $\mathbb{D}$ it describes a rate independent behavior. Homogeneity in $\mathbb{T}$ is also a prerequisite in order to describe proportional stress paths in the case of a proportional strain path. The above requirements constitute rules for the mathematical function (6.77).

6.6.1 Finite elasticity

When we apply the principles of determinism and local action in finite elasticity, the most general constitutive equation is of the form

$$\mathbb{T} = f(\mathbb{F}),$$
(6.78)

where the response function f is a non-linear tensor-valued function of the single tensor $\mathbb{F}$. By noting that in a change of frame the deformation gradient $\mathbb{F}$ transforms like a vector, the response function must satisfy the following identity, required by the objectivity principle

$$f(\mathbb{Q}\mathbb{F}) = \mathbb{Q}f(\mathbb{F})\mathbb{Q}^T. \qquad (6.79)$$

Therefore, the frame-indifference principle requires that the dependence on $\mathbb{F}$ have the form of an arbitrary function with additional dependence on $\mathbb{Q}$ as shown by equation (6.79). It can in addition be proved that in terms of the second Piola–Kirchhoff stress tensor, the response function depends only on one of the strain tensors (*i.e.*, $\mathbb{U}$ or $\mathbb{C}$ or $\mathbb{E}$) and not on rotations. The requirement of frame indifference is relaxed if it is assumed that the motion involves a small contribution of rotation in a polar decomposition of the deformation tensor, as is commonly done in the classical small displacement elasticity.

6.6.2 *Rate equations*

When the constitutive equation includes the material time derivative of the stress tensor $\dot{\mathbb{T}}$, the principle of objectivity is not straightforwardly satisfied for any arbitrary motion. The difficulty is that the material time derivative $\dot{\mathbb{T}}$ is not frame-indifferent, even though $\mathbb{T}$ is frame invariant. A possible remedy is to use the co-rotational stress rate tensor

$$\hat{\mathbb{T}} =: \dot{\mathbb{T}} - \mathbb{W}\mathbb{T} + \mathbb{T}\mathbb{W}, \qquad (6.80)$$

where $\mathbb{W}$ is the spin tensor, provided that all the other variables are frame-indifferent (as is, for example, the rate of deformation tensor $\mathbb{D}$).

6.6.3 *Finite non-linear consolidation*

A theoretical formulation and a numerical solution method for the consolidation of an elasto-plastic soil with finite deformation has been given by Carter *et al.* [13], with constitutive rate equations satisfying the objectivity principle. Here again the general conclusion is confirmed that the need to account for large deformation as well as non-linear soil behavior arises for soft materials and when the imposed load is large if compared with the soil stiffness. Both these cases are of interest in soil mechanics, when dealing with soft clay strata.

References

[1] D. AMBROSI, L. PREZIOSI: Modelling injection moulding processes with deformable porous preforms, *SIAM J. Appl. Math.* **61**, pp. 22–42, 2000.

[2] D. AMBROSI: Infiltration through deformable porous media, *ZAMM* **82**, pp. 115–124, 2002.

[3] R.J. ATKIN, R.E. CRAINE: Continuum theory of mixtures: basic theory and historical development, *Quart. J. Appl. Math.* **29**, pp. 209–244, 1976.

[4] J. BEAR: *Dynamics of Fluids in Porous Media*, Elsevier Science, Amsterdam, 1972.

[5] J. BEAR, Y. BACHMAT: *Introduction to Modeling of Transport Phenomena in Porous Media*, Kluwer Academic Publisher, Dordrecht, 1991.

[6] A. BEDFORDF, D.S. DRUMHELLER: Theory of immiscible and structured mixtures, *Int. J. Engng. Sci.* **21**, pp. 863–960, 1983.

[7] M.A. BIOT: General theory of three-dimensional consolidation, *J. Appl. Phys.* **12**, pp. 155–165, 1941.

[8] M.A. BIOT: Variational Lagrangian thermodynamics of non–isothermal finite strain mechanics of porous solids and thermo-molecular diffusion, *Int. J. Solids Struct.* **13**, pp. 579–597, 1977.

[9] R.M. BOWEN: Theory of mixtures, in: A.C. ERINGEN (ed): *Continuum physics; Volume III: Mixtures and EM Field Theories*, Academic Press, New York, 1976, pp. 1–127.

[10] R.M. BOWEN: Incompressible porous media models by the use of the theory of mixtures, *Int. J. Engng. Sci.* **18**, pp. 1129–1148, 1980.

[11] R.M. BOWEN: Compressible porous media models by the use of the theory of mixtures, *Int. J. Engng. Sci.* **20**, pp. 697–735, 1982.

[12] E. BOURGEOIS, L. DORMIEUX: Consolidation of a nonlinear poroelastic layer in finite deformations, *Eur. J. Mech. A/Solids* **5**, pp. 575–598, 1996.

[13] J.P. CARTER, J.R. BOOKER, J.C. SMALL: The analysis of finite elasto–plastic consolidation, *Int. J. Numer. Anal. Meth. Geomech.* **3**, pp. 107–129, 1979.

[14] O. COUSSY: Thermodynamics of saturated porous solids in finite deformations, *Eur. J. Mech. A/Solids* **8**, pp. 1–14, 1989.

[15] O. COUSSY: *Mechanics of Porous Continua*, Wiley, New York, 1995.

[16] C.W. CRYER: A comparison of the 3-dimensional theories of Biot and Terzaghi, *Quart. J. Mec. Appl. Math.* **16**, pp. 401–412, 1963.

[17] H. DARCY: *Les Fontaines Publiques de la Ville de Dijon*, Paris, 1856.

[18] R. DE BOER: Highlights in the historical development of the porous media theory: toward a consistent macroscopic theory, *Appl. Mech. Rev.* **49**, pp. 201–262, 1996.

[19] R. DE BOER: *Theory of Porous Media*, Springer, Berlin, 2000

[20] R. DE BOER, W. EHLERS: *Theorie der Mehrkomponentenkontinua mit Anwendung auf Bodenmechanische Problem*, Forschungsberichte aus dem Fachbereich Bauwesen, Universität-GH-Hessen **40**, 1986.

[21] W.J. DRUGAN, J.R. WILLIS: A micromechanics-based nonlocal constitutive equation and estimates of representative volume element size for elastic composites, *J. Mech. Phys. Solids* **44**, pp. 497, 1996.

[22] W. EHLERS: Constitutive equations for granular materials in geomechanical context, in: K. HUTTER (ed.): *Continuum Mechanics in Environmental Sciences and Geophysics*, CISM Courses and Lectures, Springer, Berlin **337**, pp. 313–402, 1993.

[23] R.E. GIBSON, G.L. ENGLAND, M.J.L. HUSSEY: The theory of one dimensional consolidation of saturated clays, *Geotechnique* **17**, pp. 261–273, 1967.

[24] W.Y. GU, W.M. LAI, V.C. MOW: Transport and multi–electrolytes in charged hydrated biological soft tissues, in: R. DE BOER (ed.): *Porous Media: Theory and Experiments*, Kluwer Academic Publisher, Dordrecht, 1999, pp. 143–157.

[25] D. KOLYMBAS: An outline of hypoplasticity, *Arch. Appl. Mech.* **61**, pp. 143–151, 1991.

[26] P.V. LADE, R. DE BOER: The concept of effective stress for soil, concrete and rock, *Geotechnique* **47**, pp. 61–78, 1997.

[27] R. LANCELLOTTA, L. PREZIOSI: A general nonlinear mathematical model for soil consolidation problems, *Int. J. Engng. Sci.* **35**, pp. 1045–1063, 1997.

[28] R.W. LEWIS, B.A. SCHREFLER: *The Finite Element Method in the Static and Dynamic Deformation and Consolidation of Porous Media*, Wiley, New York, 1998.

[29] H.A. LORENTZ: *The Principle of Relativity*, New York, Dover, 1952.

[30] K.Z. MARKOV: Elementary micromechanics of heterogeneous media, in: K.Z. MARKOV, L. PREZIOSI (eds.): *Heterogeneus Media: Modelling, Mathematical Methods and Simulations*, Birkhäuser, Boston, 2000, pp. 1–162.

[31] I. MÜLLER: A thermodynamic theory of mixture of fluids, *Arch. Rat. Mech. Anal.* **28**, pp. 1–39, 1968.

[32] S. NEMAT-NASSER, N. YU, M. HORI: Solids with periodically distributed cracks, *Int. J. Solids Struct.* **30**, p. 2071, 1993.

[33] S.P. NEUMAN: Theoretical derivation of Darcy's law, *Acta Mechanica* **25**, pp. 153-170, 1977.

[34] V. PANE: *Sedimentation and Consolidation of Clays*, Ph.D. thesis, University of Colorado, Boulder, 1985.

[35] K.R. RAJAGOPAL, L. TAO: *Mechanics of Mixtures*, World Scientific Publishing, Singapore, 1995.

[36] J.C. SLATTERY: Flow of viscoelastic fluids through porous media, *AIChE J.* **13**, pp. 1066–1071, 1967.

[37] K. TERZAGHI: Die berechnung der durchlässigkeitsziffer des tones aus dem verlauf der hydrodynamischen spannungserscheinungen, *Sitz. Akad. Wissen. Wien. Math.-Naturw.-Schaf. Kl.*, Abt. IIa **132**, pp. 125–138, 1923.

[38] K. TERZAGHI: *Erdbaumechanik auf Bodenphysikalischer Grundlage*, Leipzig Deuticke., 1925 [also *From Theory to Practice*, Wiley, New York, pp. 146–148, 1960].

[39] C. TRUESDELL: Sulle basi della termodinamica, *Rend. Lincei* **22**, pp. 33–38 and pp. 158–166, 1957.

[40] C. TRUESDELL: Hypo-elasticity, *J. Rational Mech. Anal.* **4**, pp. 83–133, 1965.

[41] C. TRUESDELL: Thermodynamics of diffusion, in: C. TRUESDELL: *Rational Thermodynamics*, 2nd-edition, Springer–Verlag, New York, pp. 219–236, 1984.

[42] C. TRUESDELL, W. NOLL: The non-linear field theories of mechanics, in: S. FLÜGGE (ed.): *Handbuch der Physik*, Springer-Verlag, Berlin **III/3**, 1965.

[43] C. TRUESDELL, R.A. TOUPIN: The classical field theories, in: S. FLÜGGE (ed.): *Handbuch der Physik*, Springer-Verlag, Berlin **III/1**, pp. 226–902, 1960.

[44] S. WHITAKER: The equation of motion in porous media, *Chem. Eng. Sci.* **21**, pp. 291–300, 1966.

[45] K. WILMAŃSKI: Porous media at finite strains: the new model with balance equation for porosity, *Arch. Mech.* **48**, pp. 591–628, 1996.

[46] K. WILMAŃSKI: *Mathematical theory of porous media*, *WIAS Preprint* **602**, 2000.

[47] K. WILMAŃSKI: *Note on the notion of incompressibility in theories of porous and granular materials*, *WIAS Preprint* **605**, 2000.

DAVIDE AMBROSI and LUIGI PREZIOSI
Dipartimento di Matematica
Politecnico di Torino
I-10129 Torino, ITALY
E-mail: Ambrosi@calvino.polito.it
         Preziosi@polito.it

RENATO LANCELLOTTA
Dipartimento di Ingegneria Strutturale e Geotecnica
Politecnico di Torino
I-10129 Torino, ITALY
E-mail: Lancellotta@polito.it

Chapter 7

Flow of Water in Rigid and Non-Rigid, Saturated and Unsaturated Soils

Peter A.C. Raats

ABSTRACT Flow of water in soils is studied intensely in soil physics and hydrology. The continuum theory of mixtures can be used as a framework for outlining the theory. The well-established theory for flow of water in rigid soils is reviewed briefly, with emphasis on the description and main implications of the non-linear theory proposed by Richards in 1931. For non-rigid soils, it is convenient to formulate the theory in terms of material coordinates of the solid phase. For one-dimensional, vertical flow, it is shown that, compared to the theory for rigid soils, the effect of the weight of the soil may be to reverse the sign of the gravitational force in the Richards equation. Progress with multi-dimensional deformation processes is outlined, including the complications arising from cracking. Finally, available analytical and numerical solutions of one-dimensional, equilibrium and flow problems for water in non-rigid soils are reviewed.

7.1   Introduction

The theory for movement of water in unsaturated soils, published by Richards [56] 70 years ago, is still an important starting point for the analysis of most soil physical problems. Over the last half century, the interest in finding new solutions of the Richards equation by either analytical or numerical methods was intense, particularly with regard to field situations [35, 52]. During this period we see also a widening of the scope of soil physics beyond the classical theory of Richards with further studies of various multiphase aspects (particularly, simultaneous movement of water and air, simultaneous transport of heat and moisture, flow of water and transport of solutes in structured soils), of water movement in soils subject to swelling and shrinkage and of transport of solutes in unsaturated soils. The latter two subjects became manageable by replacing the traditional spatial descriptions by material descriptions in which, respectively, the solid phase serves as the reference continuum for the water and the water serves as the reference continuum for the solutes [44, 46, 47]. Progress was driven not only

by a healthy theoretical, computational and experimental basis, but also
by productive interaction with adjoining disciplines and by challenging so-
cietal problems. The focus of this review is on the physics of soil water,
particularly water in soils subject to swelling and shrinkage.

The organization of this chapter is as follows. In Section 2 the theory
of Richards for flow of water in rigid soils is reviewed briefly. In Section 3
the soil physical theory for flow of the liquid and gaseous phases in soils
subject to swelling and shrinkage is reviewed. In Section 4 solutions of
one-dimensional flow problems are presented.

7.2   Flow of water in rigid soils

7.2.1   The Richards equation

About 70 years ago, Lorenzo A. Richards consolidated the efforts of previous generations of soil physicists (notably Franklin H. King, Charles S.
Slichter, Lyman J. Briggs, Edgar Buckingham, Willard Gardner and W.B.
Haines) by formulating a general, macroscopic theory for movement of water in rigid, unsaturated soils [56, 35]. Richards theory fits experience in
many branches of continuum mechanics: it combines the simplest possible
balance of mass, expressed in the equation of continuity, and of momentum,
expressed in Darcy's law:

$$\frac{\partial \theta_w}{\partial t} = -\nabla \cdot (\theta_w \mathbf{v}_w) - \lambda_w, \tag{7.1}$$

$$\theta_w \mathbf{v}_w = -k_w \nabla h + k_w \nabla z, \tag{7.2}$$

where t is the time, ∇ is the vector differential operator, $\mathbf{v}_w$ is the velocity
of the water, λ_w is the volumetric rate of uptake of water by plant roots
(see [43] for some details regarding this term in the mass balance of water),
k_w is the hydraulic conductivity, h is the capillary pressure head and z
is the vertical coordinate taken positive downward. The capillary pressure
head h is defined by:

$$h = \frac{p_w - p_g}{\gamma_w g} = -\frac{p^c}{\gamma_w g}, \tag{7.3}$$

where p_w and p_g are the pressures of the aqueous and gaseous phases, p^c is
the capillary pressure, γ_w is the density of the water and g is the magnitude
of the gravitational force per unit mass. In the theory of Richards, it is
assumed that the pressure of the gaseous phase is spatially uniform and
constant. The capillary pressure head h and the hydraulic conductivity
k_w are highly non-linear functions of the volumetric water content θ_w.
Moreover, the relationship $h(\theta_w)$ is hysteretic.

For future reference, three alternative forms of (7.2) are recorded also:

$$\theta_w \mathbf{v}_w = -k_w \nabla H, \tag{7.4}$$

$$\theta_w \mathbf{v}_w = -D\nabla\theta_w + k_w \nabla z, \tag{7.5}$$

$$\theta_w \mathbf{v}_w = -\nabla\phi + k_w \nabla z, \tag{7.6}$$

where $H = h + z$ is the total head, $D = k_w \mathrm{d}h/\mathrm{d}\theta_w$ is the soil water diffusivity and ϕ is the matric flux potential defined by

$$\phi - \phi_0 = \int_{h_0}^{h} k_w \mathrm{d}h = \int_{\theta_{w0}}^{\theta_w} D\mathrm{d}\theta_w, \tag{7.7}$$

where h_0 and θ_{w0} are reference values and $\theta_{w0} = \theta_w(h_0)$. The volumetric flux $\theta_w v_w$ is the sum of a matric component $-k_w\nabla h = -D\nabla\theta_w = -\nabla\phi$ and a gravitational component $k_w\nabla z$. The matric component of the volumetric flux is given by the gradient of ϕ and therefore it is appropriate to call ϕ the matric flux potential. A transformation of the type (7.7) was given around 1880 by Kirchhoff in his lectures on heat conduction [22]. For this reason, ϕ is often called the Kirchhoff potential and the transformation from h and θ_w to ϕ is referred to as the Kirchhoff transformation.

The theory of Richards can be formulated within the framework of the modern continuum theory of mixtures [79], provided that one recognizes from the outset the existence of the separate solid, liquid and gaseous phases [44, 51]. Richards' theory can also be justified on the basis of the principles of surface tension and viscous flow at the pore scale [26, 83].

7.2.2 Solutions of the Richards equation: solutions of the form $z = \zeta_{\theta_w}(\theta_w, t)$ or $z = \zeta_h(h, t)$

Several early solutions of the Richards equation are of the form [33, 36, 47]

$$z = \zeta_{\theta_w}(\theta_w, t), \qquad z = \zeta_h(h, t). \tag{7.8}$$

Examples of this are (i) solutions for steady upward and downward flows, (ii) solutions in the form of traveling, time-invariant waves, (iii) the Boltzmann solution with $t^{\frac{1}{2}}$ proportionality for horizontal absorption and (iv) Philip's series expansion in $t^{\frac{1}{2}}$ for vertical infiltration. The first of the two functional relationships expressed in (7.8) is consistent with the water content θ_w and the time t as independent variables, while the position z and the flux $\theta_w v_w$ are regarded as the dependent variables. In this context the mass balance is expressed as

$$\left(\frac{\partial z}{\partial t}\right)_{\theta_w} = \left(\frac{\partial\theta_w v_w}{\partial\theta_w}\right)_t. \tag{7.9}$$

Integration of (7.9) between θ_w and $\theta_{w\infty}$ gives

$$\theta_w v_w - \theta_{w\infty} v_{w\infty} = \frac{\partial}{\partial t}\int_{\theta_{w\infty}}^{\theta_w} z\, \mathrm{d}\theta_w. \tag{7.10}$$

This θ_w-integrated mass balance equation is the basis of two important types of approximate, iterative methods that were first introduced in the early 1970s by Parlange and by Philip and Knight. The two types use different constraints in the iterative procedures. Integration of (7.10) with respect to position z gives

$$\int\limits_0^\infty \left[\theta_w v_w - \theta_{w\infty} v_{w\infty}\right]\,\mathrm{d}z = \frac{\partial}{\partial t}\int\limits_{\theta_{w0}}^{\theta_{w\infty}} \frac{1}{2}z^2\,\mathrm{d}\theta_w = \frac{\partial}{\partial t}\int\limits_0^\infty z\left(\theta_w - \theta_{w\infty}\right)\,\mathrm{d}z. \qquad (7.11)$$

This integral moment balance is the constraint used by Parlange in an initial series of 10 papers published in *Soil Science* in the period 1971–1975 and in numerous later papers [30]. Alternatively, integration of (7.10) with respect to time t gives

$$\int\limits_0^t \left(\theta_{w0} v_{w0}\left[t\right] - \theta_{w\infty} v_{w\infty}\right)\,\mathrm{d}t = \int\limits_{\theta_{w\infty}}^{\theta_{w0}[t]} z\left[\theta_w, t\right]\,\mathrm{d}\theta_w, \qquad (7.12)$$

where the subscript 0 denotes values at the soil surface. This integral mass balance is the constraint used in the flux-concentration method of Philip and Knight [36], formulated in response to the work of Parlange. The approximate, iterative methods using either one of these two integral constraints have yielded solutions for horizontal absorption and vertical infiltration with a given water content or flux at the soil surface and without or with a surface crust.

7.2.3 Solutions of the Richards equation: analytical solutions for particular classes of soils

In the context of Richards equation, the relationships among the water content θ_w, pressure head h and hydraulic conductivity k_w define the hydraulic properties of a soil. Two major conferences on physical characterization were organized in 1989 and 1997 by the U.S. Salinity Laboratory and the University of California at Riverside [81, 82]. One can distinguish two groups of parametric expressions describing hydraulic properties:

1. a group yielding flow equations that can be solved analytically, in most cases as a result of linearization following one or more transformations;

2. a group that is favored in numerical studies and to a large extent shares flexibility with a rather sound basis in Poiseuillean flow in networks of capillaries.

In this subsection we consider the parametric expressions describing hydraulic properties yielding flow equations that can be solved analytically, in most cases as a result of linearization following one or more transformations. To the first group belong:

- the class of linear soils with the diffusivity $D = k_w \frac{\mathrm{d}h}{\mathrm{d}\theta_w}$ constant and the hydraulic conductivity k_w linear in θ_w;

  This leads to a linear Fokker–Planck equation, which can be solved relatively easily [33]. It yields useful results for integral aspects of the water balance, but is rather unreliable with respect to details of the distribution of the water content.

- the class of delta function or Green&Ampt soils with the diffusivity given by $D = \frac{1}{2}S^2 \left(\theta_{w1} - \theta_{w0}\right)^{-1} \delta \left(\theta_{w1} - \theta_w\right)$;

  This class implies discontinuities of the water content at wetting fronts and gives primarily reasonable results for one-dimensional, vertical infiltration [33, 37]. It yields very useful results for integral aspects of the water balance, but fails to give any details of the distribution of the water content.

- the class of Brooks–Corey power function soils;

  This class has turned up in recent years regularly as resulting from fractal models of pore structure. It also plays a key role in the mathematical literature on the so-called porous medium equation, *i.e.*, the non-linear diffusion equation with the diffusivity a power function of the volumetric water content, and on the corresponding special case of the Richards equation [13]. The main interest has been in so-called similarity solutions.

- the Gardner class of soils defined by $k_w = k_{w0} \exp \alpha \left(h - h_0\right)$, where α is an inverse characteristic length of the soil [10];

  For this class of soils the steady flow equation reduces to the linear equation

$$\nabla^2 \phi - \alpha \frac{\partial \phi}{\partial z} = \lambda_w. \tag{7.13}$$

  The linear equation (7.13) is the basis of an extensive literature on steady, multi-dimensional flows [36, 47].

- the class of versatile non-linear soils defined by $D = a_1 \left(\theta_w^* - \theta_w\right)^{-2}$ and $k_w = a_2 + a_3 \left(\theta_w^* - \theta_w\right) + a_4 / \left\{2 \left(\theta_w^* - \theta_w\right)\right\}$, where a_1, a_2, a_3 and a_4 are constants.

  For these hydraulic properties the Richards equation can be transformed into the Burgers equation by applying the Kirchhoff transformation and the Storm transformation. The Burgers equation can in turn be transformed into the linear diffusion equation by applying the Hopf–Cole transformation.

More details about these classes of soils and associated flow problems can be found elsewhere [52].

7.2.4 Solutions of the Richards equation: numerical solutions for particular classes of soils

In this subsection we consider the group of parametric expressions describing the soil hydraulic properties that is favored in numerical studies and to a large extent shares flexibility with a rather sound basis in Poiseuillean flow in networks of capillaries. In this group of relationships among the water content θ_w, pressure head h and hydraulic conductivity k_w, the latter is calculated from the water retention characteristic and certain assumptions concerning the geometry of the pore system [48, 49, 50]. The procedure in essence links physico-mathematical models at the Darcy and Navier–Stokes scales. By introducing a very general equation for the water retention characteristic and a rather general predictive model for the relative hydraulic conductivity, various well-known classes of soils can be regarded as subclasses of one and the same superclass, thus summarizing a widely scattered literature.

The water retention characteristic of the superclass of soils is given by

$$\lambda h = \left\{ \frac{1 - S^{b_1}}{S^{b_2}} \right\}^{b_3}, \tag{7.14}$$

where λ is the inverse characteristic length of the soil, S is the reduced water content and b_1, b_2 and b_3 are empirical parameters. The reduced water content is defined by

$$S = \frac{\theta_w - \theta_{wr}}{\theta_{ws} - \theta_{wr}}, \tag{7.15}$$

where θ_{wr} is the residual water content and θ_{ws} is the water content at saturation. Subclasses of the superclass of soils are indicated in Table 1. Equation (7.14) includes the subclasses of Brooks–Corey and of Van Genuchten as special cases.

The general predictive model for the relative hydraulic conductivity for the superclass of soils is given by

$$k_{wr} = S^{c_1} \left\{ \frac{\int_0^S (\lambda h)^{-c_2} \, dS}{\int_0^1 (\lambda h)^{-c_2} \, dS} \right\}^{c_3}, \tag{7.16}$$

where c_1, c_2 and c_3 are constants. Equation (7.16) includes the subclasses of Burdine and Mualem as special cases. Introducing the retention curve (7.14) for the superclass of soils in (7.16) gives

$$k_{wr} = S^{c_1} \left\{ I_S^{b_1} \left[(c_2 b_2 b_3 + 1)/b_1, (1 - c_2 b_3) \right] \right\}^{c_3}, \tag{7.17}$$

where $I_S^{b_1} \{\cdot, \cdot\}$ is the incomplete beta function. Important special cases are:

- Burdine: $c_1 = n + m_1 - 1$, $c_2 = 2 + m_2$ and $c_3 = 1$,
- Mualem: $c_1 = n + m_1 - 2$, $c_2 = 1 + m_2$ and $c_3 = 2$,

where m_1 and m_2 are tortuosity parameters and n is the connectivity parameter with $1 < n < 2$.

Already in the 1950s it was clear that most practical flow problems in the unsaturated zone require numerical solutions. By 1975 many finite difference solutions of one-dimensional flow problems were available, but progress with multi-dimensional problems was hampered by both the numerical methods used and the speed of the available computers. Since then efficient finite element [59] and control volume [19] methods and faster computers have made it possible to solve transient flow problems of practical spatial and temporal scale and involving complications such as (i) multi-dimensional regions that are partly and variably saturated, (ii) spatially variable soil physical properties, (iii) hysteresis of water retention characteristics, and (iv) uptake of water by plant roots. In almost all numerical models the physical properties of the soils are represented by functions included in the superclass of soils described in the preceding paragraphs.

The package of Simunek *et al.* [59] was used by De Vos *et al.* [8] to analyze a large field data set, yielding detailed information about the profile distribution of the soil physical properties and the time course and composition of the tile drain discharge. More can be found in three papers authored by De Vos, Heinen, Otten and De Vos, with several co-authors, in [82], including implementations of two different hysteresis models for the water retention characteristic in applications to greenhouse horticulture.

7.2.5   Development of experimental methods and physical characterization of soils

Experimental methods in soil physics are treated in several books [23, 9, 72]. In soil physics research and practice, generally highly specialized instrumentation is needed. Besides papers, conferences and books, companies initiated and sometimes run by soil physicists are playing a key role, especially in the USA. The close connections between the research groups and scientific instrument companies have stimulated rapid commercial development and widespread use of new methods. Improvement of already existing methods is of course also strongly stimulated by advances in other sectors of physics and engineering.

Twenty five years ago methods for measuring the water content and the water potential and its components were already available. The water content was measured not only gravimetrically, but also methods based on scattering of neutrons and adsorption of gamma rays were widely used. Quite generally, significant improvements have come from developments in electronics and data collection systems. The major instrumental development of the last 25 years has been in the realm of dielectric methods, which allow one to infer not only the water content, but also the electrolyte concentration. Tensiometry has changed from mainly water or mercury filled U-tubes or vacuum gauges to electrical transducers, allowing also the development of rapid response micro-tensiometers. With accurate pressure transducers and dataloggers, the sensitivity to disturbing factors such as diurnal temperature variations and bypass flow along tensiometer rods and tubing during intensive and/or long duration rainfall can now be determined.

Inverse methods to determine the physical properties of soils on the basis of equilibrium and steady flow date back to the beginnings of modern soil physics by Buckingham and Richards. The solutions of the Richards equation in the 1950s became the basis for the next generation of methods. In the last quarter century there has been much progress with experimental design, both for use in the laboratory and in the field. The combination of analytical or numerical solutions with optimization algorithms led to a wide variety of new methods, while the shortcomings with regard to non-uniqueness have been identified and repaired. The two Riverside conferences on physical characterization mentioned earlier provide a lot of further details [81, 82].

7.3 Flow of the liquid and gaseous phases in soils subject to swelling and shrinkage

7.3.1 Composition of soils subject to swelling and shrinkage

Let the subscripts s, w and nw denote, respectively, the solid, wetting and non-wetting phases or, where appropriate, the solid, aqueous and gaseous phases of a soil. The volume fractions θ_s, θ_w and θ_{nw} are the fractions of space occupied by phases s, w and nw. They are subject to the constraint

$$\theta_s + \theta_w + \theta_{nw} = 1. \tag{7.18}$$

The bulk mass densities ρ_s, ρ_w and ρ_{nw} and the true mass densities γ_s, γ_w and γ_{nw} are related by

$$\rho_s = \theta_s \gamma_s, \qquad \rho_w = \theta_w \gamma_w, \qquad \rho_{nw} = \theta_{nw} \gamma_{nw}. \tag{7.19}$$

The total bulk mass density ρ is defined as the sum of the bulk mass densities ρ_s, ρ_w and ρ_{nw}:

$$\rho = \rho_s + \rho_w + \rho_{nw}. \tag{7.20}$$

The pivotal role of the solid phase s suggests the introduction of the phase ratios ϑ_w and ϑ_{nw} defined by

$$\vartheta_w = \frac{\theta_w}{\theta_s}, \qquad \vartheta_{nw} = \frac{\theta_{nw}}{\theta_s}. \tag{7.21}$$

The porosity n and the void ratio e are defined by

$$n = \theta_w + \theta_{nw}, \qquad e = \vartheta_w + \vartheta_{nw}. \tag{7.22}$$

With these assumptions the composition is fully characterized by the void ratio e and the wetting phase ratio ϑ_w for unsaturated soils and by just the liquid ratio ϑ_w $(= e)$ for saturated soils.

In this paper it is assumed that the true mass densities γ_s, γ_w and γ_{nw} are constants, reflecting incompressibility of the solid, wetting and non-wetting phases or, in the context of soil physics, incompressibility of the solid and aqueous phases, and constant atmospheric pressure and corresponding constant γ_{nw} of the gaseous phase. If the non-wetting phase is a gaseous phase, then the total bulk mass density ρ is approximately given by

$$\rho \approx \rho_s + \rho_w = \frac{\frac{\gamma_s}{\gamma_w} + \vartheta_w}{1 + e}\gamma_w. \tag{7.23}$$

This expression (7.23) for ρ will be used later to determine the effect of the overburden. If the wetting phase saturates the soil, then in (7.23) the void ratio e in the denominator may be replaced by the wetting phase ratio ϑ_w.

7.3.2 Deformation and motion of the solid phase

Two approaches can be used to describe the deformation and motion of the solid phase. As labels for parcels of the solid phase one can use locations $\mathbf{x}_\kappa = \mathbf{X}_s$ in the reference configuration κ at some reference time t_κ. The spatial approach describes which parcels $\mathbf{X}_s$ of the solid phase in the course of time t are located at given locations $\mathbf{x}$. The material approach describes the places $\mathbf{x}$ occupied in the course of time t by any parcel $\mathbf{X}_s$:

$$\mathbf{x} = \mathbf{x}\left(\mathbf{X}_s, t\right). \tag{7.24}$$

Differentiating (7.24) gives the two key concepts for describing the deformation and motion of the solid phase, namely the deformation gradient tensor $\mathbf{F}_s$ and the velocity $\mathbf{v}_s$ defined by

$$\mathbf{F}_s = \left(\frac{\partial \mathbf{x}}{\partial \mathbf{X}_s}\right)_t, \qquad \mathbf{v}_s = \left(\frac{\partial \mathbf{x}}{\partial t}\right)_{\mathbf{X}_s}. \tag{7.25}$$

The spatial time derivative $(\partial \circ / \partial t)_{\mathbf{x}}$ and the material time derivative $(\partial \circ / \partial t)_{\mathbf{X}_s}$ are related by

$$\left(\frac{\partial \circ}{\partial t}\right)_{\mathbf{X}_s} = \left(\frac{\partial \circ}{\partial t}\right)_{\mathbf{x}} + \mathbf{v}_s \cdot \frac{\partial \circ}{\partial \mathbf{x}}. \tag{7.26}$$

Numerous concepts describing various aspects of the deformation can be derived from the deformation gradient tensor $\mathbf{F}_s$. In particular it can be shown that the deformation at any point can be regarded as resulting from a translation, a rigid rotation of the principal axis of strain and stretches along this axis.

Differentiating the first of (7.25) with respect to t for fixed $\mathbf{X}_s$, or the second of (7.25) with respect to $\mathbf{X}_s$ for fixed t gives

$$\left(\frac{\partial \mathbf{F}_s}{\partial t}\right)_{\mathbf{X}_{sr}} = \frac{\partial \mathbf{v}_s}{\partial x} \cdot \mathbf{F}_s. \tag{7.27}$$

Of particular interest is the trace of (7.27):

$$\nabla \cdot \mathbf{v}_s = J_s^{-1} \left(\frac{\partial J_s}{\partial t}\right)_{\mathbf{X}_s}, \tag{7.28}$$

where $J_s = \det \mathbf{F}_s$ compares the volume currently occupied by a parcel of the solid phase to the volume occupied in the reference configuration. Integration of (7.28) and noting that $J_{s\kappa} = 1$ gives

$$J_s = J_\kappa \exp \int_{t_\kappa}^t \nabla \cdot \mathbf{v}_s dt = \exp \int_{t_\kappa}^t \nabla \cdot \mathbf{v}_s dt. \tag{7.29}$$

The spatial form of the balance of mass for the solid phase is given by

$$\left(\frac{\partial \rho_s}{\partial t}\right)_{\mathbf{X}_s} = -\nabla \cdot (\rho_s \mathbf{v}_s) + \lambda_s, \tag{7.30}$$

where λ_s is the source strength. The source term may represent the decay of the solid phase due to biochemical decomposition of the solid phase in organic soils [51]. Using (7.26) and (7.28), the material form of this mass balance can be written as

$$\left(\frac{\partial (J_s \rho_s)}{\partial t}\right)_{\mathbf{X}_s} = J_s \rho_s \left(\frac{\lambda_s}{\rho_s}\right). \tag{7.31}$$

Integration of (7.31) gives

$$\rho_s = \rho_{s\kappa} J_s^{-1} \exp \left[\int_{t_\kappa}^t \left(\frac{\lambda_s}{\rho_s}\right) dt\right]. \tag{7.32}$$

Equation (7.32) gives the current density of the solid phase as the product of the density in the reference configuration, the inverse of the Jacobian of the deformation gradient tensor and a term accounting for production or decay of the material of the solid phase since the reference time t_κ. If the solid phase is non-reactive, then (7.32) reduces to Euler's equation for the mass density ρ_s or, making use of the constancy of γ_s, for the volume fraction θ_s:

$$\rho_s = \rho_{s\kappa} J_s^{-1} \quad \text{or} \quad \theta_s = \theta_{s\kappa} J_s^{-1}. \tag{7.33}$$

7.3.3 Balances of mass and momentum

The spatial forms of the balances of mass for the liquid and gaseous phases
are given by

$$\left(\frac{\partial \rho_w}{\partial t}\right)_{\mathbf{x}} = -\nabla \cdot (\rho_w \mathbf{v}_w) + \lambda_w, \tag{7.34}$$

$$\left(\frac{\partial \rho_{nw}}{\partial t}\right)_{\mathbf{x}} = -\nabla \cdot (\rho_{nw} \mathbf{v}_{nw}) + \lambda_{nw}, \tag{7.35}$$

where λ_w and λ_{nw} are the source strengths, which among others might
represent uptake by plant roots. Using (7.26), (7.28) and (7.33) these mass
balances can be written in terms of the phase ratios ϑ_w and ϑ_{nw} and relative
to the deformation and motion of the solid phase (see [53], [4], [44], [46]
and [51]:

$$\left(\frac{\partial \vartheta_w}{\partial t}\right)_{\mathbf{X}_s} = -\nabla_s \cdot \left\{\vartheta_w \mathbf{F}_s^{-1}\left(\mathbf{v}_w - \mathbf{v}_s\right)\right\} + \vartheta_w \left(\frac{\lambda_w}{\rho_w}\right), \tag{7.36}$$

$$\left(\frac{\partial \vartheta_{nw}}{\partial t}\right)_{\mathbf{X}_s} = -\nabla_s \cdot \left\{\vartheta_{nw} \mathbf{F}_s^{-1}\left(\mathbf{v}_{nw} - \mathbf{v}_s\right)\right\} + \vartheta_{nw} \left(\frac{\lambda_{nw}}{\rho_{nw}}\right). \tag{7.37}$$

In (7.36) and (7.37) the solid phase s serves as a reference in several respects:
- The composition is expressed relative to the solid phase.
- All derivatives involve the material coordinate $\mathbf{X}_s$ of the solid phase.
- $\mathbf{F}_s^{-1}$ transforms the relative velocities $(\mathbf{v}_w - \mathbf{v}_s)$ and $(\mathbf{v}_{nw} - \mathbf{v}_s)$ to the
configuration of the reference continuum at reference time t_κ.

To simplify the balances of momentum of the fluid phases, the following
assumptions are made [54, 44, 51]:
- The true pressures p_w and p_{nw} of the fluid phases give rise to the
bulk stress tensors $-p_w \mathbf{I}$ and $-p_{nw}\mathbf{I}$. The bulk shear stresses in the fluid
phases are assumed to be negligible. Operationally, the pressures p_w and
p_{nw} can be measured in a pool of the fluid connected with the soil via a
membrane permeable to the fluid, but not to the solid and the other fluid.
It is appropriate to call p_w the tensiometer pressure.
- The force exerted upon a fluid phase by the solid phase and the other
fluid phase consists of two parts:

1) Forces proportional to the pressures and the gradients of the volume
fractions: $p_w \nabla \theta_w$ and $p_{nw} \nabla \theta_{nw}$. Following earlier work by Heinrich and
Desoyer in the fifties, such terms for two-phase and three-phase media can
be motivated on the basis of an integration of the normal component of the
stresses at interfaces. Forces of exactly this form later also arose in several
other theories (see [44] for some references).

2) Drag forces proportional to the differences of the phase velocities:
$r_{sw}\left(\mathbf{v}_s - \mathbf{v}_w\right) + r_{nww}\left(\mathbf{v}_{nw} - \mathbf{v}_w\right)$ and $r_{snw}\left(\mathbf{v}_s - \mathbf{v}_{nw}\right) + r_{wnw}\left(\mathbf{v}_w - \mathbf{v}_{nw}\right)$.
- The gravitational force g is the only external body force.

192   P.A.C. Raats

- The inertial force is negligible.

Then the resulting expressions for the mass fluxes relative to the solid phase are [54, 44]

$$
\begin{pmatrix} \theta_w \left(\mathbf{v}_w - \mathbf{v}_s\right) \\ \theta_{nw} \left(\mathbf{v}_{nw} - \mathbf{v}_s\right) \end{pmatrix}
$$

$$
= \begin{pmatrix} k_{ww} & k_{nww} \\ k_{wnw} & k_{nwnw} \end{pmatrix} \begin{pmatrix} \left(\mathbf{F}_s^T\right)^{-1} \nabla_s p_w - \gamma_w \mathbf{g} \\ \left(\mathbf{F}_s^T\right)^{-1} \nabla_s p_{nw} - \gamma_{nw} \mathbf{g} \end{pmatrix}
$$

(7.38)

where the matrix of conductivities $k_{\alpha\beta}$ is related to the matrix of resistivities $r_{\alpha\beta}$ by

$$
\begin{pmatrix} k_{ww} & k_{nww} \\ k_{wnw} & k_{nwnw} \end{pmatrix} = \begin{pmatrix} \dfrac{(r_{snw}+r_{wnw})\rho_w^2}{D} & \dfrac{r_{nww}\rho_{nw}\rho_w}{D} \\ \dfrac{r_{wnw}\rho_w\rho_{nw}}{D} & \dfrac{(r_{sw}+r_{nww})\rho_{nw}^2}{D} \end{pmatrix}
$$

(7.39)

with

$$
D = r_{sw}r_{snw} + r_{sw}r_{wnw} + r_{snw}r_{nww}.
$$

(7.40)

If the drag between the liquid and gaseous phases is negligible, *i.e.*, if r_{nww} and r_{wnw} are relatively small, then the matrix of conductivities (7.39) reduces to

$$
\begin{pmatrix} k_{ww} & k_{nww} \\ k_{wnw} & k_{nwnw} \end{pmatrix} = \begin{pmatrix} \dfrac{\rho_w^2}{r_{sw}} & 0 \\ 0 & \dfrac{\rho_{nw}^2}{r_{snw}} \end{pmatrix},
$$

(7.41)

so that the flux of each of the fluid phases is dependent only on the driving forces for that phase. If in addition the resistivity r_{snw} is very small, for example because of preferential wetting of the solid phase by the liquid phase, and if furthermore the effect of the gravitational force $\mathbf{g}$ upon the gaseous phase is negligible, then the pressure p_{nw} of the gaseous phase will be uniform and under most circumstances in soil physics and hydrology be equal to the atmospheric pressure. It is then convenient to introduce the capillary pressure p^c defined by

$$
p^c = p_w - p_{nw}.
$$

(7.42)

The mass flux of the liquid phase relative to the solid is then given by

$$
\theta_w \left(\mathbf{v}_w - \mathbf{v}_s\right) = -k_w \left\{ \left(\mathbf{F}_s^T\right)^{-1} \nabla_s p^c - \mathbf{g}\gamma_w \right\}
$$

(7.43)

with the conductivity k_w given by

$$
k_w = \frac{\rho_w \rho_w}{r_{sw}}.
$$

(7.44)

If the solid phase undergoes a rigid motion, then $\mathbf{F}_s = \mathbf{I}$. If further the coordinate system is chosen to be attached to the solid phase, then (7.36) and (7.37) revert to (7.34) and (7.35) and (7.38) reduces to

$$
\begin{pmatrix} \theta_w \mathbf{v}_w \\ \theta_{nw} \mathbf{v}_{nw} \end{pmatrix} = \begin{pmatrix} k_{ww} & k_{nww} \\ k_{wnw} & k_{nwnw} \end{pmatrix} \begin{pmatrix} \nabla p_w - \gamma_w \mathbf{g} \\ \nabla p_{nw} - \gamma_{nw} \mathbf{g} \end{pmatrix}.
$$

(7.45)

Equation (7.45), in which each of the fluxes depends on both pressure gradients, was also obtained by [83] on the basis of volume averaging the force balance equations in two fluid phases at the pore scale. Equation (7.45) has rarely been applied as such. However, with the conductivity matrix given by (7.41), these expressions for the fluxes have been widely used in petroleum engineering since the 1930s. It is therefore not surprising that in soil physics a full-fledged two-fluid-phase approach has been advocated primarily by people influenced by experience in petroleum engineering [28, 27]. Effects arising from restricted access of air have been observed often in the field and the laboratory. Especially interesting effects occur if locally one of the two phases completely fills the pores [41]. In recent years, multiphase models are used widely to study the flow of liquid contaminants in aquifers and of air injected below the groundwater table for remediation purposes, so-called air-sparging [80]. In fact, drawn by funding opportunities, a lot of cross-fertilization between soil physicists and petroleum engineers studying these problems has occurred over the last 15 years (for examples see several papers in [81, 82].

Finally, let us assume that circumstances such that equations (7.43) and (7.44) apply and moreover the solid phase undergoes a rigid motion and the coordinate system is chosen to be attached to the solid phase. Then (7.43) reduces to (7.2), so that the model of Richards for flow of water in rigid soils results as a special case.

7.3.4 One-dimensional flow of the liquid phase in non-rigid soils

Equation (7.43) is the basic expression for the flux of water relative to the solid phase. In the previous section concerning rigid soils, the capillary pressure was related to the volume fraction θ_w. For non-rigid soils this idea is generalized to relationships between the liquid ratio ϑ_w, the capillary pressure p^c, the void ratio e and the envelope pressure P, which is the total vertical pressure. For the vertical direction z, the total force balance can be written as

$$\frac{\partial P}{\partial z} = \rho g = \frac{\frac{\gamma_s}{\gamma_w} + \vartheta_w}{1 + e} \gamma_w g, \tag{7.46}$$

where the expression (7.23) for the total bulk density ρ_s has been used. Note that the vertical gradient of the envelope pressure P is determined by the local distribution of the three phases and the true densities γ_s and γ_w. Integration of (7.46) gives

$$P = p_a + \gamma_w g \int_0^z \frac{\frac{\gamma_s}{\gamma_w} + \vartheta_w}{1 + e} \mathrm{d}z, \tag{7.47}$$

assuming that the air pressure p_a acts at the soil surface. Equations (7.46) and (7.47) are the spatial forms of, respectively, the differential and integral expressions of the distribution of the envelope pressure.

Perhaps it would be most natural to postulate the existence of an energy function ϵ_1, determined uniquely by ϑ_w and e, and leading to expressions for p^c and P in terms of ϑ_w and e. In fact in the context of the linear theory for saturated soils, an analogous proposal was put forward by Biot [1], who also included effects of shear strains of the solid phase, leading to expressions for the shear stresses in the solid phase. However, inspired by early work of D. Croney and J.D. Coleman and guided by G. H. Bolt, soil physicists have used another energy function ϵ_2 determined uniquely by ϑ_w and P and related to ϵ_1 by the Legendre transformation

$$\epsilon_2 = \epsilon_1 - Pe. \tag{7.48}$$

Changes of $\mathrm{d}\epsilon_2$ are related to changes of ϑ_w and P by

$$\mathrm{d}\epsilon_2 = p^c \mathrm{d}\vartheta_w - e\mathrm{d}P, \tag{7.49}$$

where

$$p^c(\vartheta_w, P) = \frac{\partial \epsilon_2(\vartheta_w, P)}{\partial \vartheta_w}, \tag{7.50}$$

$$e(\vartheta_w, P) = \frac{\partial \epsilon_2(\vartheta_w, P)}{\partial P}. \tag{7.51}$$

Changes in p^c and e can be expressed in terms of ϑ_w and P by

$$\mathrm{d}p^c = \frac{\partial p^c(\vartheta_w, P)}{\partial \vartheta_w}\mathrm{d}\vartheta_w + \frac{\partial p^c(\vartheta_w, P)}{\partial P}\mathrm{d}P, \tag{7.52}$$

$$\mathrm{d}e = \frac{\partial e(\vartheta_w, P)}{\partial \vartheta_w}\mathrm{d}\vartheta_w + \frac{\partial e(\vartheta_w, P)}{\partial P}\mathrm{d}P. \tag{7.53}$$

If it is assumed that $\mathrm{d}\epsilon_2$ is an exact differential, then two of the coefficients in (7.52) and (7.53) are related by the Maxwell relationship

$$\frac{\partial p^c(\vartheta_w, P)}{\partial P} = \frac{\partial e(\vartheta_w, P)}{\partial \vartheta_w} = \alpha(\vartheta_w, P). \tag{7.54}$$

A proper name for $\alpha(\vartheta_w, P)$ is differential or local load factor: it describes the effectiveness of the envelope pressure in affecting the capillary pressure p^c.

Equation (7.52) describes how locally changes in wetness and overburden contribute to changes in the capillary pressure. To get a global view, it is desirable to integrate (7.52) from some standard state to the local state of interest. The assumption of the existence of an energy function implies that the result of the integration should be independent of the path of the integration. It is most convenient to choose as a standard state $p^c = 0$, $P = p_a$ and $\vartheta_w = e = \infty$, $i.e.$, pure liquid at atmospheric pressure, and then first decrease ϑ_w to the desired value, while keeping $P = p_a$, and subsequently increasing P to the desired value, while keeping ϑ_w constant. The resulting expression for the capillary pressure p^c is

$$p^c\left(\vartheta_w, P\right) = p^c\left(\vartheta_w, P = p_a\right) + \widehat{\alpha}\left(\vartheta_w, P\right)\left(P - p_a\right), \qquad (7.55)$$

where the capillary pressure $p^c\left(\vartheta_w, P = p_a\right)$ of the unloaded soil is given by

$$p^c\left(\vartheta_w, P = p_a\right) = \int_{\infty}^{\vartheta_w} \left(\frac{\partial p^c}{\partial P}\right)_{P=p_a} d\vartheta_w, \qquad (7.56)$$

and the integral or average load factor is given by

$$\widehat{\alpha}\left(\vartheta_w, P\right) = \frac{1}{P - p_a} \int_{p_a}^{P} \frac{\partial p^c\left(\vartheta_w, P\right)}{\partial P} dP = \frac{1}{P - p_a} \int_{p_a}^{P} \frac{\partial e(\vartheta_w, P)}{\partial \vartheta_w} dP, \quad (7.57)$$

where the Maxwell relationship (7.54) has been used. According to (7.57), $\widehat{\alpha}\left(\vartheta_w, P\right)$ is the average slope of e versus ϑ_w between p_a and P at a particular value of ϑ_w. In (7.57) the average load factor $\widehat{\alpha}\left(\vartheta_w, P\right)$ describes the effectiveness of the total pressure P in affecting $p^c\left(\vartheta_w, P\right)$. In saturated, non-rigid soils $\vartheta_w = e$ and $\widehat{\alpha} = 1$ and (7.55) reduces to

$$p^c\left(\vartheta_w, P\right) = p^c\left(e, P = p_a\right) + \left(P - p_a\right). \qquad (7.58)$$

In soil mechanics the content of (7.58) is referred to as the principle of effective stress.

An early attempt by Philip [32] to cope with the effect of the envelope pressure upon the capillary pressure p^c ignored the role of P in the relationship $\widehat{\alpha}\left(\vartheta_w, P\right)$. The attempt was criticized by Youngs [84]. Philip [34] acknowledged the shortcoming and, using equation (7.57) derived by G.H. Bolt, in principle resolved the problem. Groenevelt and Bolt [17] proceeded to give a detailed thermodynamic discussion of (7.57) and related matters (see also [3] and [76]). The relationship between e and ϑ_w at a particular value of the envelope pressure P is called the shrinkage curve. It was first introduced by Haines [18]. Probably the best set of shrinkage curves is still that of Talsma [77]

Specializing (7.43) to vertical flow and introducing (7.52) and using (7.46) gives

$$\theta_l\left(v_{wz} - v_{sz}\right) = -k\left(\frac{\partial p^c\left(\vartheta_w, P\right)}{\partial \vartheta_w}\right)\left(\frac{\partial \vartheta_w}{\partial z}\right) + kF\gamma_w g, \qquad (7.59)$$

with

$$F = 1 - \alpha\left(\vartheta_w, P\right) \frac{\frac{\gamma_s}{\gamma_w} + \vartheta_w}{1 + e}. \qquad (7.60)$$

The load factor $\alpha\left(\vartheta_w, P\right)$ describes the effectiveness of the envelope pressure P in affecting the capillary pressure p^c. For rigid soils $\alpha\left(\vartheta_w, P\right) = 0$, so that $F = 1$ and (7.59) reduces to (7.2). The deviation of the factor F in (7.59) from unity accounts for the influence of the overburden upon the gravitational component of the flux. For saturated, non-rigid soils $\vartheta_w = e$ and the load factor $\alpha(\vartheta_w, P) = 1$, so that F reduces to

$$F = 1 - \frac{\frac{\gamma_s}{\gamma_w} + e}{1 + e} = \frac{1 - \frac{\gamma_s}{\gamma_w}}{1 + e}. \tag{7.61}$$

If $\gamma_s > \gamma_w$, as is the case for soils, then $F < 0$, so that its influence in (7.61) is to reverse the direction of the gravitational component of the flux [32]. Infiltration in non-rigid soils is analogous with capillary rise in rigid soils. Contrary to the situation for rigid soils, long-time, traveling waves in non-rigid soils arise when the liquid phase moves upward relative to the solid phase, such as occurs in sedimentation of suspensions.

7.3.5 Multi-dimensional deformation and one-dimensional flow

The drying process of a non-rigid soil may start as a combination of purely one-dimensional flow of the aqueous phase and deformation of the solid phase. This changes when cracks appear. Some attempts have been made to cope with such systems by describing them in terms of one-dimensional flow of the aqueous phase and transversely isotropic deformation of the solid phase, with the deformation gradient tensor of the solid phase given by

$$\mathbf{F}_s = \begin{pmatrix} \frac{\partial x}{\partial X_s} & 0 & 0 \\ 0 & \frac{\partial y}{\partial Y_s} & 0 \\ 0 & 0 & \frac{\partial z}{\partial Z_s} \end{pmatrix} \tag{7.62}$$

$$= \begin{pmatrix} \left(\frac{\partial z}{\partial Z_s}\right)^{\frac{1-n}{2n}} & 0 & 0 \\ 0 & \left(\frac{\partial z}{\partial Z_s}\right)^{\frac{1-n}{2n}} & 0 \\ 0 & 0 & \left(\frac{\partial z}{\partial Z_s}\right) \end{pmatrix},$$

where

$$\begin{array}{ll} n = 1 & \text{for purely axial deformation,} \\ n = \frac{1}{2} & \text{for balanced axial and lateral deformation,} \\ n = \frac{1}{3} & \text{for isotropic deformation,} \\ n = 0 & \text{for purely lateral deformation.} \end{array} \tag{7.63}$$

From (7.62) it follows that for transversely isotropic deformation

$$J_s = \det \mathbf{F}_s = \frac{\partial x}{\partial X_s}\frac{\partial y}{\partial Y_s}\frac{\partial z}{\partial Z_s} = \left(\frac{\partial z}{\partial Z_s}\right)^{\frac{1}{n}}. \tag{7.64}$$

From equation (7.64) and the Euler's equation (7.33) for the mass density ρ_s, it follows that

$$\frac{\partial z}{\partial Z_s} = \left(\frac{\rho_s}{\rho_{s\kappa}}\right)^{-n}.$$
(7.65)

Introducing (7.65) in (7.62)$_2$ gives

$$\mathbf{F}_s = \begin{pmatrix} \left(\frac{\rho_s}{\rho_{s\kappa}}\right)^{\frac{n-1}{2}} & 0 & 0 \\ 0 & \left(\frac{\rho_s}{\rho_{s\kappa}}\right)^{\frac{n-1}{2}} & 0 \\ 0 & 0 & \left(\frac{\rho_s}{\rho_{s\kappa}}\right)^{-n} \end{pmatrix}.$$
(7.66)

Using the definition of the void ratio e as the ratio of the porosity and the volume fraction of the solid phase, equations (7.65) and (7.66) can be rewritten as

$$\frac{\partial z}{\partial Z_s} = \left(\frac{1+e}{1+e_\kappa}\right)^n,$$
(7.67)

$$\mathbf{F}_s = \begin{pmatrix} \left(\frac{1+e}{1+e_\kappa}\right)^{\frac{1-n}{2}} & 0 & 0 \\ 0 & \left(\frac{1+e}{1+e_\kappa}\right)^{\frac{1-n}{2}} & 0 \\ 0 & 0 & \left(\frac{1+e}{1+e_\kappa}\right)^n \end{pmatrix}.$$
(7.68)

The expression for the deformation gradient for the class of transversely isotropic deformations in terms of the mass density was introduced by Raats [42] (see also [45] and [25]) in an analysis of axial fluid flow in swelling and shrinking rods. The r_s-factor, introduced by Rijniersce [57, 58] in a study of physical changes in the soils of IJsselmeerpolders, is the reciprocal of the parameter n introduced above. More recently, this r_s-factor has been used extensively to characterize swelling and shrinkage in clay soils [6, 7, 11, 12]. The expression for the deformation gradient for the class of transversely isotropic deformations in terms of the void ratio is due to Garnier *et al.* [11, 12]. The r_s-factor also plays a key role in the recent paper by Kim *et al.* [20].

7.4   Solutions of one-dimensional flow problems

Just as within the theory for movement of water in saturated rigid soils, analytical and quasi-analytical methods as well as numerical methods have been used to solve boundary value problems. Purely analytical solutions mainly apply to specific classes of soils, allowing transformations from non-linear to linear boundary value problems. Quasi-analytic solutions involve a varying mix of mathematical analysis, approximations and numerical evaluation of integrals and/or solutions of ordinary differential equations. Purely numerical methods have the advantage that they can deal with any class of soils and any type of boundary conditions, including temporal changes of these conditions.

7.4.1 Equilibrium and steady upward and downward flows

At equilibrium $v_{wz} = v_{sz}$ and then (7.43) implies the usual balance of pressure gradient force and gravitational force

$$\frac{\mathrm{d}p^c}{\mathrm{d}z} = \gamma_w g. \tag{7.69}$$

Setting $v_w = v_s$ in (7.59) and solving for $\frac{\mathrm{d}\vartheta_w}{\mathrm{d}z}$ gives

$$\frac{\mathrm{d}\vartheta_w}{\mathrm{d}z} = \frac{F\gamma_w g}{\frac{\partial p^c(\vartheta_w,P)}{\partial \vartheta_w}}. \tag{7.70}$$

The denominator of the right-hand side of (7.70) is positive. Therefore equation (7.70) implies the existence of three types of equilibrium profiles:
- Xeric profiles: $\frac{\mathrm{d}\vartheta_w}{\mathrm{d}z} > 0$, whenever $F > 0$;
- Pycnotatic profiles: $\frac{\mathrm{d}\vartheta_w}{\mathrm{d}z} = 0$, whenever $F = 0$;
- Hydric profiles: $\frac{\mathrm{d}\vartheta_w}{\mathrm{d}z} < 0$, whenever $F < 0$.

This classification of and names for the various types of equilibrium profiles were introduced by Philip[32]. For rigid soils, (7.70) reduces to Buckingham's expression for the equilibrium profile.

Steady upward and downward flows can be analyzed by spatially integrating equation (7.43) for p^c or equation (7.59) for ϑ_w. From (7.59) it follows that for steady flows $\frac{\mathrm{d}\vartheta_w}{\mathrm{d}z}$ is given by

$$\frac{\mathrm{d}\vartheta_w}{\mathrm{d}z} = \frac{Fk_w\gamma_w g - \theta_w\left(v_w - v_s\right)}{k_w \frac{\partial p^c(\vartheta_w,P)}{\partial \vartheta_w}}. \tag{7.71}$$

The denominator of the right-hand side of (7.71) is positive. Therefore steady flow profiles can be classified as follows:
- Xeric profiles: $\frac{\mathrm{d}\vartheta_w}{\mathrm{d}z} > 0$, whenever $F > \frac{\theta_w(v_w - v_s)}{\gamma_w g}$;
- Pycnotatic profiles: $\frac{\mathrm{d}\vartheta_w}{\mathrm{d}z} = 0$, whenever $F = \frac{\theta_w(v_w - v_s)}{\gamma_w g}$;
- Hydric profiles: $\frac{\mathrm{d}\vartheta_w}{\mathrm{d}z} < 0$, whenever $F < \frac{\theta_w(v_w - v_s)}{\gamma_w g}$.

For non-rigid soils even finding the above equilibrium and steady flow profiles proved to be quite a challenge. An early attempt to describe the equilibrium profile was made by Babcock and Overstreet in 1957, and the problem was reconsidered by Babcock in 1963, stimulated by remarks from Collis-George in 1961 and discussions with R.D. Miller [74]. An imaginative attack at both the equilibrium and steady flow profiles was made by Philip [32, 34]. Sposito [73, 74, 75] and Giráldez and Sposito [15] also worked out the implications for equilibrium profiles and steady flows. Numerical calculations of steady flow profiles were presented by Giráldez and Sposito [15], including a comparison with profiles in corresponding rigid soils. Steady flows in saturated soils were also discussed by Miller [25].

7.4.2 Analytical solutions of linearized flow problems

All analytical solutions for flows in nonrigid soils known to me are solutions of linearized flow equations for saturated soils. The simplest, but also most restrictive, linearization is based on assuming that the diffusivity $D_m = \left(k_{wm}\frac{\partial p^c(\vartheta_w,P)}{\partial \vartheta_w}\right)/(1+e)$ is constant and $k_{wm} = \frac{k_w}{1+e}$ is linear in ϑ_w. Blake and Colombera [2] used the resulting linear Fokker–Planck equation to calculate ϑ_w as a function of Z_s and t in a sedimenting column of particles. They carried out experiments on two columns of different length. A γ-ray attenuation method was used to determine the ϑ_w-profile. The measured and calculated volume of clear liquid on top agreed well. The measured and calculated ϑ_w-profiles agreed well for early times, but the eventual near equilibrium profiles did not compare so favorably.

Linearization can also be accomplished by a combination of physical simplifications and mathematical transformations. Broadbridge [5] reduced the flow equation to Burgers equation by assuming D_m to be constant and k_{wm} to be quadratic in ϑ_w. Burgers equation is a special non-linear Fokker–Planck equation that can be transformed to a linear diffusion equation by the Hopf–Cole transformation. Broadbridge derived an analytical solution for constant rate infiltration. The resulting expression for the ponding time t_p is

$$\frac{t_p}{t_{grav}} = \left(\frac{\pi}{4R*}\right)\operatorname{inverf}^2\left(\frac{i}{\sqrt{R*}}\right), \qquad (7.72)$$

where t_{grav} is the time at which gravitational effects begin to dominate capillary effects, 'inverf' is the inverse error function and $R*$ is the dimensionless infiltration rate (see Appendix of Broadbridge's paper for details regarding t_{grav} and $R*$). Broadbridge compared his expression for the ponding time with those obtained by Giráldez and Sposito [16] on the basis of quasi-analytical (see Subsection 4) and numerical solutions (see Subsection 5).

7.4.3 Solutions in the form of time-invariant traveling waves

As far as I am aware solutions in the form of time invariant traveling waves (TITWs) and related shock fronts have so far not received attention in the literature on expansive soils. The most likely reason for this is that the wetting fronts are the main TITWs widely known to soil physicists. But also in the context of rigid soils there are other situations where TITWs arise. Examples are downward flow to an upward or downward moving water table [55] and uptake of water by a growing root system [48]. The potential emergence of TITWs in a given situation is dependent of the sign of the gravitational term in the flow equation. That is why in the context of rigid soils TITWs occur with infiltration but not with capillary rise.

Equations (7.59)–(7.61) imply that the effect of the overburden may be to reverse the sign of the gravitational term. If that happens, then TITWs no longer occur with infiltration, but they may arise in situations where the liquid phase moves upward relative to the solid phase. The analogy of infiltration in non-rigid soils and capillary size in rigid soils was already pointed out by Philip [32]. It is therefore understandable that TITW large time limits do not arise in the studies of infiltration by Smiles [60], Giráldez and Sposito [16] and BroadBridge [5].

I expect TITWs may occur in the contexts of sedimentation and of upward flow from a moving water table in expansive soils. The experimental data on sedimentation of "red mud" slurry, essentially a suspension of hydrated iron oxide particles in strong sodium hydroxide solution, presented by Blake and Colombera [2] show some evidence of approach to a TITW pattern at intermediate time. They compared these data with an analysis based on a linearized, convection-dispersion equation for the moisture ratio ϑ_w formulated in terms of material coordinates (see Subsection 4.2). The linearization results from assuming $k_{wm} = \frac{k_w(\vartheta_w)}{1+\vartheta_w} = \gamma(\vartheta_w - \vartheta_w*)$ and $D_m = k_{wm}\frac{\mathrm{d}p^c}{\mathrm{d}\vartheta_w} = $ constant. To show an approach to a TITW would require $\frac{\mathrm{d}k_{wm}}{\mathrm{d}\vartheta_w}$ not to be constant but to be an increasing function of ϑ_w. Such models in fact occur widely in the literature on sedimentation. In most of that literature diffusive terms are ignored and the attention focuses on the shock waves arising from the resulting non-linear kinematic wave equation. The non-linear sedimentation problem involving interaction of the convective and diffusive terms deserves more attention. The same applies, in the context of expansive soils, for the related problem of upward movement of water with a moving water table.

7.4.4 Quasi-analytical solutions

It is often useful to regard the water ratio ϑ_w and the time t as independent variables, while the relative flux $F_w = \vartheta_w(v_w - v_s)$ and the material coordinates Z_s are regarded as the dependent variables. In this context the mass balance equation (7.36) is expressed by

$$\frac{\partial Z_s(t, \vartheta_w)}{\partial t} = \frac{\partial F_w(t, \vartheta_w)}{\partial \vartheta_w}. \tag{7.73}$$

Integration of (7.73) between ϑ_w and $\vartheta_{w\infty}$ yields

$$F_w - F_{w\infty} = \frac{\partial}{\partial t} \int\limits_{\vartheta_{w\infty}}^{\vartheta_w} Z_s \mathrm{d}\vartheta_w. \tag{7.74}$$

Integration of (7.74) with respect to the material coordinate Z_s gives

$$\int\limits_0^\infty (F_w - F_{w\infty})\,\mathrm{d}Z_s = \frac{\partial}{\partial t}\int\limits_{\vartheta_{w0}}^{\vartheta_{w\infty}} \frac{1}{2}Z_s^2 \mathrm{d}\vartheta_w = \frac{\partial}{\partial t}\int\limits_0^\infty Z_s\left(\vartheta_w - \vartheta_{w\infty}\right)\mathrm{d}Z_s. \tag{7.75}$$

Integration of (7.74) with respect to time gives

$$\int\limits_0^t \left(F_{w0}\,[t] - F_{w\infty}\right)\mathrm{d}t = \int\limits_{\vartheta_{w\infty}}^{\vartheta_{w0}[t]} Z_s \mathrm{d}\vartheta_w, \tag{7.76}$$

where the subscript 0 denotes values at the soil surface. Recall that the analogous integral equations (7.11) and (7.12) have been used by Parlange and coworkers and by Philip, Knight and coworkers, respectively, as constraints in iterative solutions in numerous papers on movement of water in rigid soils. The approximate methods, using either (7.75) or (7.76) as integral constraint, have yielded solutions for several flow problems involving slurries or expansive soils. Following this is a brief review of these results.

Smiles and Harvey [70] derived an expression for the diffusivity D_m in terms of the sorptivity:

$$D_m\left(\vartheta_{w0}\right) = \frac{\left(\dfrac{\partial S^2}{\partial \vartheta_{w\infty}} - \dfrac{rS^2}{\left(\vartheta_{w0}-\vartheta_{wn}\right)^2}\right)}{2\left(\vartheta_{w0}-\vartheta_{wn}\right)}, \tag{7.77}$$

where ϑ_{w0} and ϑ_{wn} are the surface and initial values of ϑ_w, S is the sorptivity and r is a constant that can take values $0 < r < 1$. Following critical comments by Parlange [29], Smiles [61, 62, 63, 65] explored the method further. The diffusivities where shown to be independent of the initial water content and to agree with the data from steady flow experiments. Parlange [29] pointed out that clay particles likely become reoriented during the consolidation process, leading to an anisotropic conductivity/liquid ratio relationship. Smiles [61, 62] admitted that this was so, but argued that for one-dimensional processes such effects may be implicitly accounted for, as was in fact already suggested by McNabb [24]. It appears that such effects may be more important in multi-dimensional flows.

Kirby and Smiles [21] used the method of Smiles and Harvey [70] to determine the influence of solution salt concentration upon the physical properties of bentonite suspensions. They found that both the dependencies of the capillary pressure and the hydraulic conductivity upon the liquid ratio are sensitive to the solution salt concentration, but that the capillary pressure/conductivity relationship is insensitive.

Smiles [64] used (7.76) to analyze constant rate filtration of an initially uniform slurry supported by a membrane which permits ready passage of liquid but not of solid. Smiles gave two versions of the theory, one with and one without influence of gravity. In the absence of influence of gravity, the imposed rate of filtration F_{w0} enters the solution only via the reduced material coordinate Z_s* and time $t*$ defined by

$$Z_s* = F_{w0} Z_s \qquad t* = F_{w0}^2 t. \qquad (7.78)$$

Constant pressure filtration, including the effect of membrane resistance, was analyzed by Smiles $et\ al.$ [67]. In the absence of influence from gravity, there is in this problem the interesting feature that the membrane conductance α enters the solution only via the reduced material coordinate Z_s** and time $t**$ defined by

$$Z_s** = \alpha Z_s \qquad t** = \alpha^2 t. \qquad (7.79)$$

Note that the forms of (7.78) and (7.79) are similar.

Smiles $et\ al.$ [66] also analyzed gravity drainage of a column of bentonite, to the top of which additional effluent is applied continuously at a constant rate, and from the bottom of which the liquid phase escapes through a membrane which prevents escape of the solid phase. Among the results are the spatial distributions of the liquid ratio and the filtration rate, both as functions of time.

The quasi-analytical approach was used by Giráldez and Sposito [16] to analyze infiltration in swelling soils, thereby extending earlier work for rigid soils by Parlange and Smith. The analysis gave equations for the ponding time and the postponding infiltration rate. Ponding times for swelling soils are shorter than those for non-swelling analogs. Postponding infiltration rates in swelling soils approach zero instead of becoming equal to the hydraulic conductivity, as in rigid soils. The results agreed with those from a numerical model. Among the results is a generalization to swelling soils of a three-parameter infiltration equation proposed by Parlange $et\ al.$ [31].

7.4.5 Numerical solutions

Stroosnijder [78] and Giráldez [14] were the first to use numerical methods in studies of swelling soils. Stroosnijder developed a CSMP program to analyze infiltration of ponded water into a swelling soil. From related experiments Stroosnijder concluded that hysteresis and retarded equilibration were complicating factors. Retarded equilibration was included in the program and led to an S-shaped graph of the cumulative infiltration as a function of the square root of time.

Giráldez [14] used a Crank–Nicolson implicit scheme to analyze drainage to a water table in a non-rigid soil. In comparison with a corresponding rigid soil, the drainage from the non-rigid soil was slow and cumulatively small. A similar numerical method was used by Giráldez and Sposito [16] to analyze infiltration with at the soil surface a prescribed rate of infiltration up to the time t_p at which ponding occurs and thereafter, *i.e.*, for $t > t_p$, a constant capillary pressure p^c and a corresponding water ratio ϑ_w. For rates of infiltration less than ten times the hydraulic conductivity corresponding to ϑ_w, the ponding time t_p for the swelling soil was found to be smaller for the nonrigid soil than for the corresponding rigid soil.

7.5   Concluding remark

Philip [35] expressed regret about the apartness of the worlds of soil physics and soil mechanics. This has not changed drastically in the last quarter century. Few soil physicists keep up with developments in soil mechanics as presented in this book. In soil mechanics, the Richards equation shows up only sporadically and the work on swelling soils by soil physicists is not widely known. Perhaps the most encouraging is that now a small number of people with knowledge and experience in both disciplines work on the fundamental aspects of soil tillage operations and on water and wind erosion problems. Water erosion is closely intertwined with in- and exfiltration, while wind erosion is influenced by cohesion induced by soil water. Occasionally there are also contributions to geomorphology, such as the recent work on flow around solid object and air filled cavities, flow in stratified media, unstable flow. I hope this review will stimulate further interactions.

Acknowledgments: I would like to thank Dr. Kees Rappoldt for introducing me to OzTeX, a Macintosh implementation of TeX.

References

[1] M.A. Biot: General theory of three-dimensional consolidation, *J. Appl. Phys.* **12**, pp. 155–164, 1941.

[2] J.R. Blake, P.M. Colombera: Sedimentation: a comparison between theory and experiment, *Chem. Eng. Sci.* **32**, pp. 221–228, 1977.

[3] G.H. Bolt, S. Iwata, A.J. Peck, P.A.C. Raats, A.A. Rode, G.Vachaud, A.D. Voronin: Soil physics terminology, *Bull. Int. Soc. Soil Sci.* **49**, pp. 26–35, 1976.

[4] R. BOWEN: Compressible porous media models by use of the theory of mixtures, *Int. J. Engng. Sci.* **20**, pp. 697–735, 1982.

[5] P. BROADBRIDGE: Infiltration in saturated swelling soils and slurries: exact solutions for constant supply rate, *Soil Sci.* **149**, pp. 13–22, 1990.

[6] J.J.B. BRONSWIJK: Shrinkage geometry of a heavy clay soil at various stresses, *Soil Sci. Soc. Am. J.* **54**, pp. 1500–1502, 1990.

[7] J.J.B. BRONSWIJK, J.J. EVERS-VERMEER: Shrinkage of Dutch clay soil aggregates, *Neth. J. Agric. Sci.* **38**, pp. 175–194, 1990.

[8] J.A. DE VOS, D.L.R. HESTERBERG, P.A.C. RAATS: Water flow and nitrate leaching in a layered silt loam, *Soil Sci. Soc. Am. J.* **64**, pp. 517–527, 2000.

[9] C. DIRKSEN: *Soil Physics Measurements*, Catena Verlag, Reiskirchen, 2000.

[10] W.R. GARDNER: Some steady-state solutions of the unsaturated moisture flow equation with application to evaporation from a water table, *Soil Sci.* **85**, pp. 228–232, 1958.

[11] P. GARNIER, E. PERRIER, R. ANGULO JARAMILLO, P. BAVEYE: Numerical model of 3-dimensional anisotropic deformation and 1-dimensional water flow in swelling soils, *Soil Sci.* **162**, pp. 410–420, 1997.

[12] P. GARNIER, M. RIEU, P. BOIVIN, M. VAUCLIN, P. BAVEYE: Determining the hydraulic properties of a swelling soil from a transient evaporation experiment, *Soil Sci. Soc. Am. J.* **61**, pp. 1555–1563, 1997.

[13] B.H. GILDING: Qualitative mathematical analysis of the Richards equation, *Transp. Porous Media* **5**, pp. 561–566, 1991.

[14] J.V. GIRÁLDEZ: *The Theory of Infiltration and Drainage in Swelling Soils*, Ph.D. thesis, University of California, Riverside (Univ. Microfilms, Ann Arbor, Michigan, 1976, Abstract DC J 77–11772).

[15] J.V. GIRÁLDEZ, G. SPOSITO: Moisture profiles during steady vertical flows in soils, *Water Resour. Res.* **14**, pp. 314–318, 1978.

[16] J.V. GIRÁLDEZ, G. SPOSITO: Infiltration in swelling soils, *Water Resour. Res.* **21**, pp. 33–44, 1985.

[17] P.H. GROENEVELT, G.H. BOLT: Water retention in soil, *Soil Sci.* **113**, pp. 238–245, 1972.

[18] W.B. HAINES: The volume changes associated with variations of water content in soil, *J. Agric. Sci. Cambridge* **13**, pp. 296–311, 1923.

[19] M. Heinen, P. de Willigen: FUSSIM2: A two-dimensional simulation model for water flow, solute transport and root uptake of water and nutrients in partly unsaturated porous media, *Quantitative Approaches in Systems Analysis* **20**, DLO Research Institute for Agrobiology and Soil Fertility and the C.T. de Wit Graduate School for Production Ecology, Wageningen, The Netherlands, 1998.

[20] D.J. Kim, R. Angulo-Jaramillo, M. Vauclin, J. Feyen, S.I. Choi: Modelling of soil deformation and water flow in swelling soil, *Geoderma* **92**, pp. 217–238, 1999.

[21] J.M. Kirby, D.E. Smiles: Hydraulic conductivity of aqueous bentonite suspensions, *Aust. J. Soil Res.* **26**, pp. 561–574, 1988.

[22] G. Kirchhoff: *Vorlesungen über die Theorie der Wärme, Herausgegeben von M. Planck*, Teubner, Leipzig, 1894, p. 13.

[23] A. Klute (ed.): Methods of soil analysis; part 1: physical and mineralogical methods, *Agronomy Monograph* **9**, 2nd edition, American Society of Agronomy, Madison, Wisconsin, USA, 1986.

[24] A. McNabb: A mathematical treatment of one-dimensional soil consolidation, *Q. Appl. Math.* **17**, pp. 337–347, 1960.

[25] E.E. Miller: Physics of swelling and cracking soils, *J. Colloid Interface Sci.* **52**, pp. 434–443, 1975

[26] E.E. Miller, R.D. Miller: Physical theory of capillary flow phenomena, *J. Appl. Phys.* **27**, pp. 324–332, 1956.

[27] D.B. McWorther, F. Marinelli: Theory of soil-water flow, in: R.W. Skaggs, J. van Schilfgaarde (eds.): Agricultural drainage, *Agronomy Monograph* **38**, American Society of Agronomy, Crop and Soil Science Societies of America, Madison, Wisconsin, USA, 2000.

[28] H.J. Morel-Seytoux: Multiphase flows in porous media, in P. Novak (ed.): *Developments in Hydraulic Engineering* **4**, Elsevier, London, 1983, pp. 103–174.

[29] J.-Y. Parlange: A note on the moisture diffusivity of saturated swelling systems from desorption experiments, *Soil Sci.* **120**, pp. 156–158, 1975.

[30] J.-Y. Parlange: Water transport in soils, *Ann. Rev. Fluid Mech.* **12**, pp. 77–102, 1980.

[31] J.-Y. Parlange, I. Lisle, R.D. Braddock, R.E. Smith: The three-parameter infiltration equation, *Soil Sci.* **133**, pp. 337–341, 1982.

206   P.A.C. Raats

[32] J.R. PHILIP: Hydrostatics and hydrodynamics in swelling soils, *Water Resour. Res.* **5**, pp. 1070–1077, 1969.

[33] J.R. PHILIP: Theory of infiltration, *Adv. Hydrosci.* **5**, pp. 215–296, 1969.

[34] J.R. PHILIP: Reply, *Water Resour. Res.* **6**, pp. 1248–1251, 1970.

[35] J.R. PHILIP: Fifty years progress in soil physics, *Geoderma* **12**, pp. 265–280, 1974.

[36] J.R. PHILIP: Quasianalytic and analytic approaches to unsaturated flow, in: W.L. STEFFEN, O.T. DENMEAD (eds.): Flow and transport in the natural environment: advances and applications, *Proc. Int. Symp. on Flow and Transport in the Natural Environment, 1987, Canberra, Australia*, Springer-Verlag, Berlin, 1988, pp. 30–47.

[37] J.R. PHILIP: How to avoid free boundary problems, in: K.H. HOFFMAN, J. SPREKELS (eds.): Free boundary problems: theory and applications, *Research Notes in Mathematics* **185**, Longman, London, 1990, pp. 193–207.

[38] J.R. PHILIP: Flow and volume change in soils and other porous media, in: T.K. KARALIS (ed.): *Mechanics of Swelling*, Springer, Berlin, 1992, pp. 3–32.

[39] J.R. PHILIP: Phenomenological approach to flow and volume change in soils and other media, *Appl. Mech. Rev.* **48**, pp. 650–658, 1995.

[40] J.R. PHILIP, D.E. SMILES: Macroscopic analysis of the behaviour of colloidal suspensions, *Adv. Colloid Interface Sci.* **17**, pp. 83–103, 1982.

[41] J.R. PHILIP, C.J. VAN DUIJN: Redistribution with air diffusion, *Water Resour. Res.* **35**, pp. 2295–2300, 1999.

[42] P.A.C. RAATS: Axial fluid flow in swelling and shrinking porous rods, *Abstracts 40th Annual Meeting of the Society of Rheology*, 1969, p. 13.

[43] P.A.C. RAATS: The distribution of the uptake of water by plants: inference from hydraulic and salinity data, *Proc. AGRIMED Seminar on the Movement of Water and Salts as a Function of the Properties of the Soil under Localized Irrigation, 1979, Bologna, Italy*, Istituto d'Agronomia, Università di Bologna, Italy, 1982, pp. 35–46.

[44] P.A.C. RAATS: Applications of the theory of mixtures in soil science, in: C. TRUESDELL: *Rational Thermodynamics*, 2nd-edition, Springer-Verlag, New York, Appendix 5D, 1984, pp. 326–343.

[45] P.A.C. RAATS: Mechanics of cracking soils, in: J. BOUMA, P.A.C. RAATS (eds.): *Proc. ISSS Symp. on Water and Solute Movement in Heavy Clay Soils. ILRI publication* **37**, International Institute for Land Reclamation and Improvement, Wageningen, The Netherlands, 1984, pp. 23–38.

[46] P.A.C. RAATS: Applications of the theory of mixtures in soil science, *Math. Modelling* **9**, pp. 849–856, 1987.

[47] P.A.C. RAATS: Quasianalytic and analytic approaches to unsaturated flow: commentary, in: W.L. STEFFEN, O.T. DENMEAD (eds.): Flow and transport in the natural environment: advances and applications, *Proc. Int. Symp. on Flow and Transport in the Natural Environment, 1987, Canberra, Australia*, Springer-Verlag, Berlin, 1988, pp. 48–58.

[48] P.A.C. RAATS: Characteristic lengths and times associated with processes in the root zone, in: D. HILLEL, D.E. ELRICK (eds): *Scaling in Soil Physics: Principles and Applications*, Soil Science Society of America, Madison, Wisconsin, 1990, pp. 59–72.

[49] P.A.C. RAATS: On the roles of characteristic lengths and times in soil physical processes, in: *Proc. 14th Int. Congr. Soil Sci., 1990, Kyoto, Japan* 1990, pp. 202–207.

[50] P.A.C. RAATS: A superclass of soils, in: M.TH. VAN GENUCHTEN, F.J. LEIJ, L.J. LUND (eds.): *Proc. Int. Workshop on Indirect Methods for Estimating the Hydraulic Properties of Unsaturated Soils, 1989, Riverside, California*, University of California, Riverside, 1992, pp. 45–51.

[51] P.A.C. RAATS: Spatial and material description of some processes in rigid and non-rigid saturated and unsaturated soils, in: J.-F. THIMUS, Y. ABOUSLEIMAN, A.H.-D. CHENG, O. COUSSY, E. DETOURNAY (eds.): Poromechanics - A tribute to Maurice A. Biot, *Proceedings of the Biot Conference on Poromechanics, 1998, Louvain-la-Neuve, Belgium*, Balkema, Rotterdam, 1998, pp. 135–140.

[52] P.A.C. RAATS: Developments in soil water physics since the mid 1960s, *Geoderma* **100**, pp. 355–387, 2001.

[53] P.A.C. RAATS, A. KLUTE: Transport in soils: the balance of mass, *Soil Sci. Soc. Am. Proc.* **32**, pp. 161–166, 1968.

[54] P.A.C. RAATS, A. KLUTE: Transport in soils: the balance of momentum, *Soil Sci. Soc. Am. Proc.* **32**, pp. 452–456, 1968.

[55] P.A.C. RAATS, W.R. GARDNER: Movement of water in the unsaturated zone near a water table, *Agronomy Monograph* **17**, American Society of Agronomy, Madison, Wisconsin, 1974, pp. 311–357.

[56] L.A. RICHARDS: Capillary conduction of liquids through porous mediums, *Physics* **1**, pp. 318–333, 1931.

[57] K. RIJNIERSCE: *A Simulation Model for Physical Soil Ripening in the IJsselmeerpolders*, Lelystad, The Netherlands, 1983.

[58] K. RIJNIERSCE: Crack formation in newly reclaimed sediments in the IJsselmeerpolders, in: J. BOUMA, P.A.C. RAATS (eds.): *Proc. Symp. on Water and Solute Movement in Heavy Clay Soils, ILRI Publication* **37**, International Institute for Land Reclamation and Improvement, Wageningen, The Netherlands, 1984, pp. 59–62.

[59] J. SIMUNEK, T. VOGEL, M.TH. VAN GENUCHTEN: HYDRUS-2D, Simulating water flow and solute transport in two-dimensional variably saturated media, version 1.2, *Research Report* **132**, U.S. Salinity Laboratory, Riverside, U.S.A, 1996.

[60] D.E. SMILES: Infiltration into a swelling soil, *Soil Sci.* **117**, pp. 140–147, 1974.

[61] D.E. SMILES: On the validity of the theory of flow in saturated swelling materials, *Austr. J. Soil Res.* **14**, pp. 389–395, 1976.

[62] D.E. SMILES: Sedimentation and filtration equilibria, *Separation Sci.* **11**, pp. 1–16, 1976.

[63] D.E. SMILES: Further comments on estimating the moisture diffusivity of saturated swelling materials using sorptivity data, *Soil Sci.* **124**, pp. 125–126, 1977.

[64] D.E. SMILES: Constant rate filtration of bentonite, *Chem. Eng. Sci.* **33**, pp. 1355–1361, 1978.

[65] D.E. SMILES: Transient- and steady-flow experiments testing theory of water flow in saturated bentonite, *Soil Sci. Soc. Am. J.* **42**, pp. 11–14, 1978.

[66] D.E. SMILES, J.H. KNIGHT, T.X.T. NGUYEN-HOAN: Gravity filtration with accretion of slurry at constant rate, *Separation Sci. Techn.* **14**, pp. 175–192, 1979.

[67] D.E. SMILES, P.A.C. RAATS, J.H. KNIGHT: Constant pressure filtration: the effect of a filter membrane, *Chem. Eng. Sci.* **37**, pp. 707–714, 1982.

[68] D.E. SMILES: Principles of constant pressure filtration, in: N.P. CHEREMISINOFF (ed.): *Encyclopedia of Fluid Mechanics; Vol. 5: Slurry Flow Technology*, Gulf Publ. Co., Houston, Texas, USA, 1986, pp. 791–824.

[69] D.E. SMILES: Material coordinates and solute movement in consolidating clay, *Chem. Eng. Sci.* **55**, pp. 773–781, 2000.

[70] D.E. SMILES, A.G. HARVEY: Measurement of moisture diffusivity of wet swelling materials, *Soil Sci.* **116**, pp. 391–399, 1973.

[71] D.E. SMILES, P.A.C. RAATS, J.H. KNIGHT: Constant pressure filtration: the effect of a filter membrane, *Chem. Eng. Sci.* **37**, pp. 707–714, 1982.

[72] K.A. SMITH, C.E. MULLINS (eds.): *Soil and Environmental Analysis: Physical Methods*, 2nd-edition, Marcel Dekker, New York, 2000.

[73] G. SPOSITO: A thermodynamic integral equation for the equilibrium moisture profile in swelling soil, *Water Resour. Res.* **11**, pp. 499-500, 1975

[74] G. SPOSITO: On the differential equation for the equilibrium moisture profile in swelling soil, *Soil Sci. Soc. Am. Proc.* **39**, pp. 1053–1056, 1975.

[75] G. SPOSITO: Steady vertical flows in swelling soils, *Water Resour. Res.* **11**, pp. 461–464, 1975

[76] G. SPOSITO: *The Thermodynamics of Soil Solutions*, Clarendon Press, Oxford, 1981.

[77] T. TALSMA: A note on shrinkage behaviour of a clay paste under various loads, *Aust. J. Soil Res.* **15**, pp. 275–277, 1977.

[78] L. STROOSNIJDER: Infiltratie en herverdeling van water in grond (Infiltration and redistribution of water in soils). *Verslagen van Landbouwkundige Onderzoekingen (Agricultural Research Reports)* **847**, 1976.

[79] C. TRUESDELL, R.A. TOUPIN: The classical field theories, in: S. FLÜGGE (ed.): *Handbuch der Physik*, Springer-Verlag, Berlin **III/1**, pp. 226-902, 1960.

[80] R. VAN DIJKE: *Multi-Phase Flow Modeling of Soil Contamination and Soil Remediation*, Ph.D. thesis, Wageningen Agricultural University, 1997.

[81] M.TH. VAN GENUCHTEN, F.J. LEIJ, L.J. LUND (eds.): *Proc. Int. Workshop on Indirect Methods for Estimating the Hydraulic Properties of Unsaturated Soils, 1989, Riverside, California*, U.S. Salinity Laboratory USDA–ARS and Department of Soil and Environmental Sciences, University of California, Riverside, 1992.

[82] M.TH. VAN GENUCHTEN, F.J. LEIJ, L.J. LUND (eds.): *Proc. Int. Workshop on Characterization and Measurement of the Hydraulic Properties of Unsaturated Porous Media, 1997, Riverside, California,* U.S. Salinity Laboratory USDA–ARS and Department of Soil and Environmental Sciences, University of California, Riverside, 1999.

[83] S. WHITAKER: Flow in porous media II: the governing equations for immiscible, two-phase flow, *Transp. Porous Media* **1**, pp. 105–125, 1986.

[84] E.G. YOUNGS, G.D. TOWNER: Comments on "Hydrostatics and hydrodynamics in swelling soils", *Water Resour. Res.* **6**, pp. 1246–1247, 1970.

PETER A.C. RAATS
Paaskamp, 16
NL-9301 KL Roden, THE NETHERLANDS
E-mail: Pac.Raats@home.nl

Chapter 8

Mass Exchange, Diffusion and Large Deformations of Poroelastic Materials

Krzysztof Wilmański [1]

ABSTRACT The paper contains a review of fundamental equations of the two component thermoporoelastic materials with the balance equation of porosity. By exploiting the second law of thermodynamics restricted to small deviations from thermodynamical equilibrium, we prove that there exists no thermodiffusional coupling of components through intrinsic parts of fluxes. Certainly such a coupling is still present due to convective contributions. Simultaneously we show that classical partial dynamical compatibility conditions on material interfaces cannot hold. For boundary conditions on permeable boundaries to hold true, it must be required that global balance equations contain at least surface sources of momentum, entropy and porosity. We show as well that the requirement of local thermodynamical equilibrium on permeable interfaces yields the continuity of absolute temperature. It means that temperature becomes a measurable physical field in porous materials undergoing processes with small deviations from thermodynamical equilibria. This result allows us to extend models of mass exchange in poroelastic materials from adsorption isothermal processes to chemical reactions, and phase transformations. Details of the latter problems are not discussed in this paper.

8.1   Introduction

The paper is devoted to the presentation of basic properties of the thermo-dynamical model of thermoporoelastic materials which I have developed during the last decade. A good deal of material contained in this work has been already published elsewhere, and I quote it here again to make the paper selfcontained and new contributions understandable. The presentation of one chosen model of porous materials does not mean, of course, that there exists any qualification for various models appearing in the literature. An appropriate one must be always chosen as a best fit for the purpose. For

---

[1] Dedicated to Professor Kolumban Hutter on occasion of his 60th birthday

instance, the model presented in this paper is particularily well suited to describe wave propagation in multicomponent systems, as well as large deformations of the skeleton. It is much too complex for applications to most consolidation problems. A model in which one assumes the incompressibility of components frequently used in soil mechanics cannot describe all modes of acoustic and surface waves, but it describes very well various instabilities in granular geotechnical materials such as piping. A model based on Darcy's law with rigid skeleton describes very well flows of fluid components (reaction-diffusion equations) but it cannot describe consolidation processes and acoustic waves. One can find multiple such examples.

Due to the above limitation of the contents the references are chosen in a very subjective manner and reflect solely results for one particular approach.

The general part of the present considerations is devoted to a two–component system consisting of an elastic skeleton (a solid component) and of the ideal fluid. Deformations and kinematics of both components are related to a reference configuration of the skeleton. This is called the Lagrangian description of motion [9]. The main new elements of the model presented in this work are contained in the exploitation of the second law of thermodynamics which yields quite explicit relations for fluxes of the balance equations under the assumption of small deviations from the thermodynamical equilibrium state. We do not make an assumption on a relation between partial heat and entropy fluxes which has been made in the thermodynamical analysis of a multicomponent system in [11, 15]. In addition we present an analysis of conditions on interfaces material with respect to the skeleton. This analysis allows us to interpret the temperature in the classical way for processes satisfying the above assumption on small deviations. This means that we can effectively construct boundary conditions for heat conduction problems.

In Section 4 we review briefly results on adsorption processes coupled to the diffusion. This problem indicates limitations of contemporary modeling of mass exchange in porous materials which is related to the assumption that processes are isothermal. Results on non-isothermal models presented in this work allow us to extend the description to processes in which we have to incorporate the latent heats of phase transformations and heats of chemical reactions.

The paper is organized in the following way. Sections 2 and 3 contain a development of the general thermodynamical two-component model. Technical considerations connected with the exploitation of the second law of thermodynamics are covered in the Appendix. Section 4 is devoted to modeling of adsorption. Section 5 contains an analysis of the structure of conditions on interfaces material with respect to the skeleton. In particular we present sufficient conditions for the continuity of absolute temperature on such an interface.

8.2   Balance equations in Lagrangian description

Large deformations of the skeleton of porous materials yield the necessity of Lagrangian description of motion. This has been proposed in a series of works [9, 10, 11] and some details can be found in the book [12]. In this Section I present only some main features of this description.

We consider a two-component porous medium described as a continuum. The motion of the skeleton is assumed to be given by a diffeomorphism

$$\mathbf{f}^S(\cdot,\cdot) : B \times T \to \mathbb{R}^3, \tag{8.1}$$

where B is a reference configuration of the skeleton, $B \subset \mathbb{R}^3$ and T is the time interval. The deformation gradient and the partial velocity of the skeleton are defined by the relations

$$\mathbf{F}^S = \operatorname{Grad} \mathbf{f}^S, \qquad \acute{\mathbf{x}}^S = \frac{\partial \mathbf{f}^S}{\partial t}, \tag{8.2}$$

and they are assumed to be continuous almost everywhere in B. The motion of the fluid component is assumed to be given by a partial velocity field

$$\acute{\mathbf{x}}^F : B \times T \to \mathbb{V}^3, \tag{8.3}$$

where $\mathbb{V}^3$ is a three-dimensional vector space. The partial fluid velocity is assumed to be continuous almost everywhere in B. Material domains of the skeleton $\mathcal{P} \subset B$ are assumed to satisfy usual conditions of continuum mechanics which we shall not quote here. Certainly they also do not depend on time, and each member of their class $\mathbb{M}^S$ is called S-material. On the other hand, material domains of the fluid $\mathcal{P} \subset B$ do depend on time, and their kinematics is described by the Lagrangian velocity field

$$\acute{\mathbf{X}}^F := \mathbf{F}^{S-1}(\acute{\mathbf{x}}^F - \acute{\mathbf{x}}^S). \tag{8.4}$$

The members of their class $\mathbb{M}^F$ are called F-material.

The set of fields characterizing temperature dependent processes of motion in porous media is of the form

$$\left\{ \rho^S, \rho^F, \mathbf{f}^S, \acute{\mathbf{x}}^F, n, \Theta \right\}, \tag{8.5}$$

where ρ^S, ρ^F are partial mass densities in the reference configuration B, n is the porosity and Θ is the absolute temperature. We return later very briefly to the problem of systems with multiple temperatures.

In the case of porous media, whose heterogeneity is limited to an interface Σ dividing the reference configuration into two subdomains, $B^+, B^-, c\ell B^+ \cup c\ell B^- = c\ell B, c\ell B^+ \cap c\ell B^- = \Sigma$, where $c\ell$ denotes the closure of domains in which the porous medium may have different material properties, we have

214   K. Wilmański

the following set of balance equations corresponding to fields (8.5):

– partial mass balance:

$$\forall \mathcal{P} - \text{S-material}: \quad \frac{d}{dt}\int_{\mathcal{P}} \rho^S dv = \int_{\mathcal{P}} \hat{\rho}^S dV, \tag{8.6}$$

$$\forall \mathcal{P} - \text{F-material}: \quad \frac{d}{dt}\int_{\mathcal{P}} \rho^F dv = \int_{\mathcal{P}} \hat{\rho}^F dV, \quad \hat{\rho}^S + \hat{\rho}^F = 0, \tag{8.7}$$

– partial momentum balance:

$\forall \mathcal{P} - \text{S-material}:$

$$\frac{d}{dt}\int_{\mathcal{P}} \rho^S \acute{\mathbf{x}}^S dV = \oint_{\partial\mathcal{P}} \mathbf{P}^S \mathbf{N} dA + \int_{\mathcal{P}} \hat{\mathbf{p}}^S dV + \oint_{\mathcal{P}\cap\Sigma} \hat{\mathbf{p}}^S_{surf} dA, \tag{8.8}$$

$\forall \mathcal{P} - \text{F-material}:$

$$\frac{d}{dt}\int_{\mathcal{P}} \rho^F \acute{\mathbf{x}}^F dV = \oint_{\partial\mathcal{P}} \mathbf{P}^F \mathbf{N} dA + \int_{\mathcal{P}} \hat{\mathbf{p}}^F dV + \oint_{\mathcal{P}\cap\Sigma} \hat{\mathbf{p}}^F_{surf} dA,$$

$$\hat{\mathbf{p}}^S + \hat{\mathbf{p}}^F = 0, \qquad \hat{\mathbf{p}}^S_{surf} + \hat{\mathbf{p}}^F_{surf} = 0, \tag{8.9}$$

– partial energy balance:

$\forall \mathcal{P} - \text{S-material}:$

$$\frac{d}{dt}\int_{\mathcal{P}} \rho^S \left(\varepsilon^S + \tfrac{1}{2}\acute{x}^{S2}\right) dV + \oint_{\partial\mathcal{P}} \mathbf{Q}^S \cdot \mathbf{N} dA = \oint_{\partial\mathcal{P}} (\mathbf{P}^S \mathbf{N}) \cdot \acute{\mathbf{x}}^S dA,$$

$$\tag{8.10}$$

$\forall \mathcal{P} - \text{F-material}:$

$$\frac{d}{dt}\int_{\mathcal{P}} \rho^F \left(\varepsilon^F + \tfrac{1}{2}\acute{x}^{F2}\right) dV + \oint_{\partial\mathcal{P}} \mathbf{Q}^F \cdot \mathbf{N} dA = \oint_{\partial\mathcal{P}} (\mathbf{P}^F \mathbf{N}) \cdot \acute{\mathbf{x}}^F dA,$$

– balance of porosity:

$\forall \mathcal{P} - \text{S-material}:$

$$\frac{d}{dt}\int_{\mathcal{P}} n dV + \oint_{\partial\mathcal{P}} \mathbf{J} \cdot \mathbf{N} dA = \int_{\mathcal{P}} \hat{n} dV + \int_{\mathcal{P}\cap\Sigma} \hat{n}_{surf} dA. \tag{8.11}$$

The sources of mass $\hat{\rho}^S, \hat{\rho}^F$, the volume sources of momentum $\hat{\mathbf{p}}^S, \hat{\mathbf{p}}^F$ and the surface sources of momentum $\hat{\mathbf{p}}^S_{surf}, \hat{\mathbf{p}}^F_{surf}$ are assumed to satisfy the local conservation laws (8.7), (8.9). This condition can be weakened which is not essential for considerations of this work. We justify the necessity of the presence of momentum surface sources and porosity surface source on the interface in Section 5.

The partial Piola–Kirchhoff stress tensors are denoted by $\mathbf{P}^S, \mathbf{P}^F$, the heat flux vectors are $\mathbf{Q}^S, \mathbf{Q}^F$. $\hat{n}, \hat{n}_{surf}$ denote the volume source, and the surface source of porosity, respectively. The flux of porosity is denoted by $\mathbf{J}$. $\mathbf{N}$ is a unit vector orthogonal to the surface ∂P.

The local form of these equations in $B\backslash\Sigma$ is

$$\frac{\partial \rho^S}{\partial t} = \hat{\rho}^S, \qquad \frac{\partial \rho^F}{\partial t} + \operatorname{Div}\,(\rho^F \acute{\mathbf{X}}^F) = \hat{\rho}^F,$$

$$\frac{\partial \rho^S \acute{\mathbf{x}}^S}{\partial t} = \operatorname{Div}\,\mathbf{P}^S + \hat{\mathbf{p}}^S,$$

$$\frac{\partial \rho^F \acute{\mathbf{x}}^F}{\partial t} + \operatorname{Div}\,\left(\rho^F \acute{\mathbf{x}}^F \otimes \acute{\mathbf{X}}^F - \mathbf{P}^F \right) = \hat{\mathbf{p}}^F,$$

$$\frac{\partial \rho^S (\varepsilon^S + \frac{1}{2} \acute{x}^{S2})}{\partial t} + \operatorname{Div}\,\left(\mathbf{Q}^S - \mathbf{P}^{ST} \acute{\mathbf{x}}^S \right) = 0, \tag{8.12}$$

$$\frac{\partial \rho^F (\varepsilon^F + \frac{1}{2} \acute{x}^{F2})}{\partial t} + \operatorname{Div}\,\left(\rho^F (\varepsilon^F + \tfrac{1}{2} \acute{x}^{F2}) \acute{\mathbf{X}}^F + \mathbf{Q}^F - \mathbf{P}^{FT} \acute{\mathbf{x}}^F \right) = 0,$$

$$\frac{\partial n}{\partial t} + \operatorname{Div}\,\mathbf{J} = \hat{n},$$

where $\acute{\mathbf{X}}^F$ denotes the Lagrangian relative velocity (see (8.4)). We use these equations to construct field equations for thermoporoelastic materials.

8.3 Thermodynamics of thermoporoelastic materials

In order to close the system (8.12) and obtain field equations, and boundary conditions for fields (8.5), we need constitutive relations for the constitutive quantities

$$\mathcal{Z} \quad := \quad \left\{ \hat{\rho}^S, \hat{\rho}^F, \mathbf{P}^S, \mathbf{P}^F, \hat{\mathbf{p}}^S, \hat{\mathbf{p}}^S_{surf}, \hat{\mathbf{p}}^F, \hat{\mathbf{p}}^F_{surf}, \right. \tag{8.13}$$

$$\left. \varepsilon^S, \varepsilon^F, \mathbf{Q}^S, \mathbf{Q}^F, \mathbf{J}, \hat{n}, \hat{n}_{surf} \right\}.$$

Certainly, constitutive relations for sources are not all independent due to conservation laws.

We assume the quantities $\mathcal{Z}$ to be differentiable functions of the constitutive variables

$$\mathcal{C} := \left\{ \rho^S, \rho^F, \mathbf{F}^S, \acute{\mathbf{X}}^F, n, \Theta, \mathbf{G} \right\}, \qquad \mathbf{G} := \operatorname{Grad}\,\Theta, \tag{8.14}$$

i.e.,

$$\mathcal{Z} = \mathcal{Z}(\mathcal{C}). \tag{8.15}$$

As we see further we need additional fields of microstructural variables in order to describe processes of mass exchange. We introduce them in the next section as they do not influence basic consequences of the second law of thermodynamics which we proceed to present.

Any solution of field equations which follows from (8.12) by the substitution of (8.15) we call a thermodynamical process. As we consider solely the case of a common temperature for the solid and the fluid we use the energy balance in the bulk form which follows by adding equations $(8.12)_5$ and $(8.12)_6$.

The second law is assumed to be constructed in the same way as the balance equations of Section 2. We assume an existence of non-trivial fields of partial entropies η^S, η^F, and their fluxes $\mathbf{H}^S, \mathbf{H}^F$, such that

$$\forall \mathcal{P} - \text{S-material} :$$
$$\frac{d}{dt} \int_{\mathcal{P}} \rho^S \eta^S dV + \int_{\mathcal{P}} \text{Div } \mathbf{H}^S dV = \int_{\mathcal{P}} \hat{\eta}^S dV + \int_{\mathcal{P} \cap \Sigma} \hat{\eta}^S_{surf} dv,$$

$$(8.16)$$

$$\forall \mathcal{P} - \text{F-material} :$$
$$\frac{d}{dt} \int_{\mathcal{P}} \rho^F \eta^F dV + \int_{\mathcal{P}} \text{Div } \mathbf{H}^F dV = \int_{\mathcal{P}} \hat{\eta}^F dV + \int_{\mathcal{P} \cap \Sigma} \hat{\eta}^F_{surf} dv,$$

$$\eta^S = \eta^S(\mathcal{C}), \ \eta^F = \eta^F(\mathcal{C}), \ \mathbf{H}^S = \mathbf{H}^S(\mathcal{C}), \ \mathbf{H}^F = \mathbf{H}^F(\mathcal{C}). \qquad (8.17)$$

It is assumed that at each point $X \in B\backslash\Sigma$ the inequality

$$\hat{\eta}^S + \hat{\eta}^F \geq 0, \qquad (8.18)$$

holds for all solutions of field equations.

By means of balance equations (8.16) it can be written in the local form

$$\frac{\partial}{\partial t}\left(\rho^S \eta^S + \rho^F \eta^F\right) + \text{Div}\left(\rho^F \eta^F \acute{\mathbf{X}}^F + \mathbf{H}^S - \mathbf{H}^F\right) \geq 0. \qquad (8.19)$$

This entropy inequality yields thermodynamical admissibility conditions which we discuss in the Appendix. For our further considerations we limit our attention to the model describing small deviations from the state of thermodynamical equilibrium. This state is defined within the present model as such for which the following conditions hold:

$$\mathbf{G}|_E = 0, \quad \hat{\rho}^S\big|_E = 0, \quad \hat{\mathbf{p}}^S\big|_E = 0, \quad \hat{n}|_E = 0. \qquad (8.20)$$

Then, as we show in the Appendix, the following relations hold true:

$$\acute{\mathbf{X}}\big|_E = 0, \quad \Delta|_E = 0, \quad \Delta := n - n_E, \quad n_E = n_E\left(\frac{\rho^F}{\rho^S}\right), \qquad (8.21)$$

and the basic constitutive relations are as follows. The intrinsic heat flux in both components defined by the sum of partial fluxes is independent of the relative velocity $\acute{\mathbf{X}}^F$, and of the change of porosity Δ, and it has the form

$$
\begin{aligned}
\mathbf{Q}^S + \mathbf{Q}^F &= -K_\Theta \mathbf{G} = \Theta\left(\mathbf{H}^S + \mathbf{H}^F\right), &(8.22)\\
K_\Theta &= K_\Theta\left(\mathcal{C}_E\right), \quad \mathcal{C}_E := \left\{\rho^F, \rho^S, \mathbf{F}^S, \Theta\right\}
\end{aligned}
$$

while the constitutive relation for the flux of porosity simplifies to a single constant

$$
\mathbf{J} = \varphi J^S \acute{\mathbf{X}}^F, \quad \varphi = \text{const.} \tag{8.23}
$$

This constant is determined for a particular initial state of the porous medium, which means it may still be parametrically dependent on an initial porosity. This was indicated in earlier works on this model where it was argued that $\varphi \approx n_E$ for the constant equilibrium porosity n_E.

Under the restriction of processes to a small neighbourhood of the thermodynamical equilibrium, coupling through partial Piola–Kirchhoff stress tensors reduces solely to coupling through the dynamical change of porosity Δ:

$$
\begin{aligned}
\mathbf{P}^S &= \rho^S \frac{\partial \psi^S}{\partial \mathbf{F}^S} - \Theta \Lambda_1^n \varphi J^S \Delta \mathbf{F}^{S-T}, &(8.24)\\[2mm]
\mathbf{P}^F &= \left(-p^F + \Theta \Lambda_1^n \varphi \Delta\right) J^S \mathbf{F}^{S-T}, \quad p^F := \rho^{F2} \frac{\partial \psi^F}{\partial \rho^F} J^{S-1},
\end{aligned}
$$

where

$$
\begin{aligned}
\psi^F &: = \varepsilon^F - \Theta \eta^F = \psi^F\left(\rho^F J^{S-1}, \Theta, \Delta\right), \\
\psi^S &: = \varepsilon^S - \Theta \eta^S = \psi^S\left(\rho^S, \mathbf{F}^S, \Theta, \Delta\right), &(8.25)\\
\Lambda_1^n &: = -\frac{1}{\Theta} \frac{\partial}{\partial \Delta}\left(\rho^S \psi^S + \rho^F \psi^F\right)\bigg|_E.
\end{aligned}
$$

The free energies ψ^S, ψ^F contain only two contributions. One is independent of Δ, the other one is quadratic in Δ and, in addition,

$$
\eta^F = -\frac{\partial \psi^F}{\partial \Theta}. \tag{8.26}
$$

The sources are given by the relations

$$
\hat{\mathbf{p}}^S = \pi\left(\acute{\mathbf{x}}^F - \acute{\mathbf{x}}^S\right) + \hat{\rho}^S \acute{\mathbf{x}}^S,
$$

$$
\hat{\rho}^S = R\left(\psi^F + \frac{p^F}{\rho^F J^{S-1}} - \psi^S - \rho^S \frac{\partial \psi^S}{\partial \rho^S}\right), \tag{8.27}
$$

where the coefficients π, R may still depend on all equilibrium constitutive variables $\mathcal{C}_E$. Obviously, the formula for mass sources contains a difference of functions recalling the chemical potentials of the fluid $\psi^F + \frac{p^F}{\rho^F J^{S-1}}$ and of the skeleton $\psi^S + \rho^S \frac{\partial \psi^S}{\partial \rho^S}$. However, the second contribution to the potential of the skeleton does not coincide with the partial pressure (see $(8.24)_1$) as ρ^S and $\mathbf{F}^S$ are independent.

Finally the following dissipation inequality must hold:

$$
\frac{1}{\Theta} K_\Theta \mathbf{G} \cdot \mathbf{G} + \pi \left(\acute{\mathbf{x}}^F - \acute{\mathbf{x}}^S \right) \cdot \left(\acute{\mathbf{x}}^F - \acute{\mathbf{x}}^S \right)
$$
$$
+ \frac{\partial}{\partial \Delta} \left(\rho^S \psi^S + \rho^F \psi^F \right) \Big|_E \frac{1}{\tau} \Delta^2 \qquad (8.28)
$$
$$
+ R \left(\psi^F + \frac{p^F}{\rho^F J^{S-1}} - \psi^S - \rho^S \frac{\partial \psi^S}{\partial \rho^S} \right)^2 \geq 0.
$$

This completes the general thermodynamical construction of the two-component thermoporoelastic model.

In more general cases of multicomponent systems, only partial results on thermodynamical admissibility are available [13, 15].

Further in this work we use as well the Eulerian description. The local balance equations and the thermodynamical results presented above have in this description the following form in a generic point $\mathbf{x} \in \mathbf{f}^S (B, t)$:

mass balance

$$
\frac{\partial \rho_t^S}{\partial t} + \operatorname{div} \left(\rho_t^S \mathbf{v}^S \right) = \hat{\rho}_t^S, \qquad \frac{\partial \rho_t^F}{\partial t} + \operatorname{div} \left(\rho_t^F \mathbf{v}^F \right) = -\hat{\rho}_t^S,
$$
$$
\rho_t^S := \rho^S J^{S-1}, \quad \rho_t^F := \rho^F J^{S-1}, \quad \hat{\rho}_t^S := \hat{\rho}^S J^{S-1}, \qquad (8.29)
$$
$$
\mathbf{v}^S := \acute{\mathbf{x}}^S \left(\mathbf{f}^{S-1} (\mathbf{x}, t), t \right), \quad \mathbf{v}^F := \acute{\mathbf{x}}^F \left(\mathbf{f}^{S-1} (\mathbf{x}, t), t \right),
$$

and the operator div $(\cdot)$, as well as grad $(\cdot)$ in the following relations, concerns the Eulerian differentiation with respect to $\mathbf{x}$,

momentum balance

$$
\rho_t^S \left(\frac{\partial \mathbf{v}^S}{\partial t} + \mathbf{v}^S \cdot \operatorname{grad} \mathbf{v}^S \right) = \operatorname{div} \mathbf{T}^S + \hat{\mathbf{p}}_t^S - \hat{\rho}_t^S \mathbf{v}^S,
$$
$$
\mathbf{T}^S := J^{S-1} \mathbf{P}^S \mathbf{F}^{ST}, \quad \hat{\mathbf{p}}_t^S := J^{S-1} \hat{\mathbf{p}}^S,
$$
$$
\rho_t^F \left(\frac{\partial \mathbf{v}^F}{\partial t} + \mathbf{v}^F \cdot \operatorname{grad} \mathbf{v}^F \right) = \operatorname{div} \mathbf{T}^F - \hat{\mathbf{p}}_t^S + \hat{\rho}_t^S \mathbf{v}^F, \qquad (8.30)
$$
$$
\mathbf{T}^F := J^{S-1} \mathbf{P}^F \mathbf{F}^{ST},
$$

energy balance

$$\rho_t^S \left(\frac{\partial \varepsilon^S}{\partial t} + \mathbf{v}^S \cdot \operatorname{grad} \varepsilon^S \right) + \rho_t^F \left(\frac{\partial \varepsilon^F}{\partial t} + \mathbf{v}^F \cdot \operatorname{grad} \varepsilon^F \right)$$

$$+ \operatorname{div} \left(\mathbf{q}^S + \mathbf{q}^F \right) = \mathbf{T}^S \cdot \operatorname{grad} \mathbf{v}^S + \mathbf{T}^F \cdot \operatorname{grad} \mathbf{v}^F \qquad (8.31)$$

$$+ \hat{\rho}_t^S \left(\varepsilon^F - \varepsilon^S - \frac{1}{2} \left(\mathbf{v}^F - \mathbf{v}^S \right) \cdot \left(\mathbf{v}^F - \mathbf{v}^S \right) \right)$$

$$+ \left(\hat{\mathbf{p}}_t^S - \hat{\rho}_t^S \mathbf{v}^S \right) \cdot \left(\mathbf{v}^F - \mathbf{v}^S \right),$$

$$\mathbf{q}^S := J^{S-1} \mathbf{F}^S \mathbf{Q}^S, \quad \mathbf{q}^F := J^{S-1} \mathbf{F}^S \mathbf{Q}^F,$$

balance of porosity

$$\frac{\partial \Delta}{\partial t} + \mathbf{v}^S \cdot \operatorname{grad} \Delta + J^S \operatorname{div} \left(\varphi \left(\mathbf{v}^F - \mathbf{v}^S \right) \right) = \hat{n}. \qquad (8.32)$$

Apart from the mass sources we need solely linear constitutive laws, and these have the form:

partial Cauchy stress tensors

$$\mathbf{T}^S = \mathbf{T}_0^S + \lambda^S \mathbf{e}^S \cdot \mathbf{11} + 2\mu^S \mathbf{e}^S + \beta \Delta \mathbf{1},$$

$$\mathbf{T}^F = \left(-p^F - \beta \Delta \right) \mathbf{1}, \quad p^F = p_0^F + \kappa \left(\rho_t^F - \rho_0^F \right),$$

$$\beta := \left. \frac{\partial}{\partial \Delta} \left(\rho_t^S \psi^S + \rho_t^F \psi^F \right) \right|_{\Delta=0} \varphi, \qquad (8.33)$$

where $\mathbf{T}_0^S, p_0^F, \rho_0^F$ denote reference values of the Cauchy stress in the skeleton, partial pressure in the fluid, and the partial mass density of the fluid, respectively, λ^S, μ^S, κ are Lamé parameters of the skeleton, and the compressibility parameter of the fluid, respectively, and they may still be dependent on a reference porosity n_0; the small deformation of the skeleton $\mathbf{e}^S$ is

$$\mathbf{e}^S := \frac{1}{2} \left(\mathbf{1} - \mathbf{F}^{S-T} \mathbf{F}^{S-1} \right), \qquad (8.34)$$

$$\|\mathbf{e}^S\| := \max \left(\left| \lambda^{(1)} \right|, \left| \lambda^{(2)} \right|, \left| \lambda^{(3)} \right| \right), \quad \|\mathbf{e}^S\| \ll 1,$$

$\lambda^{(a)}, a = 1, 2, 3$ being the eigenvalues (principal stretches) of $\mathbf{e}^S$,

internal energies

$$\varepsilon^S = \varepsilon^S \left(\rho_t^S, \mathbf{e}^S, \Theta, \Delta \right), \qquad (8.35)$$

$$\varepsilon^F = \varepsilon^F \left(\rho_t^F, \Theta, \Delta \right),$$

where the dependence on Δ is even, and at most quadratic,

the intrinsic heat flux

$$\mathbf{q}^S + \mathbf{q}^F = -\aleph \operatorname{grad} \Theta, \quad \aleph \approx K_\Theta, \qquad (8.36)$$

the porosity source and the equilibrium porosity

$$\hat{n} = -\frac{\Delta}{\tau}, \quad n_E = n_E\left(\frac{\rho_t^F}{\rho_t^S}\right), \quad e.g. \quad n_E = n_0\frac{\rho_t^F}{\rho_0^F}\frac{\rho_0^S}{\rho_t^S}. \tag{8.37}$$

We skip here easy proofs of the above relations.

8.4   Mass exchange, adsorption

Macroscopic processes of mass exchange between components of mixtures of fluids and solids belong to one of the three fundamental classes: phase changes, chemical reactions or adsorption/desorption processes. Within the first two classes the exchange of mass is accompanied by thermal effects due to the presence of a latent heat of reaction. The processes of the last class can be considered to be isothermal, for instance for a small concentration of adsorbate.

We skip here the presentation of phase changes. Let us only mention that theories of both diffusionless phase changes as well as those with diffusion (*e.g.*, phase field theories) have recently experienced strong development.

Continuum models of diffusion processes with mass exchange are developed very well for mixtures of fluids. There is very little done for porous materials. Some work was done on combustion problems, and most of the results are based on the classical model of Goodman and Cowin [6] (see, *e.g.*, [5]). Difficulties are connected with the coupling of diffusion and heat conduction. Particularly in processes in which one has to account for multiple temperatures there is barely any progress at all.

In this work we limit our attention to adsorption processes and present a construction of the mass source contribution to mass balance equations of a three-component continuous model of porous materials.

Adsorption belongs to the most important practical problems within theories of porous and granular materials. This is connected primarily with a very large internal surface per unit volume in such materials on which the mass exchange takes place. For example in sandstone it reaches the value of $1.5 \times 10^5 \frac{m^2}{m^3}$ in comparison with $6\frac{m^2}{m^3}$ for its external surface. This property is used in many technological processes. For instance in the growth of SiC single crystals by sublimation, the vapour of silicium flows through a porous graphite wall in which it forms various carbite connections. A charcoal granular material is also used in gas masks. Lungs, many filters and chemical reactors are made of porous materials for the same reason.

The model of such a mass exchange between a fluid component and a solid in porous and granular materials is based on the classical work of Langmuir (*e.g.*, see the review in [1]). In the original works of Langmuir the theory of adsorption was limited to flat solid surfaces interacting with a gas. However for porous materials whose pores are large (their diameter is greater than approximately 500 Å = 50 nm) one can still rely on the assumption that the influence of the curvature of the surface is small.

On the microscopic level of description of porous and granular materials we rely on the assumption that particles of the adsorbate change their kinematics from fluid to solid due to a weak van der Waals interaction with internal surfaces of the skeleton (a solid component of the system). The transfer of particles from the fluid component to the internal surface of the solid depends on a partial pressure of the fluid adsorbate, on an area of this surface, and on a number of available *bare sites* on this surface. The physical interpretation of the latter depends on the nature of adsorption processes on internal surfaces. On the macroscopic level (*i.e.*, averaged over the *representative elementary volume (REV)* of a porous or granular material) the normalized fraction of these sites per unit volume is denoted by $1 - x$, *i.e.*, x is the fraction of *occupied sites*. If the area of the internal surface contained in the representative elementary volume is denoted by f_{int}, and the mass of adsorbate per unit area of the internal surface by m^A, then the amount of mass which is already adsorbed in the representative elementary volume is equal to the product $m^A x f_{int}$.

Let us denote by V the volume of the representative elementary volume. Then the amount of mass of adsorbate transferred in unit time from the liquid phase to the solid skeleton is given by the balance relation

$$\hat{\rho}_t^A = -m^A \frac{d\,(xy)}{dt}, \qquad y := \frac{f_{int}}{V}, \tag{8.38}$$

where $\hat{\rho}_t^A$ denotes the intensity of mass source per unit time and unit macroscopic volume in the current configuration.

In order to construct the model we have to specify the rates in this relation.

For $\frac{dx}{dt}$ we assume that changes of the fraction x are described by the Langmuir relation

$$\frac{dx}{dt} = a\,(1 - x)\,p^A - bxe^{-\frac{E_b}{k\Theta}}, \tag{8.39}$$

where p^A denotes the partial pressure of the adsorbate in the fluid phase, E_b is the energy barrier for particles adsorbed on the solid surface due to the Van der Waals interaction forces, and it is assumed to be constant, a and b are material parameters which within the present model may depend solely on the temperature, k is the Boltzmann constant and Θ is the absolute temperature. In the case of full phase equilibrium we obtain from the equation (8.39) the following relation for the fraction of occupied sites:

$$x = x_L := \frac{\frac{p^A}{p_0}}{1 + \frac{p^A}{p_0}}, \qquad p_0 := \frac{b}{a}e^{-\frac{E_b}{k\Theta}}, \tag{8.40}$$

which defines the so-called Langmuir isotherm. It begins in the origin $\frac{p^A}{p_0} = 0$ with the zero value of occupied sites and saturates at the value

222   K. Wilmański

1 for $\frac{p^A}{p_0} \to \infty$. At any given partial pressure p^A the fraction x is uniquely determined, and it may change its value if we vary the pressure. This corresponds to a slow transition from one thermodynamical equilibrium to another one. In reality such processes are conducted through nonequilibrium states which are described by the rate equation (8.39) and are connected with the dissipation.

In the mass source $(8.38)_1$ we have also another contribution connected with the change of the internal surface. Consequently we must formulate a relation for the rate $\frac{dy}{dt}$. We make the assumption that changes of the internal surface are coupled with dissipative changes of the porosity n which in turn describe relaxation processes of semimacroscopic changes of the volume of skeleton. This seems to be appropriate in processes of small deformations of the skeleton with accompanying small changes of the equilibrium porosity n_E. Then their influence on changes of internal surface can be neglected as being of a higher order than dissipative changes.

First of all let us notice, for sufficiently smooth internal surfaces of porous and granular materials with a random geometry of pore spaces, a change of an average characteristic linear dimension of the internal surface, and this of pores in the elementary representative volume can be assumed to be proportional: $\delta f_{int}^{\frac{1}{2}} \sim \delta \left(nV \right)^{\frac{1}{3}}$. Simultaneously dissipative changes of the porosity are given by a source $\hat{n}$ which describes the intensity of these changes per unit time and volume of the porous material. Bearing the above assumption in mind we obtain immediately

$$\frac{1}{y}\frac{dy}{dt} = \varsigma \frac{\hat{n}}{n}, \tag{8.41}$$

where the proportionality factor ς is assumed to be constant for the purpose of this work.

Obviously in a thermodynamical phase equilibrium $\hat{n} \equiv 0$, and the equilibrium fraction x is connected with the partial pressure p^A through the relation (8.40). Then the mass source (8.38) vanishes identically. The behavior of the continuous model based on the above assumptions has been checked on a simple bench-mark homogeneous problem [16]. It was found that results are indeed qualitatively in agreement with observations.

We present here the set of field equations which cover a much more extensive class of problems. In particular we can describe couplings of adsorption and diffusion and we can as well incorporate boundary conditions on permeable boundaries which are characteristic for the majority of practical problems.

We use the Eulerian description of the system in which mass densities are referred to the current configuration. Then for the mass density of the skeleton, the fluid carrier of the adsorbate and the adsorbate in the liquid state, we have for $\mathbf{x} \in \mathbf{f}^S \left(B, t \right)$, $t \in T$,

$$\rho_t^S := \rho^S J^{S-1}, \quad \rho_t^F := \rho^F J^{S-1}, \quad \rho_t^A := \rho^A J^{S-1}. \tag{8.42}$$

We consider solely isothermal processes. According to these remarks we have to determine the fields

$$\left\{\rho_t^S, \rho_t^L, c, \mathbf{v}^S, \mathbf{e}^S, \mathbf{v}^F \equiv \mathbf{v}^A, n, x, y\right\}, \quad \rho_t^L := \rho_t^F + \rho_t^A, \tag{8.43}$$

where the concentration c is defined by the relation

$$c := \frac{\rho_t^A}{\rho_t^F + \rho_t^A} \ll 1. \tag{8.44}$$

Inspection of the list (8.43) reveals that the model contains, in addition to usual fields describing multicomponent systems, three microstructural fields: Δ, x, y. The first one describes changes of the microstructural geometry, and the remaining two exchange of mass related to both energetic properties of the microstructure (the number of occupied sites x) and the geometry (the fraction of the internal surface y).

The velocity of the third component does not appear because the adsorbate in the fluid phase moves with the same velocity as the other fluid component. Therefore we use only two momentum balance equations, for the skeleton and for both fluid components together.

Field equations follow from three mass balance equations, two momentum balance equations, the balance equation of porosity, integrability condition for the deformation of the skeleton and two evolution equations for two additional microstructural variables. They have the form:

mass balance

$$\frac{\partial \rho_t^S}{\partial t} + \mathrm{div}\ \left(\rho_t^S \mathbf{v}^S\right) = -\rho_t^L \hat{c}, \quad \frac{\partial \rho_t^L}{\partial t} + \mathrm{div}\ \left(\rho_t^L \mathbf{v}^F\right) = \rho_t^L \hat{c},$$

$$\frac{\partial c}{\partial t} + \mathbf{v}^F \cdot \mathrm{grad}\ c = (1 - c)\,\hat{c}, \quad \hat{c} := \frac{\hat{\rho}_t^A}{\rho_t^L} = -\frac{m^A}{\rho_0^L}\frac{d\,(xy)}{dt}, \tag{8.45}$$

momentum balance

$$\frac{\partial \rho_t^L \mathbf{v}^F}{\partial t} + \mathrm{div}\ \left(\rho_t^L \mathbf{v}^F \otimes \mathbf{v}^F + p^L \mathbf{1}\right) + \pi\left(\mathbf{v}^F - \mathbf{v}^S\right) = 0,$$

$$\rho_t^S \frac{\partial \mathbf{v}^S}{\partial t} = \mathrm{div}\ \mathbf{T}^S + \pi\left(\mathbf{v}^F - \mathbf{v}^S\right), \tag{8.46}$$

porosity balance

$$\frac{\partial \Delta}{\partial t} + \varphi \mathrm{div}\ \left(\mathbf{v}^F - \mathbf{v}^S\right) = -\frac{\Delta}{\tau},$$

where

$$\mathbf{T}^S = \mathbf{T}_0^S + \lambda^S \mathrm{tr}\mathbf{e}^S \mathbf{1} + 2\mu^S \mathbf{e}^S + \beta \Delta \mathbf{1}, \tag{8.47}$$

$$p^L = p_0^L + \kappa\left(\rho_t^L - \rho_0^L\right) + \beta\Delta, \quad p^F = (1 - c)\,p^L, \quad p^A = cp^L,$$

with material parameters $\varphi, \lambda^S, \mu^S, \kappa, \beta, \pi$ being constant. They depend parametrically on the constant initial porosity n_0. In addition we have:

integrability condition

$$\frac{\partial \mathbf{e}^S}{\partial t} = \text{symgrad } \mathbf{v}^S, \qquad (8.48)$$

evolution equations for microstructural variables

$$\frac{d \ln \frac{y}{y_0}}{dt} = -\varsigma \frac{\Delta}{n_E}, \qquad y(t = 0) = y_0 \equiv \frac{f_{int}(t = 0)}{V},$$

$$\frac{dx}{dt} = \frac{1}{\tau_{ad}} \left[(1 - x) \frac{cp^L}{p_0} - x \right], \qquad x(t = 0) = \frac{\frac{c_0 p_0^L}{p_0}}{1 + \frac{c_0 p_0^L}{p_0}}, \qquad (8.49)$$

$$\tau_{ad} := \frac{1}{b} e^{\frac{E_b}{kT}}, \qquad c_0 := c(t = 0).$$

Again the material parameters $\varsigma, p_0, \tau_{ad}$ are assumed to be constant.

General results for this system of equations have not been obtained as yet. However some important particular problems have been solved under the assumptions of negligible accelerations and a negligible explicit time dependence of porosity. Their discussion can be found in the Ph.D. thesis of B. Albers [1] and subsequent publications [2, 3, 4]. We quote here solely the most important conclusions of these works.

Investigation of a one-dimensional flow of an ideal liquid through a poroelastic linear material has shown that the rate of adsorption depends on the magnitude of the relative velocity. This dependence is non-monotonic. The rate is small for either small or very large relative velocities and there appears a maximum of the rate at an intermediate velocity. Both position of this maximum as well as its amplitude depend on the time lapse from the beginning of the adsorption process. It has been also found out that an influence of changes of internal surface is limited to a very small neighbourhood of the initial instant of time. This is understandable as the relaxation time of porosity is much smaller than that of adsorption. Simultaneously it has been confirmed that an intensity of adsorption processes coupled to diffusion depends on the surface permeability which controls the relative velocity in the system. In the work in progress, similar results seem to follow from a numerical analysis of a two-dimensional problem.

8.5   Interfaces, ideal walls, boundary conditions

8.5.1   Introduction

Properties of interfaces in multicomponent systems with different kinematics of components are much more involved than those following from dynamical compatibility conditions of the usual continuum thermodynamics. This is related to the existence of boundary layers in transition regions between a porous body and a neighbouring system (*e.g.*, a fluid component flowing through a permeable boundary of the porous body to the exterior or another porous body with, maybe, a different number of components which is the case if it is, for instance, not fully saturated). Boundary layers are replaced in the present model by singular surfaces and these, as a consequence of these properties, must possess a structure of its own replacing gradients of fields in transition regions. This is the reason for introducing surface sources on material surfaces (interfaces of the skeleton) as we indicated in Section 2. We proceed to improve this motivation and to investigate consequences of such improved conditions on the construction of thermodynamical properties of fields and boundary value problems.

One such problem appears in a physical interpretation of temperature. We limit our attention to a single temperature field common for all components, as the problem for systems with multiple temperatures does not have a solution as yet.

The classical thermodynamical argument concerning the interpretation of temperature is as follows. If we bring together two thermodynamical systems each of them being in a state of thermodynamical equilibrium, and the contact surface admits solely a non-mechanical flux of energy between them (*i.e.*, the mechanical working of one system on the other is not allowed), then we say that these two systems are in thermodynamical equilibrium with each other if this non-mechanical flux vanishes. By constructing equivalence classes of such systems we introduce an empirical temperature as a scalar-valued function on the set of all systems, which is the same for systems in thermodynamical equilibrium with each other. The classical considerations of the integrability of the Gibbs equation lead then to the notion of an absolute temperature as a special choice of an empirical temperature. This argument is transferred to systems in which solely local thermodynamical equilibria appear. However, we can indeed consider local equilibria on interfaces if we can prove the continuity of the temperature in globally non-equilibrium processes. Such an argument is based in single-component systems on dynamical compatibility conditions. Namely on a material surface of such a system the global energy and entropy balances yield continuity of the normal component of the heat flux and of the entropy flux. Consequently, if these two fluxes are related to each other by a classical proportionality relation with the proportionality factor being equal to the inverse of the absolute temperature, then it follows that the temperature must be continuous as well. Consequently, if one of the systems is identified with a thermometer, we can measure the temperature by the contact through the interface and we can control the temperature on the boundary, if we want to construct the boundary value problem for heat conduction.

In the case of multi-component systems, permeable interfaces are not material for some components and, consequently, partial heat and entropy fluxes are not continuous. The question arises if we can still use the classical argument on the continuity of the temperature and, consequently, if we can construct boundary value problems in terms of the temperature for heat conduction in such systems. We proceed to investigate this question.

8.5.2 Compatibility conditions on an interface

We consider a smooth orientable surface Σ material with respect to the skeleton, $i.e.$,

$$\Psi(\mathbf{X}) = 0, \qquad \mathbf{X} \in B, \qquad \mathbf{N} := \frac{\mathrm{Grad}\ \Psi}{|\mathrm{Grad}\ \Psi|}, \tag{8.50}$$

where $\mathbf{N}$ is the unit normal vector specifying the positive and negative sides of the surface Σ. In its current configuration, this surface is described by the equation

$$\psi(\mathbf{x}, t) \quad := \quad \Psi\left(\mathbf{f}^{S-1}(\mathbf{x}, t)\right) = 0, \quad x \in \mathbf{f}^{S}(B, t),$$

$$i.e., \qquad \mathbf{n} \quad := \quad \frac{\mathrm{grad}\ \psi}{|\mathrm{grad}\ \psi|} = \frac{\mathbf{F}^{S-T}\mathbf{N}}{|\mathbf{F}^{S-T}\mathbf{N}|}, \quad u = \acute{\mathbf{x}}^{S} \cdot \mathbf{n}, \tag{8.51}$$

with $\mathbf{n}$ being the unit normal vector and u the normal speed of propagation of the image $\sigma := \mathbf{f}^{S}(\Sigma, t)$. The vector $\mathbf{n}$ is well defined due to the relation

$$\left[\!\left[J^{S}\mathbf{F}^{S-T}\mathbf{N} \right]\!\right] = 0, \qquad \left[\!\left[\ldots \right]\!\right] := (\ldots)^{+} - (\ldots)^{-}, \tag{8.52}$$

which follows from the smoothness assumption. The brackets $(\ldots)^{+}, (\ldots)^{-}$ denote the positive and negative finite limits on the surface Σ.

Let us consider the balance equations reduced to this surface.

Mass balance
According to (8.6) in the absence of mass sources the jump of the mass density of skeleton $\left[\!\left[\rho^{S} \right]\!\right]$ is not limited by the balance equations and the jump of the mass density of the fluid must fulfil the condition

$$\forall_{\mathbf{X}\in\Sigma}\left[\!\left[\rho^{F}\acute{\mathbf{X}}^{F} \right]\!\right] \cdot \mathbf{N} = 0 \implies \forall_{x\in\sigma}\left[\!\left[\rho_{t}^{F}(\mathbf{v}^{F} - \mathbf{v}^{S}) \right]\!\right] \cdot \mathbf{n} = 0. \tag{8.53}$$

The latter relation in the current configuration shows that the mass flow of the fluid through the interface is continuous. The interface does not contain sinks.

Momentum balance
Due to the presence of surface sources in (8.8), we obtain in $\mathbf{X} \in \Sigma$

$$\left[\!\left[\mathbf{P}^{S}\mathbf{N} \right]\!\right] - \hat{\mathbf{p}}_{surf}^{S} = 0, \quad \left(\rho^{F}\acute{\mathbf{X}}^{F} \cdot \mathbf{N} \right)\left[\!\left[\acute{\mathbf{x}}^{F} \right]\!\right] = \left[\!\left[\mathbf{P}^{F} \right]\!\right]\mathbf{N} + \hat{\mathbf{p}}_{surf}^{F}, \tag{8.54}$$

or, in the Eulerian description, for $\mathbf{x} \in \sigma$,

$$
\begin{aligned}
\left[\!\left[\mathbf{T}^S \mathbf{n} \right]\!\right] + \hat{\mathsf{p}}^S_{surf} &= 0, \\
\hat{\mathsf{p}}^S_{surf} &:= J^{S-1} \left| \mathbf{F}^{S-T} \mathbf{N} \right|^{-1} \hat{\mathbf{p}}^S_{surf}, \\
\left(\rho^F_t (\mathbf{v}^F - \mathbf{v}^S) \cdot \mathbf{n} \right) \left[\!\left[\mathbf{v}^F \right]\!\right] &= -\left[\!\left[\mathsf{p}^F \right]\!\right] \mathbf{n} + \hat{\mathsf{p}}^F_{surf}, \\
\hat{\mathsf{p}}^F_{surf} &:= J^{S-1} \left| \mathbf{F}^{S-T} \mathbf{N} \right|^{-1} \hat{\mathbf{p}}^F_{surf},
\end{aligned}
\tag{8.55}
$$

where we have used the constitutive assumption that the fluid is ideal (see (8.24) and $(8.33)_2$), *i.e.*,

$$
\mathbf{P}^F = J^S \mathbf{T}^F \mathbf{F}^{S-T} \equiv -J^S \mathsf{p}^F \mathbf{F}^{S-T}, \quad \mathsf{p}^F := p^F + \beta\Delta.
\tag{8.56}
$$

Relation $(8.55)_2$ motivates the necessity of the surface sources of momentum. It has been argued (*e.g.*, [15, 14, 7, 1]) that the boundary conditions on permeable boundaries of a skeleton should follow from the bulk momentum balance

$$
\left(\rho^F \acute{\mathbf{X}}^F \cdot \mathbf{N} \right) \left[\!\left[\acute{\mathbf{x}}^F \right]\!\right] - \left[\!\left[\mathbf{P}^F + \mathbf{P}^S \right]\!\right] \cdot \mathbf{N} \Big|_\Sigma = 0
\tag{8.57}
$$

and from the flow condition for the fluid

$$
\begin{aligned}
\rho^F \acute{\mathbf{X}}^F \cdot \mathbf{N} \Big|_\Sigma &= \alpha_0 \left[\!\left[J^{S-1} \frac{\mathbf{P}^F \cdot \mathbf{F}^S}{n} \right]\!\right] \Big|_\Sigma \\
\implies \rho^F_t \left(\mathbf{v}^F - \mathbf{v}^S \right) \cdot \mathbf{n} \Big|_\sigma &= \alpha \left(p^{F-} - \frac{n^-}{n^+} p^{F+} \right) \Big|_\sigma, \\
\alpha &:= \frac{3\alpha_0}{n^-} \left(\mathbf{N} \cdot \mathbf{C}^{S-1} \mathbf{N} \right)^{-1/2},
\end{aligned}
\tag{8.58}
$$

$$
\acute{\mathbf{X}}^F - \acute{\mathbf{X}}^F \cdot \mathbf{N}\mathbf{N} \Big|_\Sigma = 0,
\tag{8.59}
$$

where α is the so-called surface permeability coefficient. The condition (8.59) is characteristic for ideal fluids and, if needed, can be replaced by a Beavers–Joseph type of condition for the slip motion. In such a case the constitutive law for the partial stress $\mathbf{T}^F$ must be modified in order to include shear stresses (*e.g.*, due to viscosity of the fluid component). Condition (8.58) states that the amount of fluid mass which flows through a permeable boundary is driven by the discontinuity of the pressure. It has been assumed that for relatively slow processes the pore pressure can be described by the simple relation $p = \frac{p^F}{n}$. Relation $(8.58)_2$ can be easily motivated on theoretical grounds. If one assumes that in a thin transition layer near the interface Σ a simple Darcy law holds true,

$$-n \operatorname{grad} \frac{p^F}{n} - \pi(\mathbf{v}^F - \mathbf{v}^S) = 0, \tag{8.60}$$

then

$$\frac{\rho_t^F \langle n \rangle}{\pi L} \left[\!\left[\frac{p^F}{n} \right]\!\right] \approx \rho_t^F (\mathbf{v}^F - \mathbf{v}^S) \cdot \mathbf{n}, \quad \langle n \rangle := \tfrac{1}{2}(n^+ + n^-), \tag{8.61}$$

where L is the thickness of the boundary layer. Relation (8.61) coincides, of course, with $(8.58)_2$, if $\alpha := \frac{\rho_t^F \langle n \rangle}{\pi L}$.

In the case of thermodynamical equilibrium we have $\acute{\mathbf{X}}^F \cdot \mathbf{N}\big|_{\Sigma} = 0$ and relations $(8.55)_2$ and $(8.58)_2$ imply

$$\left[p^F \right] + \hat{\mathbf{p}}_{surf}^F \cdot \mathbf{n} \bigg|_{\sigma} = 0, \qquad \left[\!\left[\frac{p^F}{n} \right]\!\right]\bigg|_{\sigma} = 0. \tag{8.62}$$

Consequently, if the source $\hat{\mathbf{p}}_{surf}^F \cdot \mathbf{n}\big|_{\sigma}$ were zero, the porosity n had to be continuous. This, certainly, cannot be the case.

The presence of the surface source of momentum can be easily understood in semi-microscopical terms. Various values of the surface permeability coefficient α yield a different distribution of the total load between solid and fluid components which is exerted by subbodies on each other through the interface Σ. Hence the partial pressure p^F cannot be continuous on Σ.

Energy balance

Bearing the global balance equations (8.10) in mind, we obtain, for $\mathbf{X} \in \Sigma$,

$$\left[\mathbf{Q}^S \right] \cdot \mathbf{N}\bigg|_{\Sigma} = \left[\mathbf{P}^S \right] \mathbf{N} \cdot \acute{\mathbf{x}}^S \bigg|_{\Sigma} \equiv -\hat{\mathbf{p}}_{surf}^S \cdot \acute{\mathbf{X}}^S \bigg|_{\Sigma}, \tag{8.63}$$

$$\rho^F \acute{\mathbf{X}}^F \cdot \mathbf{N} \left[\!\left[\varepsilon^F + \tfrac{1}{2}\acute{\mathbf{x}}^{F2} \right]\!\right] + \left[\mathbf{Q}^F \right] \cdot \mathbf{N}\bigg|_{\Sigma} = \left[\!\left[(\mathbf{P}^F \mathbf{N}) \cdot \acute{\mathbf{X}}^F \right]\!\right]\bigg|_{\Sigma}.$$

The second condition can be easily transformed to the form

$$\rho^F \acute{\mathbf{X}}^F \cdot \mathbf{N} \left[\!\left[\varepsilon^F + \tfrac{1}{2}(\acute{\mathbf{x}}^F - \acute{\mathbf{x}}^S) \cdot (\acute{\mathbf{x}}^F - \acute{\mathbf{x}}^S) \right]\!\right]\bigg|_{\Sigma} + \left[\mathbf{Q}^F \right] \cdot \mathbf{N}\bigg|_{\Sigma}$$
$$\tag{8.64}$$
$$= \left[\!\left[(\mathbf{P}^F \mathbf{N}) \cdot (\acute{\mathbf{x}}^F - \acute{\mathbf{x}}^S) \right]\!\right]\bigg|_{\Sigma} - \hat{\mathbf{p}}_{surf}^F \cdot \acute{\mathbf{x}}^F \bigg|_{\Sigma},$$

where the momentum condition $(8.54)_2$ has been used. Consequently the bulk energy transport through the interface Σ can be written in the form

$$\left[\!\left[\mathbf{Q}^S + \mathbf{Q}^F \right]\!\right] \cdot \mathbf{N} \Big|_{\Sigma} \tag{8.65}$$

$$= -\rho^F \acute{\mathbf{X}}^F \cdot \mathbf{N} \left[\!\left[\varepsilon^F - J^S \frac{p^F}{\rho^F} + \tfrac{1}{2}(\acute{\mathbf{x}}^F - \acute{\mathbf{x}}^S) \cdot (\acute{\mathbf{x}}^F - \acute{\mathbf{x}}^S) \right]\!\right] \Big|_{\Sigma} ,$$

where the relation (8.56) has been applied. It is clear that the heat flux $(\mathbf{Q}^S + \mathbf{Q}^F) \cdot \mathbf{N}$ is not continuous on permeable boundaries.

Entropy balance

The global partial entropy balance equations

$$\forall P - \text{S-material}:$$
$$\frac{d}{dt} \int_{\mathcal{P}} \rho^S \eta^S dV + \oint_{\partial \mathcal{P}} \mathbf{H}^S \cdot \mathbf{N} dA = \int_{\mathcal{P}} \hat{\eta}^S dV + \oint_{\partial \mathcal{P}} \hat{\eta}^S_{surf} dA,$$
$$\tag{8.66}$$
$$\forall P - \text{F-material}:$$
$$\frac{d}{dt} \int_{\mathcal{P}} \rho^F \eta^F dV + \oint_{\partial \mathcal{P}} \mathbf{H}^F \cdot \mathbf{N} dA = \int_{\mathcal{P}} \hat{\eta}^F dV + \oint_{\partial \mathcal{P}} \hat{\eta}^F_{surf} dA$$

yield for the interface

$$\left[\!\left[\mathbf{H}^S \right]\!\right] \cdot \mathbf{N}|_{\Sigma} = \hat{\eta}^S_{surf}, \tag{8.67}$$

$$\rho^F \acute{\mathbf{X}}^F \cdot \mathbf{N} \left[\!\left[\eta^F \right]\!\right] \Big|_{\Sigma} + \left[\!\left[\mathbf{H}^F \right]\!\right] \cdot \mathbf{N}|_{\Sigma} = \hat{\eta}^F_{surf}.$$

Hence the intrinsic bulk transport of the entropy through the interface satisfies the relation

$$\left[\!\left[\mathbf{H}^S + \mathbf{H}^F + \rho^F \eta^F \acute{\mathbf{X}}^F \right]\!\right] \cdot \mathbf{N} \Big|_{\Sigma} = \hat{\eta}^S_{surf} + \hat{\eta}^F_{surf}. \tag{8.68}$$

We combine this result with the relation (8.65) for the intrinsic bulk transport of energy. Bearing relation $(8.22)_1$ in mind we obtain

$$\rho^F \acute{\mathbf{X}}^F \cdot \mathbf{N} \left[\!\left[\varepsilon^F - J^S \frac{p^F}{\rho^F} - \Theta \eta^F \right]\!\right] \Big|_{\Sigma} \tag{8.69}$$

$$= -[\Theta] \left\langle \mathbf{H}^S + \mathbf{H}^F + \rho^F \eta^F \acute{\mathbf{X}}^F \right\rangle \cdot \mathbf{N} \Big|_{\Sigma} + \hat{\eta}^S_{surf} + \hat{\eta}^F_{surf} ,$$

where $2 \langle \cdots \rangle = (\cdots)^+ + (\cdots)^-$, *i.e.*, it is an average value on the interface, and we neglected the quadratic contribution of the relative velocity. This is justified as the relation $(8.22)_1$ was derived under the assumption of a small deviation from the state of thermodynamical equilibrium.

In classical thermodynamics the problem of continuity of the absolute temperature is considered on the so-called *ideal walls* (see Müller [8]). The existence of ideal walls is required if we want temperature to be a measurable quantity[2]. Then entropy productions on such a surface are zero. If we make this assumption for the interface Σ, then the absolute temperature Θ is continuous on this surface if it is either impermable or if the Gibbs free energy of the fluid component (chemical potential) is continuous:

$$\left[\!\left[\mu^F\right]\!\right] = 0, \quad \mu^F := \varepsilon^F - J^S \frac{p^F}{\rho^F} - \Theta \eta^F. \tag{8.70}$$

This condition seems to be plausible because the density of the true Gibbs free energy of the fluid component μ^{FR} is approximately equal to μ^F, due to the relation between the true mass density ρ^{FR}, and the partial mass density ρ^F: $\rho^F = n\rho^{FR}$. Hence the assumption on a local thermodynamical equilibrium yielding the continuity of μ^{FR} leads to the continuity of μ^F.

The above considerations show that processes arbitrarily deviating from the state of thermodynamical equilibrium yield problems with the operational definition of temperature. In such processes one cannot expect that surface entropy sources vanish. They are most likely of the second order in non-equilibrium variables and, consequently, remain in the jump condition. The requirement of continuity of the true chemical potential is not fulfilled either because one has to account for convective contributions in both energy and entropy jump conditions.

We complete the considerations for interfaces material with respect to the skeleton with the analysis of porosity equation. From (8.11) we obtain easily

$$\left[\!\left[\varphi J^S \acute{\mathbf{X}}^F \cdot \mathbf{N}\right]\!\right] \equiv \rho^F \acute{\mathbf{X}}^F \cdot \mathbf{N} \left[\!\left[\frac{\varphi}{\rho^F J^{S-1}}\right]\!\right] = \hat{n}_{surf}, \tag{8.71}$$

where relation (8.23) has been used.

Obviously the above relation could not be satisfied on an interface between two different porous materials for which $[[\varphi]] \neq 0$, if the initial porosity of both bodies was different and the surface source of porosity was zero. Note that quantities appearing on the left-hand side are all specified either by the initial conditions or by a solution of field equations. On the other hand field equations do not contain contributions of $\hat{n}_{surf}$. Consequently relation (8.71) can be considered to be the definition of this source.

Let us mention in passing that surface sources $\hat{\mathbf{p}}^S_{surf}, \hat{n}_{surf}$ are not needed for consistency of the model if the surface is not material, *e.g.*, in the case of shock waves. In those cases the usual dynamical compatibility conditions yielding Rankine–Hugoniot conditions preserve their validity. The presence

---

[2]Another example of such a wall for the transport of mass rather than energy is the *semipermeable membrane* of the mixture of fluids on which the chemical potential is continuous.

of sources is strictly related to a material change of microstructure on an interface between two different porous materials.

8.6   Conclusions

New results presented in this work concern two topics: a relation between partial fluxes of heat and entropy following from the second law of thermodynamics, and relations on permeable interfaces separating a porous material from a single component system or a different porous material.

We have shown that the assumption on small deviations from thermodynamical equilibrium, *i.e.*,

$$\max\left\{\|\mathbf{G}\|, \left\|\acute{\mathbf{X}}^{F}\right\|, \|\Delta\|\right\} \ll 1, \tag{8.72}$$

where the norms are chosen as supremum norms on $B \times T$, yields an explicit answer to the first question in the form

$$\mathbf{H}^{S} + \mathbf{H}^{F} = \frac{\mathbf{Q}^{S} + \mathbf{Q}^{F}}{\Theta}, \quad \mathbf{J} = \varphi J^{S}\acute{\mathbf{X}}^{F}, \tag{8.73}$$

where $\mathbf{Q}^{S} + \mathbf{Q}^{F}$ is independent of $\acute{\mathbf{X}}^{F}$ and $\mathbf{J}$ is independent of $\mathbf{G}$.

Under the same assumption and under the condition of local equilibrium of the fluid component on interfaces (*i.e.*, the continuity of the chemical potential of the fluid component), we have shown that the absolute temperature is continuous on such interfaces.

These results allow us to extend the model of poroelastic materials which has been investigated in earlier contributions to non-isothermal processes.

Apart from these two important results we have shown that the couplings between two components reduce also in a considerable manner under the condition of small deviations from the state of thermodynamical equilibrium. Namely we have shown that partial Cauchy stresses in the skeleton cannot depend on the current mass density of the fluid and, *vice versa*, partial Cauchy stresses in the fluid depend solely on the current partial mass density of the fluid and on the deviation of porosity Δ, but not on deformations of the skeleton. It means that in the linear simplified version of the model we do not obtain Biot's multi-component model of porous materials. However, in spite of some claims in the literature, this difference has solely a quantitative influence on properties of weak discontinuity waves, but it does not influence either the number of modes or their basic properties. This was in a way expected if one inspected carefully the analysis of wave propagation in mixtures of fluids. In the case of so-called ideal mixtures (no interaction terms in partial free energies) the number of modes remains the same as in the case of interacting mixtures and only the speeds of propagation change a little.

Finally let us note that there is an indication that mass exchange processes yield their own contributions to stresses in the skeleton independent of the deformation. Namely in contrast to fluid, for which the definition of the chemical potential contains the partial pressure divided by the mass density $\frac{p^F}{\rho^F J^{S-1}}$, the chemical potential coupled to the mass source for the skeleton contains the contribution $\rho^S \frac{\partial \psi^S}{\partial \rho^S}$, rather than $-\frac{\text{tr}\mathbf{T}^S}{3\rho^S J^{S-1}}$ which would be a usual partial pressure contribution in the skeleton. It means that the presence of mass exchange yields additional stress effects in the skeleton which would appear even in the case of lack of deformations of the skeleton.

8.7 Appendix: Evaluation of the entropy inequality

In this Appendix we evaluate solutions of the local entropy inequality. As usual in thermodynamics, field equations are considered to be constraints imposed on the class of smooth solutions of the inequality. These constraint conditions are eliminated by Lagrange multipliers [8, 12]. Hence we have for all sufficiently smooth fields (8.5)

$$\rho^S \frac{\partial \eta^S}{\partial t} + \rho^F \frac{\partial \eta^F}{\partial t} + \rho^F \acute{\mathbf{X}}^F \cdot \text{Grad } \eta^F + \text{Div } \left(\mathbf{H}^S + \mathbf{H}^F \right)$$

$$-\Lambda^S \frac{\partial \rho^S}{\partial t} - \Lambda^F \left(\frac{\partial \rho^F}{\partial t} + \acute{\mathbf{X}}^F \cdot \text{Grad } \rho^F + \rho^F \text{Div } \acute{\mathbf{X}}^F \right)$$

$$-\boldsymbol{\Lambda}^{v^F} \cdot \left(\rho^F \left(\frac{\partial \acute{\mathbf{x}}^F}{\partial t} + \acute{\mathbf{X}}^F \cdot \text{Grad } \acute{x}^F \right) - \text{Div } \mathbf{P}^F \right)$$

$$-\Lambda^\varepsilon \left(\rho^S \frac{\partial \varepsilon^S}{\partial t} + \rho^F \left(\frac{\partial \varepsilon^F}{\partial t} + \acute{\mathbf{X}}^F \cdot \text{Grad } \varepsilon^F \right) \right.$$

$$\left. + \text{Div } \left(\mathbf{Q}^S + \mathbf{Q}^F \right) - \mathbf{P}^S \cdot \text{Grad } \acute{x}^S - \mathbf{P}^F \cdot \text{Grad } \acute{x}^F \right) \qquad \text{(A.1)}$$

$$-\Lambda^n \left(\frac{\partial n}{\partial t} + \text{Div } \mathbf{J} \right) - \boldsymbol{\Lambda}^F \cdot \left(\frac{\partial \mathbf{F}^S}{\partial t} - \text{Grad } \acute{x}^S \right)$$

$$+ \left(\boldsymbol{\Lambda}^{v^S} - \boldsymbol{\Lambda}^{v^F} \right) \cdot \left(\hat{\mathbf{p}}^S - \hat{\rho}^S \acute{x}^S \right) - \boldsymbol{\Lambda}^{v^S} \cdot \left(\rho^S \frac{\partial \acute{\mathbf{x}}^S}{\partial t} - \text{Div } \mathbf{P}^S \right)$$

$$-\Lambda^\varepsilon \left(-\hat{\mathbf{p}}^S \cdot \left(\acute{x}^F - \acute{x}^S \right) + \hat{\rho}^S \left(\varepsilon^S - \frac{1}{2}\acute{x}^{S2} - \varepsilon^F + \frac{1}{2}\acute{x}^{F2} \right) \right)$$

$$+ \hat{\rho}^S \left(\eta^S - \eta^F \right) + \hat{\rho}^S \left(\Lambda^S - \Lambda^F \right) + \Lambda^n \hat{n} \geq 0,$$

where the multipliers $\Lambda^S, \Lambda^F, \boldsymbol{\Lambda}^F, \boldsymbol{\Lambda}^{v^S}, \boldsymbol{\Lambda}^{v^F}, \Lambda^\varepsilon, \Lambda^n$ are constitutive functions continuously differentiable with respect to constitutive variables almost everywhere on the domain $B \times T$.

We have replaced the field of motion of the skeleton $\mathbf{f}^S$ by the field of deformation gradient $\mathbf{F}^S$ and the field of velocity $\acute{x}^S$. Then the new fields must satisfy the following compatibility conditions in almost all points of the domain B:

$$\frac{\partial \mathbf{F}^S}{\partial t} = \operatorname{Grad} \acute{\mathbf{x}}^S, \quad \operatorname{Grad} \mathbf{F}^S = \left(\operatorname{Grad} \mathbf{F}^S\right)^{\overset{23}{T}}. \tag{A.2}$$

We account for the first condition in the same way as we do in the case of all other field equations, while the second one shall be directly substituted in thermodynamical relations.

It is easy to see that application of constitutive relations (8.15) yields a linearity of the above inequality with respect to the following derivatives:

time derivatives

$$\left\{ \frac{\partial \rho^S}{\partial t}, \frac{\partial \rho^F}{\partial t}, \frac{\partial \mathbf{F}^S}{\partial t}, \frac{\partial n}{\partial t}, \frac{\partial \Theta}{\partial t}, \frac{\partial \mathbf{G}}{\partial t}, \frac{\partial \acute{\mathbf{x}}^S}{\partial t}, \frac{\partial \acute{\mathbf{x}}^F}{\partial t} \right\}, \tag{A.3}$$

spatial derivatives

$$\left\{ \operatorname{Grad} \rho^S, \operatorname{Grad} \rho^F, \operatorname{Grad} \mathbf{F}^S, \operatorname{Grad} n, \operatorname{Grad} \mathbf{G}, \operatorname{Grad} \acute{\mathbf{x}}^S, \operatorname{Grad} \acute{\mathbf{x}}^F \right\}. \tag{A.4}$$

This means that coefficients of these derivatives must vanish identically, and we obtain the following set of relations determining Lagrange multipliers:

$$\Lambda^S = \rho^S \left(\frac{\partial \eta^S}{\partial \rho^S} - \Lambda^\varepsilon \frac{\partial \varepsilon^S}{\partial \rho^S} \right) + \rho^F \left(\frac{\partial \eta^F}{\partial \rho^S} - \Lambda^\varepsilon \frac{\partial \varepsilon^F}{\partial \rho^S} \right),$$

$$\Lambda^F = \rho^S \left(\frac{\partial \eta^S}{\partial \rho^F} - \Lambda^\varepsilon \frac{\partial \varepsilon^S}{\partial \rho^F} \right) + \rho^F \left(\frac{\partial \eta^F}{\partial \rho^F} - \Lambda^\varepsilon \frac{\partial \varepsilon^F}{\partial \rho^F} \right),$$

$$\mathbf{\Lambda}^F = \rho^S \left(\frac{\partial \eta^S}{\partial \mathbf{F}^S} - \Lambda^\varepsilon \frac{\partial \varepsilon^S}{\partial \mathbf{F}^S} \right) + \rho^F \left(\frac{\partial \eta^F}{\partial \mathbf{F}^S} - \Lambda^\varepsilon \frac{\partial \varepsilon^F}{\partial \mathbf{F}^S} \right),$$

$$\Lambda^n = \rho^S \left(\frac{\partial \eta^S}{\partial n} - \Lambda^\varepsilon \frac{\partial \varepsilon^S}{\partial n} \right) + \rho^F \left(\frac{\partial \eta^F}{\partial n} - \Lambda^\varepsilon \frac{\partial \varepsilon^F}{\partial n} \right), \tag{A.5}$$

$$0 = \rho^S \left(\frac{\partial \eta^S}{\partial \Theta} - \Lambda^\varepsilon \frac{\partial \varepsilon^S}{\partial \Theta} \right) + \rho^F \left(\frac{\partial \eta^F}{\partial \Theta} - \Lambda^\varepsilon \frac{\partial \varepsilon^F}{\partial \Theta} \right),$$

$$0 = \rho^S \left(\frac{\partial \eta^S}{\partial \mathbf{G}} - \Lambda^\varepsilon \frac{\partial \varepsilon^S}{\partial \mathbf{G}} \right) + \rho^F \left(\frac{\partial \eta^F}{\partial \mathbf{G}} - \Lambda^\varepsilon \frac{\partial \varepsilon^F}{\partial \mathbf{G}} \right),$$

$$\rho^S \mathbf{\Lambda}^{v^S} = XS - \mathbf{F}^{S-T} \left\{ \rho^S \left(\frac{\partial \eta^S}{\partial \acute{\mathbf{X}}^F} - \Lambda^\varepsilon \frac{\partial \varepsilon^S}{\partial \acute{\mathbf{X}}^F} \right) \right.$$
$$\left. + \rho^F \left(\frac{\partial \eta^F}{\partial \acute{\mathbf{X}}^F} - \Lambda^\varepsilon \frac{\partial \varepsilon^F}{\partial \acute{\mathbf{X}}^F} \right) \right\},$$

$$\rho^F \mathbf{\Lambda}^{v^F} = \mathbf{F}^{S-T} \left\{ \rho^S \left(\frac{\partial \eta^S}{\partial \acute{\mathbf{X}}^F} - \Lambda^\varepsilon \frac{\partial \varepsilon^S}{\partial \acute{\mathbf{X}}^F} \right) \right.$$
$$\left. + \rho^F \left(\frac{\partial \eta^F}{\partial \acute{\mathbf{X}}^F} - \Lambda^\varepsilon \frac{\partial \varepsilon^F}{\partial \acute{\mathbf{X}}^F} \right) \right\},$$

in the case of coefficients of time derivatives, and the identities limiting constitutive relations

$$\frac{\partial \left(\mathbf{H}^S + \mathbf{H}^F\right)}{\partial \rho^S} - \Lambda^\varepsilon \frac{\partial \left(\mathbf{Q}^S + \mathbf{Q}^F\right)}{\partial \rho^S} - \Lambda^n \frac{\partial \mathbf{J}}{\partial \rho^S}$$

$$+ \rho^F \acute{\mathbf{X}}^F \left(\frac{\partial \eta^F}{\partial \rho^S} - \Lambda^\varepsilon \frac{\partial \varepsilon^F}{\partial \rho^S}\right) + \left(\frac{\partial \mathbf{P}^S}{\partial \rho^S}\right)^T \mathbf{\Lambda}^{v^S} + \left(\frac{\partial \mathbf{P}^F}{\partial \rho^S}\right)^T \mathbf{\Lambda}^{v^F} = 0,$$

$$\frac{\partial \left(\mathbf{H}^S + \mathbf{H}^F\right)}{\partial \rho^F} - \Lambda^\varepsilon \frac{\partial \left(\mathbf{Q}^S + \mathbf{Q}^F\right)}{\partial \rho^F} - \Lambda^n \frac{\partial \mathbf{J}}{\partial \rho^F} - \Lambda^F \acute{\mathbf{X}}^F$$

$$+ \rho^F \acute{\mathbf{X}}^F \left(\frac{\partial \eta^F}{\partial \rho^F} - \Lambda^\varepsilon \frac{\partial \varepsilon^F}{\partial \rho^F}\right) + \left(\frac{\partial \mathbf{P}^S}{\partial \rho^F}\right)^T \mathbf{\Lambda}^{v^S} + \left(\frac{\partial \mathbf{P}^F}{\partial \rho^F}\right)^T \mathbf{\Lambda}^{v^F} = 0,$$

$$\underset{\text{sym}}{\overset{23}{}} \left\{ \left[\frac{\partial \left(\mathbf{H}^S + \mathbf{H}^F\right)}{\partial \mathbf{F}^S}\right]^{\overset{12}{T}} - \Lambda^\varepsilon \left[\frac{\partial \left(\mathbf{Q}^S + \mathbf{Q}^F\right)}{\partial \mathbf{F}^S}\right]^{\overset{12}{T}} - \Lambda^n \left(\frac{\partial \mathbf{J}}{\partial \mathbf{F}^S}\right)^{\overset{12}{T}} \right.$$

$$+ \rho^F \Lambda^F \mathbf{F}^{S-T} \otimes \acute{\mathbf{X}}^F + \rho^F \left(\frac{\partial \eta^F}{\partial \mathbf{F}^S} - \Lambda^\varepsilon \frac{\partial \varepsilon^F}{\partial \mathbf{F}^S}\right) \otimes \acute{\mathbf{X}}^F \qquad (A.6)$$

$$\left. + \left[\left(\frac{\partial \mathbf{P}^S}{\partial \mathbf{F}^S}\right)^{\overset{14}{T}} \mathbf{\Lambda}^{v^S}\right]^{\overset{12}{T}} + \left[\left(\frac{\partial \mathbf{P}^F}{\partial \mathbf{F}^S}\right)^{\overset{14}{T}} \mathbf{\Lambda}^{v^F}\right]^{\overset{12}{T}} \right\} = 0,$$

$$\frac{\partial \left(\mathbf{H}^S + \mathbf{H}^F\right)}{\partial n} - \Lambda^\varepsilon \frac{\partial \left(\mathbf{Q}^S + \mathbf{Q}^F\right)}{\partial n} - \Lambda^n \frac{\partial \mathbf{J}}{\partial n}$$

$$+ \rho^F \acute{\mathbf{X}}^F \left(\frac{\partial \eta^F}{\partial n} - \Lambda^\varepsilon \frac{\partial \varepsilon^F}{\partial n}\right) + \left(\frac{\partial \mathbf{P}^S}{\partial n}\right)^T \mathbf{\Lambda}^{v^S} + \left(\frac{\partial \mathbf{P}^F}{\partial n}\right)^T \mathbf{\Lambda}^{v^F} = 0,$$

$$\underset{\text{sym}}{} \left\{ \frac{\partial \left(\mathbf{H}^S + \mathbf{H}^F\right)}{\partial \mathbf{G}} - \Lambda^\varepsilon \frac{\partial \left(\mathbf{Q}^S + \mathbf{Q}^F\right)}{\partial \mathbf{G}} - \Lambda^n \frac{\partial \mathbf{J}}{\partial \mathbf{G}} \right.$$

$$\left. + \rho^F \left(\frac{\partial \eta^F}{\partial \mathbf{G}} - \Lambda^\varepsilon \frac{\partial \varepsilon^F}{\partial \mathbf{G}}\right) \otimes \acute{\mathbf{X}}^F + \left(\frac{\partial \mathbf{P}^S}{\partial \mathbf{G}}\right)^{\overset{13}{T}} \mathbf{\Lambda}^{v^S} + \left(\frac{\partial \mathbf{P}^F}{\partial \mathbf{G}}\right)^{\overset{13}{T}} \mathbf{\Lambda}^{v^F} \right\} = 0,$$

as well as implicit relations for partial stresses

$$
\Lambda^\varepsilon \mathbf{P}^S = \mathbf{F}^{S-T} \left\{ \left[\frac{\partial \left(\mathbf{H}^S + \mathbf{H}^F \right)}{\partial \acute{\mathbf{X}}^F} \right]^{\overset{12}{T}} - \Lambda^\varepsilon \left[\frac{\partial \left(\mathbf{Q}^S + \mathbf{Q}^F \right)}{\partial \acute{\mathbf{X}}^F} \right]^{\overset{12}{T}} \right.
$$

$$
- \Lambda^n \left(\frac{\partial \mathbf{J}}{\partial \acute{\mathbf{X}}^F} \right)^{\overset{12}{T}} - \rho^F \Lambda^F \mathbf{1} + \rho^F \left(\frac{\partial \eta^F}{\partial \acute{\mathbf{X}}^F} - \Lambda^\varepsilon \frac{\partial \varepsilon^F}{\partial \acute{\mathbf{X}}^F} \right) \otimes \acute{\mathbf{X}}^F \quad (A.7)
$$

$$
\left. + \left(\frac{\partial \mathbf{P}^S}{\partial \acute{\mathbf{X}}^F} \right)^{\overset{13}{T}} \boldsymbol{\Lambda}^{vS} + \left(\frac{\partial \mathbf{P}^F}{\partial \acute{\mathbf{X}}^F} \right)^{\overset{13}{T}} \boldsymbol{\Lambda}^{vF} \right\} - \boldsymbol{\Lambda}^F,
$$

$$
\Lambda^\varepsilon \mathbf{P}^F =
$$

$$
= \mathbf{F}^{S-T} \left\{ \rho^F \Lambda^F \mathbf{1} - \left[\frac{\partial \left(\mathbf{H}^S + \mathbf{H}^F \right)}{\partial \acute{\mathbf{X}}^F} \right]^{\overset{12}{T}} + \Lambda^\varepsilon \left[\frac{\partial \left(\mathbf{Q}^S + \mathbf{Q}^F \right)}{\partial \acute{\mathbf{X}}^F} \right]^{\overset{12}{T}} \right.
$$

$$
+ \Lambda^n \left(\frac{\partial \mathbf{J}}{\partial \acute{\mathbf{X}}^F} \right)^{\overset{12}{T}} - \rho^F \left(\frac{\partial \eta^F}{\partial \acute{\mathbf{X}}^F} - \Lambda^\varepsilon \frac{\partial \varepsilon^F}{\partial \acute{\mathbf{X}}^F} \right) \otimes \acute{\mathbf{X}}^F \quad (A.8)
$$

$$
\left. - \left(\frac{\partial \mathbf{P}^S}{\partial \acute{\mathbf{X}}^F} \right)^{\overset{13}{T}} \boldsymbol{\Lambda}^{vS} - \left(\frac{\partial \mathbf{P}^F}{\partial \acute{\mathbf{X}}^F} \right)^{\overset{13}{T}} \boldsymbol{\Lambda}^{vF} \right\},
$$

in the case of coefficients of spatial derivatives. There remains the residual inequality which determines the dissipation D:

$$
D := \left\{ \frac{\partial \left(\mathbf{H}^S + \mathbf{H}^F \right)}{\partial \Theta} - \Lambda^\varepsilon \frac{\partial \left(\mathbf{Q}^S + \mathbf{Q}^F \right)}{\partial \Theta} - \Lambda^n \frac{\partial \mathbf{J}}{\partial \Theta} \right.
$$

$$
+ \rho^F \acute{\mathbf{X}}^F \left(\frac{\partial \eta^F}{\partial \Theta} - \Lambda^\varepsilon \frac{\partial \varepsilon^F}{\partial \Theta} \right) + \left(\frac{\partial \mathbf{P}^S}{\partial \Theta} \right)^T \boldsymbol{\Lambda}^{vS} + \left(\frac{\partial \mathbf{P}^F}{\partial \Theta} \right)^T \boldsymbol{\Lambda}^{vF} \left. \right\} \cdot \mathbf{G}
$$

$$
+ \hat{\rho}^S \left\{ \eta^S + \Lambda^S - \Lambda^\varepsilon \left(\varepsilon^S - \frac{1}{2} \acute{x}^{S2} \right) - \eta^F - \Lambda^F + \Lambda^\varepsilon \left(\varepsilon^F - \frac{1}{2} \acute{x}^{F2} \right) \right\}
$$

$$
+ \left(\boldsymbol{\Lambda}^{vS} - \boldsymbol{\Lambda}^{vF} \right) \cdot \left(\hat{\mathbf{p}}^S - \hat{\rho}^S \acute{\mathbf{x}}^S \right) + \boldsymbol{\Lambda}^{vF} \cdot \hat{\rho}^S \left(\acute{\mathbf{x}}^F - \acute{\mathbf{x}}^S \right)
$$

$$
+ \Lambda^\varepsilon \hat{\mathbf{p}}^S \cdot \left(\acute{\mathbf{x}}^F - \acute{\mathbf{x}}^S \right) + \Lambda^n \hat{n} \geq 0. \quad (A.9)
$$

We have made use of local balance equations for sources.

Apparently the dissipation D has its minimum in the state in which it is zero. Such a state is called the state of thermodynamical equilibrium. It appears if all three sources $\hat{\rho}^S, \hat{\mathbf{p}}^S, \hat{n}$ and the temperature gradient $\mathbf{G}$ are zero. For sufficiently smooth constitutive relations, it means that D

must be at least of the second order with respect to deviations from this state, and these are described by the gradient $\mathbf{G}$ again, the difference of chemical potentials of both components whose prototype appears as a co-efficient of the mass source in (A.9), the relative velocity represented by the Lagrangian velocity $\acute{\mathbf{X}}^F \equiv \mathbf{F}^{S-1}\left(\acute{\mathbf{x}}^F - \acute{\mathbf{x}}^S\right)$ and by the deviation of the porosity n from its equilibrium value n_E:

$$\Delta := n - n_E, \quad n_E = n_E\left(\mathcal{C}_E\right), \quad \mathcal{C}_E := \left\{\rho^F, \rho^S, \mathbf{F}^S, \Theta\right\}. \tag{A.10}$$

The constitutive relation for the equilibrium porosity n_E contains solely those constitutive variables whose values in the state of equilibrium are different from zero.

This structure indicates simplifications for processes with small deviations from the state of thermodynamical equilibrium.

Substitution of the definition (A.10) in the balance equation of porosity $(8.12)_7$ yields

$$\frac{\partial n_E}{\partial t} = 0, \quad \frac{\partial \Delta}{\partial t} + \operatorname{Div} \mathbf{J} = \hat{n}, \tag{A.11}$$

provided the deviation Δ is small. We have used the fact that the source of porosity must vanish in the state of thermodynamical equilibrium.

Bearing the constitutive relation for n_E in mind we obtain

$$\left(\frac{\partial n_E}{\partial \rho^S} - \frac{\partial n_E}{\partial \rho^F}\right)\hat{\rho}^S + \frac{\partial n_E}{\partial \mathbf{F}^S} \cdot \frac{\partial \mathbf{F}^S}{\partial t} + \frac{\partial n_E}{\partial \Theta}\frac{\partial \Theta}{\partial t} = 0, \tag{A.12}$$

where we have applied mass balance equations in the case $\acute{\mathbf{X}}^F = 0$ (equilibrium!).

The first term in this relation vanishes in equilibrium and, consequently, n_E can be an arbitrary function of mass densities. On the other hand, neither $\frac{\partial \mathbf{F}^S}{\partial t}$ nor $\frac{\partial \Theta}{\partial t}$ are identically zero in equilibrium and they may have locally arbitrary values. Consequently their coefficients, being independent of those derivatives, must be identically zero and we finally obtain

$$n_E = n_E\left(\rho^S, \rho^F\right). \tag{A.13}$$

Dimensional analysis leads then to the conclusion that the equilibrium porosity n_E is solely a function of the fraction $\frac{\rho^F}{\rho^S}$.

Let us return to the problem of deviations from equilibrium. By means of the definition (A.9) we can specify the assumption that these deviations are small:

1. All constitutive quantities appearing in field equations must be at most linear functions of the constitutive variables $\mathbf{G}$, $\acute{\mathbf{X}}^F$ and Δ.

2. The dissipation D may contain at most quadratic contributions of the constitutive variables $\mathbf{G}$, $\acute{\mathbf{X}}^F$ and Δ. Partial energies and entropies do

not depend on $\mathbf{G}$ and $\acute{\mathbf{X}}^F$, but they may contain a quadratic contribution of Δ. This exception is related to the structure of dissipation due to the relaxation of porosity.

We use also the assumption that the system is isotropic.

Inspection of the dissipation inequality shows immediately that under these assumptions both $\hat{n}$ and Λ^n must be linear homogeneous functions of Δ, $i.e.$,

$$\hat{n} = -\frac{\Delta}{\tau}, \quad \Lambda^n = \Lambda_1^n \Delta, \quad \tau = \tau\left(\mathcal{C}_E\right), \quad \Lambda_1^n = \Lambda_1^n\left(\mathcal{C}_E\right). \tag{A.14}$$

The form of these relations has been chosen for convenience in the further analysis.

Simultaneously, due to relations (A.5), we obtain

$$\Lambda^{v^S} = \Lambda^{v^F} = 0. \tag{A.15}$$

Bearing relations $(A.5)_4$ and $(A.14)_2$ in mind we see that the contribution of Δ to energies and entropies must be quadratic, and such that multipliers Λ^S, Λ^F and Λ^ε are independent of Δ.

The above assumption yields as well the following representations for vector fluxes

$$\mathbf{Q}^S + \mathbf{Q}^F = -K_\Theta\left(\mathcal{C}_E\right)\mathbf{G} + K_v\left(\mathcal{C}_E\right)\acute{\mathbf{X}}^F,$$

$$\mathbf{J} = \Phi_\Theta\left(\mathcal{C}_E\right)\mathbf{G} + \Phi_v\left(\mathcal{C}_E\right)\acute{\mathbf{X}}^F,$$

$$\mathbf{H}^S + \mathbf{H}^F = -H_\Theta\left(\mathcal{C}_E\right)\mathbf{G} + H_v\left(\mathcal{C}_E\right)\acute{\mathbf{X}}^F. \tag{A.16}$$

Substitution of these relations in (A.6) yields the following set of identities:

1.

$$\rho^F\left(\frac{\partial \eta^F}{\partial \rho^S} - \Lambda^\varepsilon \frac{\partial \varepsilon^F}{\partial \rho^S}\right) = -\frac{\partial H_v}{\partial \rho^S} + \Lambda^\varepsilon \frac{\partial K_v}{\partial \rho^S} + \Lambda_1^n \Delta \frac{\partial \Phi_v}{\partial \rho^S},$$

$$-\frac{\partial H_\Theta}{\partial \rho^S} + \Lambda^\varepsilon \frac{\partial K_\Theta}{\partial \rho^S} - \Lambda_1^n \Delta \frac{\partial \Phi_\Theta}{\partial \rho^S} = 0, \tag{A.16$_1$}$$

2.

$$-\rho^S\left(\frac{\partial \eta^S}{\partial \rho^F} - \Lambda^\varepsilon \frac{\partial \varepsilon^S}{\partial \rho^F}\right) = -\frac{\partial H_v}{\partial \rho^F} + \Lambda^\varepsilon \frac{\partial K_v}{\partial \rho^F} + \Lambda_1^n \Delta \frac{\partial \Phi_v}{\partial \rho^F},$$

$$-\frac{\partial H_\Theta}{\partial \rho^F} + \Lambda^\varepsilon \frac{\partial K_\Theta}{\partial \rho^F} - \Lambda_1^n \Delta \frac{\partial \Phi_\Theta}{\partial \rho^F} = 0, \tag{A.16$_2$}$$

238   K. Wilmański

3.

$$\operatorname*{sym}^{23}\left\{\rho^F\left(\frac{\partial\eta^F}{\partial\mathbf{F}^S}-\Lambda^\varepsilon\frac{\partial\varepsilon^F}{\partial\mathbf{F}^S}+\Lambda^F\mathbf{F}^{S-T}\right)\right.$$

$$-\left(H_v-\Lambda^\varepsilon K_v-\Lambda_1^n\Delta\Phi_v\right)\mathbf{F}^{S-T}$$

$$\left.+\left(\frac{\partial H_v}{\partial\mathbf{F}^S}-\Lambda^\varepsilon\frac{\partial K_v}{\partial\mathbf{F}^S}-\Lambda_1^n\Delta\frac{\partial\Phi_v}{\partial\mathbf{F}^S}\right)\right\}\otimes\acute{\mathbf{X}}^F \qquad (A.16_3)$$

$$+\operatorname*{sym}^{23}\left\{-\frac{\partial H_\Theta}{\partial\mathbf{F}^S}+\Lambda^\varepsilon\frac{\partial K_\Theta}{\partial\mathbf{F}^S}-\Lambda_1^n\Delta\frac{\partial\Phi_\Theta}{\partial\mathbf{F}^S}\right\}\otimes\mathbf{G}=0\ ,$$

4.

$$-H_\Theta+\Lambda^\varepsilon K_\Theta-\Lambda_1^n\Delta\Phi_\Theta=0; \qquad (A.16_4)$$

with the relation containing derivatives with respect to n of thermal components of fluxes identically satisfied. According to the above assumption we account only for the first order contributions with respect to $\acute{\mathbf{X}}^F$, $\mathbf{G}$ and Δ. This concerns also identities following from $(A.16_3)$ after simplifying with respect to $\acute{\mathbf{X}}^F$ and $\mathbf{G}$.

The relations for stresses become now true constitutive relations

$$\Lambda^\varepsilon\mathbf{P}^S=-\Lambda^F-\left(\rho^F\Lambda^F-H_v+\Lambda^\varepsilon K_v+\Delta\Lambda_1^n\Phi_v\right)\mathbf{F}^{S-T},$$
$$\Lambda^\varepsilon\mathbf{P}^F=\left(\rho^F\Lambda^F-H_v+\Lambda^\varepsilon K_v+\Delta\Lambda_1^n\Phi_v\right)\mathbf{F}^{S-T} \qquad (A.17)$$

and the residual inequality has the form

$$\left[\rho^F\left(\frac{\partial\eta^F}{\partial\Theta}-\Lambda^\varepsilon\frac{\partial\varepsilon^F}{\partial\Theta}\right)+\left(\frac{\partial H_v}{\partial\Theta}-\Lambda^\varepsilon\frac{\partial K_v}{\partial\Theta}-\Lambda_1^n\Delta\frac{\partial\Phi_v}{\partial\Theta}\right)\right]\acute{\mathbf{X}}^F\cdot\mathbf{G}$$

$$-\left(\frac{\partial H_\Theta}{\partial\Theta}-\Lambda^\varepsilon\frac{\partial K_\Theta}{\partial\Theta}+\Lambda_1^n\Delta\frac{\partial\Phi_\Theta}{\partial\Theta}\right)\mathbf{G}\cdot\mathbf{G}+\Lambda^\varepsilon\left(\hat{\mathbf{p}}^S-\hat{\rho}^S\acute{\mathbf{x}}^S\right)\cdot\left(\acute{\mathbf{x}}^F-\acute{\mathbf{x}}^S\right)$$

$$-\frac{1}{\tau}\Lambda_1^n\Delta^2+\hat{\rho}^S\left[\left(\eta^S+\Lambda^S-\Lambda^\varepsilon\varepsilon^S\right)-\left(\eta^F+\Lambda^F-\Lambda^\varepsilon\varepsilon^F\right)\right.$$

$$\left.-\frac{1}{2}\Lambda^\varepsilon\left(\acute{\mathbf{x}}^F-\acute{\mathbf{x}}^S\right)\cdot\left(\acute{\mathbf{x}}^F-\acute{\mathbf{x}}^S\right)\right]\geq 0. \qquad (A.18)$$

It is clear that the approximation made above cannot admit mixed terms of this form as the first contribution to the above inequality. Hence we have in addition

$$\rho^F\left(\frac{\partial\eta^F}{\partial\Theta}-\Lambda^\varepsilon\frac{\partial\varepsilon^F}{\partial\Theta}\right)+\left(\frac{\partial H_v}{\partial\Theta}-\Lambda^\varepsilon\frac{\partial K_v}{\partial\Theta}-\Lambda_1^n\Delta\frac{\partial\Phi_v}{\partial\Theta}\right)=0. \qquad (A.19)$$

The relation $(A.16_4)$ indicates that $\Phi_\Theta=0$. The same relation together with $(A.16_{1,2,3})$ yields that Λ^ε depends solely on the temperature Θ. Then the classical argument for the state of equilibrium gives

$$\Lambda^\varepsilon=\frac{1}{\Theta}. \qquad (A.20)$$

The structure of the multiplier Λ^n given by (A.14) indicates that the partial energies and entropies may contain solely terms independent of Δ or quadratic in Δ.

Now it is convenient to introduce the notation

$$\psi^F = \varepsilon^F - \Theta\eta^F, \quad \psi^S = \varepsilon^S - \Theta\eta^S, \tag{A.21}$$

where ψ^F and ψ^S are, of course, the prototypes of Helmholtz free energies. These are constitutive quantities as well. We substitute them in the above relations after presenting another simplification.

Namely it is easy to show that it is compatible with the above thermodynamical structure to require the following condition to hold:

$$H_v = \frac{1}{\Theta}K_v. \tag{A.22}$$

Then

$$\frac{\partial\psi^S}{\partial\rho^F} = 0,$$

$$\frac{\partial\psi^F}{\partial\rho^S} = 0, \quad \frac{\partial\psi^F}{\partial\mathbf{F}^S} = -\rho^F\frac{\partial\psi^F}{\partial\rho^F}\mathbf{F}^{S-T},$$

$$\frac{\partial\Phi_v}{\partial\rho^S} = 0, \quad \frac{\partial\Phi_v}{\partial\rho^F} = 0, \quad \frac{\partial\Phi_v}{\partial\mathbf{F}^S} = \Phi_v\mathbf{F}^{S-T}, \quad \frac{\partial\Phi_v}{\partial\Theta} = 0,$$

$i.e.^3,$

$$\psi^F = \psi^F\left(\rho^F J^{S-1}, \Theta, \Delta\right), \quad \psi^S = \psi^S\left(\rho^S, \mathbf{F}^S, \Theta, \Delta\right),$$

$$\Phi_v = \varphi J^S, \quad \varphi = \text{const} \tag{A.23}$$

and

$$K_v = \Theta\rho^F\left(\frac{\partial\psi^F}{\partial\Theta} + \eta^F\right). \tag{A.24}$$

In thermodynamics of single component materials the right-hand side of this relation is identically zero. It may also be the case in the present model but the second law of thermodynamics does not impose this condition.

Let us summarize the results. Apart from the constitutive relations (A.23) for free energies which should be in addition quadratic with respect to Δ, we have

$$\mathbf{Q}^S + \mathbf{Q}^F = \Theta\left(\mathbf{H}^S + \mathbf{H}^F\right) = -K_\Theta\mathbf{G} + K_v\acute{\mathbf{X}}^F, \tag{A.25}$$

---

[3]The analogous result has been obtained for isothermal conditions in my earlier works (*e.g.*, [15]) under the assumption of constant equilibrium porosity n_E. This assumption is not made in this work and, consequently, the result for φ should be understood as a dependence on an initial value of the equilibrium porosity. This is a consequence of the assumption that processes deviate little from the thermodynamical equilibrium.

240   K. Wilmański

with K_v given by (A.24). Simultaneously

$$\mathbf{J} = \varphi J^S \mathbf{\acute{X}}^F, \quad \varphi = \text{const.} \tag{A.26}$$

Partial stresses are given by the relations

$$\mathbf{P}^S = \rho^S \frac{\partial \psi^S}{\partial \mathbf{F}^S} - \Theta \Lambda_1^n \varphi J^S \Delta \mathbf{F}^{S-T}, \tag{A.27}$$

$$\mathbf{P}^F = \left(-p^F + \Theta \Lambda_1^n \varphi \Delta\right) J^S \mathbf{F}^{S-T}, \quad p^F := \rho^{F2} \frac{\partial \psi^F}{\partial \rho^F} J^{S-1},$$

$$\Lambda_1^n = -\frac{1}{\Theta} \left(\rho^S \frac{\partial \psi^S}{\partial \Delta} + \rho^F \frac{\partial \psi^F}{\partial \Delta} \right). \tag{A.28}$$

The residual inequality defining the dissipation has the form

$$\frac{1}{\Theta} K_\Theta \mathbf{G} \cdot \mathbf{G} + \left(\hat{\mathbf{p}}^S - \hat{\rho}^S \mathbf{\acute{x}}^S\right) \cdot \left(\mathbf{\acute{x}}^F - \mathbf{\acute{x}}^S\right) + \left(\rho^S \frac{\partial \psi^S}{\partial \Delta} + \rho^F \frac{\partial \psi^F}{\partial \Delta} \right) \frac{1}{\tau} \Delta^2$$

$$+ \hat{\rho}^S \left(\psi^F + \frac{p^F}{\rho^F J^{S-1}} - \psi^S - \rho^S \frac{\partial \psi^S}{\partial \rho^S} \right) \geq 0, \tag{A.29}$$

where due to the assumption on small deviations from equilibrium the quadratic contribution of the relative velocity to the mass source was neglected.

As usual in linear non-equilibrium thermodynamics we assume sources to be proportional to their conjugated forces and, consequently, due to the isotropy we have

$$\left(\hat{\mathbf{p}}^S - \hat{\rho}^S \mathbf{\acute{x}}^S\right) = \pi \left(\mathbf{\acute{x}}^F - \mathbf{\acute{x}}^S\right),$$

$$\hat{\rho}^S = R \left(\psi^F + \frac{p^F}{\rho^F J^{S-1}} - \psi^S - \rho^S \frac{\partial \psi^S}{\partial \rho^S} \right), \tag{A.30}$$

where π and R are phenomenological coefficients.

This completes consideration of the local thermodynamical admissibility.

Acknowledgments: This paper was written during my visiting professorship with Professor Gianfranco Capriz of Università di Pisa, Department of Mathematics in October 2000. I would like to express my gratitude to the National Group of Mathematical Physics (Italy) for the grant for this stay and to CNR di Pisa San Cataldo, and particularly to Eng. Maurizio Brocato, for their hospitality.

References

[1] B. ALBERS: *Makroskopische Beschreibung von Adsorptions-Diffusions-Vorgängen in Porösen Körpern*. Ph.D. thesis, TU-Berlin, Logos-Verlag, Berlin, 2000.

[2] B. ALBERS: Coupling of adsorption and diffusion in porous and granular materials. A 1-D example of the boundary value problem, *Arch. Appl. Mech.* **70**, pp. 519–531, 2000.

[3] B. ALBERS: Pollution transport as example for adsorption/diffusion processes in porous materials, in: P.E. O'DONOGHUE, J.N. FLAVIN (eds.): *Symposium on Trends in the Applications of Mathematics to Mechanics*, Elsevier, 2000, pp. 27–34.

[4] B. ALBERS: On adsorption and diffusion in porous media, *ZAMM* **81**, pp. 683–690, 2001.

[5] P.B. BUTLER, M.F. LEMBECK, H. KRIER: Modeling of shock development and transition to detonation initiated by burning in porous propellant beds, *Combust. Flame* **46**, pp. 75–93, 1982.

[6] M.A. GOODMAN, S.C. COWIN: A continuum theory of granular materials, *Arch. Rat. Mech, Anal.* **44**, pp. 249–266, 1972.

[7] W. KEMPA: On the description of the consolidation phenomenon by means of a two-component continuum, *Arch. Mech.* **49**, pp. 893–917, 1997.

[8] I. MÜLLER, *Thermodynamics*, Pitman, New York, 1985.

[9] K. WILMAŃSKI: Lagrangian model of two-phase porous material, *J. Non-Equilibrium Thermodyn.* **20**, pp. 50–77, 1995.

[10] K. WILMAŃSKI: Porous media at finite strains. The new model with the balance equation of porosity, *Arch. Mech.* **48**, pp. 591–628, 1996.

[11] K. WILMAŃSKI: A thermodynamic model of compressible porous materials with the balance equation of porosity, *Transp. Porous Media* **32**, pp. 21–47, 1998.

[12] K. WILMAŃSKI, *Thermomechanics of Continua*, Springer, Heidelberg, 1998.

[13] K. WILMAŃSKI: Toward an extended thermodynamics of porous and granular materials, in: G. IOOSS, O. GUÈS, A. NOURI (eds.): *Trends in Applications of Mathematics to Mechanics*, Monographs and Surveys in Pure and Applied Mathematics, Chapman & Hall/CRC, Boca Raton **106**, pp. 147–160, 2000.

[14] K. WILMAŃSKI: Mathematical theory of porous media – Lecture notes of XXV summer school on mathematical physics, Ravello 2000, *WIAS-Preprint* **602**, 2000.

[15] K. WILMAŃSKI: Thermodynamics of multicomponent continua, in: J. MAJEWSKI, R. TEISSEYRE (eds.): *Earthquake Thermodynamics and Phase Transformations in the Earth's Interior*, Academic Press, San Diego, 2001, pp. 567–654.

[16] K. WILMAŃSKI: On a homogeneous adsorption in porous materials, *ZAMM* **81**, pp. 119–124, 2001.

KRZYSZTOF WILMAŃSKI
Weierstrass Institute for Applied Analysis and Stochastics
Mohrenstrasse, 39
D-10117 Berlin, GERMANY

E-mail: Wilmanski@wias-berlin.de

Part III

Numerical Simulations

Chapter 9

Continuum and Numerical Simulation of Porous Materials in Science and Technology

Wolfgang Ehlers

ABSTRACT Continuum mechanics of porous materials touches all kinds of problems arising from the necessity to successfully describe the behaviour of geomaterials such as saturated, partially saturated or empty porous solids. Geomaterials as well as further porous media like concrete, sinter materials, polymeric and metallic foams, living tissues, etc., basically fall into the category of multiphasic materials, which can be described within the framework of a macroscopic continuum mechanical approach by use of the well-founded theory of porous media (TPM).
Based on the concept of non-polar and micropolar materials, the present contribution outlines the continuum mechanical foundations of the TPM including the necessary set of constitutive equations for the description of elasto-plastic and elasto-viscoplastic frictional geomaterials with or without a viscous pore content. Furthermore, the discretization of the governing field equations and the basic numerical tools including time- and space-adaptive strategies are presented and included into the finite element tool PANDAS. Finally, a number of numerical examples exhibits the wide range of applications of the TPM approach to saturated and partially saturated problems like fluid flow in porous media, the consolidation problem and localization phenomena like the well-known biaxial experiment or the base failure problem.

9.1   Introduction

Basically, saturated, partially saturated and empty porous solid materials as for instance geomaterials like soil, rock or concrete as well as sinter materials, polymeric and metallic foams, living tissues, *etc.*, can be described by use of two generally different strategies, the micromechanical and the macromechanical approaches. Proceeding from the micromechanical or the microscopic approach, respectively, the behaviour of the individual microaggregates like, *e.g.*, the grains of granular soils is described on their own domain taking into account both the grain interaction and, at the internal boundaries between the different materials, the interaction of the soil with the pore-fluids (*e.g.*, pore water and pore-gas). On the other hand, the macromechanical or the macroscopic approach, respectively, either proceeds from a real or from a virtual averaging process, where both the porous solid and the pore-fluids are defined throughout the whole domain in the sense of superimposed continua.

In the present article, the macroscopic approach is directly applied, thus assuming the *a priori* existence of homogenized substructures. This procedure is known to be embedded in the well-founded framework of the theory of porous media (TPM), which is based on two major columns, the classical theory of mixtures (theory of heterogeneously composed continua with internal constraints) and the concept of volume fractions. The fundamentals of the TPM including the development of material models, the comparison of the TPM with other approaches to multiphasic materials, *e.g.*, *Biot*'s approach or general averaging theories, as well as recent developments can be taken from the relevant literature. The interested reader is referred, for instance, to the work by Biot [1, 2], Bowen [4, 5, 6], Coussy [8], de Boer [10, 11], de Boer & Ehlers [12, 13], Ehlers [19, 20, 22], Ehlers & Kubik [29], Hassanizadeh & Gray [40, 41] or Truesdell & Toupin [55].

However, apart from the general necessity of computing strongly coupled solid-fluid problems in geomechanical engineering like, *e.g.*, the consolidation problem under saturated and unsaturated conditions, there is a further considerable interest in the investigation of shear band localization phenomena occurring in fluid-saturated as well as in partially saturated or in empty porous materials. As a consequence of localization phenomena like, *e.g.*, the onset of shear bands during the well-known biaxial experiment, the numerical solution of the governing field equations generally reveals an ill-posed problem, when, *e.g.*, in the case of an empty skeleton material, the type of the governing differential equation changes from elliptic to hyperbolic. In the numerical computations, this is seen by the fact that the shear band width strongly depends on the finite element discretization through the mesh size and the mesh orientation. For example, each mesh refinement leads to a decrease of the shear band width until one obtains (ideally) a singular surface. To overcome this unphysical behaviour, different regularization strategies can be taken into consideration. In the case of single phase materials, extended continuum formulations with internal length scales are widely used, as there are for instance micropolar continua in the sense of the *Cosserat* brothers, general non-local continua or strain-gradient models. For a broad review of these formulations, the reader is referred to the work by Eringen [36], Ehlers & Volk [33, 34], de Borst [15], Steinmann[52] and Volk [57]. An excellent survey on the different strategies can be found, *e.g.*, in the Ph.D. thesis by Brinkgreve [7].

In contrast to a variation of the basic formulation of the continuum model, variations of the material properties towards the inclusion of viscosity effects are also known to lead to a convenient regularization of the shear band problem. In the saturated case of porous materials, there occurs the possibility to furthermore use the naturally included fluid viscosity as an additional regularization tool [34, 50]. However, it was pointed out by Ehlers & Volk [34] that the fluid viscosity effect is not dominant compared, *e.g.*, to the effect of skeleton micropolarity. Restricting the consideration to saturated and empty micropolar elasto-plastic frictional porous solid materials, it was furthermore shown in [34, 57] that the additional micropolar degrees of freedom are generally inactive except for the localization zones, where, in the case of granular soils, the grains are rolling upon each other. Following this, micropolarity is a regularization tool which, on the one hand, describes the natural behaviour of granular materials and, on the other hand, does not affect the basic material properties in the non-localized zones. Furthermore, the physical shear band width was shown to be governed by the included *Cosserat* parameters, *e.g.*, through the internal length scale. By use of viscoplasticity or fluid viscosity as a regularization tool, the shear band width correlates with the included viscosity parameters and can thus only be fitted to experimental results by changing the basic material properties through a variation of these values. This procedure, however it regularizes conveniently, often leads to an unphysical description of the material response.

The present contribution concentrates on two major topics, namely, the continuum mechanical and the numerical treatment of porous materials with respect to both scientific and technological purposes. After some preliminary remarks on the motivation and the goal of the presentation, the continuum mechanical foundation outlines, in the framework of the purely mechanical approach, a general macroscopic description of triphasic materials consisting of a deformable soil matrix saturated by two pore-fluids, the pore-water and the pore-gas. This model is appropriate to describe unsaturated soils as well as saturated material by reducing the triphasic model to a binary one of soil and pore-water. Of course, the empty material is naturally included into the description by either proceeding from a binary model of soil and pore-gas or, if the mechanical effect of the pore-gas is negligible, by simply disregarding the pore content. In addition, the soil skeleton is treated both as a standard continuum and as a *Cosserat* material. Once the continuum mechanical framework is given, the constitutive setting of saturated and unsaturated soil is outlined in the frame of the geometrically linear approach, where the soil skeleton is described as an elasto-plastic or an elasto-viscoplastic cohesive-frictional material, whereas the pore-content is treated in the sense of viscous pore-fluids. Furthermore, since the compressibility of the matrix material (material compressibility) is generally much smaller than the compressibility of the porous bulk material (bulk compressibility), the skeleton is assumed to be materially incompressible. Proceeding from non-polar materials (standard formulation), the elastic response of the soil is described by an elasticity law of *Hooke*an type relating the symmetric stress tensor to the elastic strains. In addition, the plastic or the viscoplastic response is based on a convenient yield condition to bound the elastic domain together with a non-associated flow rule for the plastic strain rates. In contrast to the usual metal plasticity models, the yield condition of frictional materials is governed, in the deviatoric plane of the principal stress space, by a triangular shape with rounded corners and, in the hydrostatic plane, by a closed and non-symmetric drop-like shape [20, 21]. In addition to the yield function, a plastic potential is introduced to govern the non-associated flow rule. However, when micropolarity comes into play (extended micropolar formulation), the stress and strain tensors are no longer symmetric. Furthermore, the set of constitutive equations has to be extended by an additional elasticity law for the couple stress tensor and a flow rule for the plastic rate of curvature tensor, where, in this context, it is important to note that the flow rule for the plastic rate of curvature tensor is not independent of the flow rule for the plastic strain rate, since these quantities are coupled through the micropolar compatibility condition [34].

Apart from the continuum mechanical description of saturated and empty elasto-plastic and elasto-viscoplastic continua, the numerical solution of initial boundary-value problems reveals that both the *a priori* time step size in the frame of an implicit integration scheme as well as the *a priori* mesh size in the frame of a finite element discretization may not be appropriate to describe shear band phenomena accurately. Following this, time-adaptive as well as space-adaptive strategies are applied. In particular, singly diagonally implicit *Runge–Kutta* (SDIRK) methods are used in the time domain, whereas the space domain is governed by both remeshing and hierarchical adaptive schemes. The basic procedure used in the present article is based on one-step time integration methods with an embedded time step control, where the solution at time t_{n+1} only depends on the solution at time t_n. This choice is of essential importance with respect to space-adaptive methods (refinements as well as coarsenings), since the transfer of the numerical solution thus only includes two meshes. Concerning details of this procedure, the reader is referred to [18, 26, 35]. In the space domain, h-adaptive methods are applied and based on a modification of the *Zienkiewicz–Zhu* error indicator [58]. In the case of saturated nonpolar elasto-plastic skeleton materials, smoothed values of the L_2-norms of characteristic quantities are compared to the respective discrete values, thus leading to a convenient and successfully applicable method to refine and to coarsen the local mesh size [23, 26, 27]. In particular, the effective solid stresses are considered to represent the elastic behaviour, the plastic part of the strain state is considered to represent the accumulated plasticity and the seepage velocities of the liquid and gas constituents are additionally considered to represent the pore-fluid viscosities. When micropolarity comes into play, it seems not to be necessary to either change or extend the set of mechanical quantities incorporated into the error indicator. This statement can easily be understood by consideration of the micropolar compatibility condition [34] exhibiting the curvature tensor of micropolar materials not to be an independent quantity but a unique function of the *Cosserat* strain tensor which itself contains the stretch and the continuum rotation as well as the additional independent micropolar rotation. As a result, it is concluded that, firstly, the plastic curvature tensor of micropolar materials is not independent from the plastic *Cosserat* strain and, secondly, that the micropolar couple stresses can be given as a function of the effective solid stresses if one considers the constitutive equations of both quantities together with the micropolar compatibility condition.

Based on the above procedure, the article finally contains a lot of numerical examples of strongly coupled solid-fluid problems like the investigation of the pore-fluid flow in different soil situations as well as the saturated and the unsaturated consolidation and different localization problems. The examples, computed with the aid of the finite element tool PANDAS[1], are chosen carefully to not only demonstrate the validity of the presented approach, but also to exhibit the convenience of the presented numerical tools. The contribution closes with some concluding remarks summarizing the main results and presenting an outlook on further work in the field.

9.2 Governing equations

9.2.1 Kinematics

By use of the theory of porous media (TPM), fully saturated, partially saturated and empty porous solid skeleton materials are considered as a mixture of k immiscible constituents φ^α with particles X^α ($\alpha = S$: solid skeleton; $\alpha = \beta$: $k - 1$ pore-fluids). Following this, partially saturated and empty skeleton materials are treated as special cases of saturated media, where, in the partially saturated case, the pore content is assumed to consist of a pore-liquid (pore-water) and a pore-gas (air) and, in the empty case, the pore content is either assumed to consist of an overall pore-gas or, if the mechanical effect is of minor importance, it may also be neglected. In general, each spatial point $\mathbf{x}$ of the current configuration is at any time t simultaneously occupied by material points X^α of all constituents φ^α (superimposed continua). These particles proceed from different reference positions $\mathbf{X}_\alpha$ at time t_0, *cf.* Figure 9.1. Thus, each constituent is assigned an individual motion function

$$\mathbf{x} = \chi_\alpha(\mathbf{X}_\alpha, t). \tag{9.1}$$

The volume fractions $n^\alpha = n^\alpha(\mathbf{x}, t)$ are defined as the local ratios of the

---

[1]Porous media Adaptive Nonlinear finite element solver based on Differential Algebraic Systems

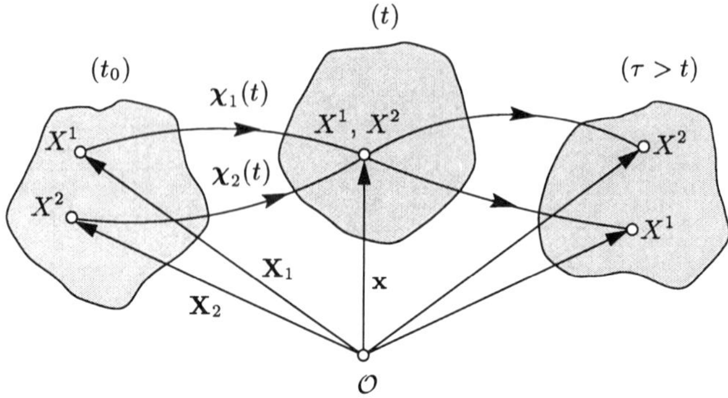

FIGURE 9.1. Motion of a binary mixture.

constituent volumes v^α with respect to the bulk volume v. Thus, in the absence of any vacant space, the saturation condition yields

$$\sum_{\alpha=1}^{k} n^\alpha = 1. \tag{9.2}$$

Associated with each constituent φ^α is an effective (realistic) or material density $\rho^{\alpha R}$ and a partial or bulk density ρ^α. The effective density $\rho^{\alpha R}$ defines the local mass of φ^α per unit of v^α, whereas the bulk density ρ^α exhibits the same mass per unit of v. Thus, the density functions are related by

$$\rho^\alpha = n^\alpha \rho^{\alpha R}. \tag{9.3}$$

Proceeding from (9.3), it is obvious that the property of material incompressibility of any constituent φ^α (defined by $\rho^{\alpha R} = \text{const.}$) is not equivalent to a global incompressibility of this constituent, since the partial density functions can still change through changes in the volume fractions n^α.

In a standard *Lagrangea*n setting, the primary kinematic variable of the skeleton material is the displacement vector

$$\mathbf{u}_S = \mathbf{x} - \mathbf{X}_S, \tag{9.4}$$

whereas the pore-fluids, in a modified *Euleri*an formulation, are governed by their seepage velocities $\mathbf{w}_\beta$ defined as the difference between the pore-fluid velocities $\mathbf{v}_\beta$ and the skeleton velocity $(\mathbf{u}_S)'_S$:

$$\mathbf{w}_\beta = \mathbf{v}_\beta - (\mathbf{u}_S)'_S. \tag{9.5}$$

Therein, $(\,\cdot\,)'_\alpha = \frac{\partial(\,\cdot\,)}{\partial t} + \text{grad}\,(\,\cdot\,)\cdot \overset{\prime}{\mathbf{x}}_\alpha$ is the material time derivative following the motion of the φ^α-th constituent.

From (9.1) and (9.4), the material deformation gradient and the displacement gradient of the solid skeleton yield

$$\mathbf{F}_S = \mathrm{Grad}_S\,\mathbf{x}\,, \qquad \mathbf{H}_S = \mathrm{Grad}_S\,\mathbf{u}_S\,. \tag{9.6}$$

In these relations, the operator "$\mathrm{Grad}_S\,(\,\cdot\,)$" defines the partial derivative of $(\,\cdot\,)$ with respect to the reference position $\mathbf{X}_S$ of φ^S. In the framework of the geometrically linear approach, the symmetric part of the displacement gradient yields the linear *Lagrange*an strain $\boldsymbol{\varepsilon}_S$, whereas the skew-symmetric part stores the information of the continuum rotation vector $\boldsymbol{\varphi}_S$. Thus,

$$
\begin{aligned}
\mathbf{H}_{S\,\mathrm{sym}} &= \tfrac{1}{2}\,(\mathbf{H}_S + \mathbf{H}_S^T) &&= \boldsymbol{\varepsilon}_S\,, \\
\mathbf{H}_{S\,\mathrm{skw}} &= \tfrac{1}{2}\,(\mathbf{H}_S - \mathbf{H}_S^T) &&= -\overset{3}{\mathbf{E}}\,\boldsymbol{\varphi}_S\,.
\end{aligned}
\tag{9.7}
$$

Therein, $\mathbf{I}$ and $\overset{3}{\mathbf{E}}$ are the fundamental tensors of second order (identity tensor) and third order (*Ricci* permutation tensor), $(\,\cdot\,)^T$ is the transpose of $(\,\cdot\,)$. Given $\mathbf{H}_{S\,\mathrm{skw}}$ as a function of $\boldsymbol{\varphi}_S$ through $(9.7)_2$, one alternatively obtains $\boldsymbol{\varphi}_S$ as the axial vector of $\mathbf{H}_{S\,\mathrm{skw}}$ via

$$\boldsymbol{\varphi}_S = -\tfrac{1}{2}\,\overset{3}{\mathbf{E}}\,(\mathbf{H}_{S\,\mathrm{skw}})\,. \tag{9.8}$$

Extending the geometrically linear description of the solid skeleton towards micropolar materials, a second primary kinematic skeleton variable occurs, namely the independent micropolar rotation $\overset{*}{\boldsymbol{\varphi}}_S$. Following this, the total average grain rotation $\bar{\boldsymbol{\varphi}}_S$ results from the sum of the continuum rotation and the additional micropolar rotation. Thus,

$$\bar{\boldsymbol{\varphi}}_S = \boldsymbol{\varphi}_S + \overset{*}{\boldsymbol{\varphi}}_S\,. \tag{9.9}$$

As a consequence of the introduction of $\bar{\boldsymbol{\varphi}}_S$, one obtains the linear *Cosserat* strain tensor $\bar{\boldsymbol{\varepsilon}}_S$ and the linear curvature tensor $\bar{\boldsymbol{\kappa}}_S$:

$$\bar{\boldsymbol{\varepsilon}}_S = \mathbf{H}_S + \overset{3}{\mathbf{E}}\,\bar{\boldsymbol{\varphi}}_S\,, \qquad \bar{\boldsymbol{\kappa}}_S = \mathrm{Grad}_S\,\bar{\boldsymbol{\varphi}}_S\,. \tag{9.10}$$

With the aid of $(9.7)_2$ and (9.9), the symmetric and skew-symmetric parts of the *Cosserat* strain yield

$$
\begin{aligned}
\bar{\boldsymbol{\varepsilon}}_{S\,\mathrm{sym}} &= \tfrac{1}{2}\,(\mathbf{H}_S + \mathbf{H}_S^T) &&= \boldsymbol{\varepsilon}_S\,, \\
\bar{\boldsymbol{\varepsilon}}_{S\,\mathrm{skw}} &= \tfrac{1}{2}\,(\mathbf{H}_S - \mathbf{H}_S^T) + \overset{3}{\mathbf{E}}\,\bar{\boldsymbol{\varphi}}_S &&= \overset{3}{\mathbf{E}}\,\overset{*}{\boldsymbol{\varphi}}_S\,.
\end{aligned}
\tag{9.11}
$$

Furthermore, the *Cosserat* strain and the linear curvature tensor are related to each other by the micropolar compatibility condition, which can be obtained from $(9.6)_2$ and (9.10) by use of the *Schwarz*ian exchangeability rule of partial derivatives:

$$\mathrm{Grad}_S\,\bar{\boldsymbol{\varepsilon}}_S - \mathrm{Grad}_S^{\overset{23}{T}}\,\bar{\boldsymbol{\varepsilon}}_S = (\overset{3}{\mathbf{E}}\,\bar{\boldsymbol{\kappa}}_S)^{\underline{3}} - (\overset{3}{\mathbf{E}}\,\bar{\boldsymbol{\kappa}}_S)^{\underline{3}\,\overset{23}{T}}\,. \tag{9.12}$$

Therein, the transpositions $(\,\cdot\,)^{\overset{ik}{T}}$ indicate an exchange of the i-th and k-th basis systems included into the tensor basis of higher order tensors. The additional superscript $(\,\cdot\,)^{\underline{n}}$ defines the included contraction $(\,\cdot\,)$ to yield a tensor of n-th order. In a simple index notation, the above compatibility condition has firstly been proposed by *Nowacki* in 1969 [48]. In general, (9.12) consists of 27 scalar equations. However, since there are only 9 independent equations incorporated into (9.12), this relation can be solved with respect to $\bar{\kappa}_S$ by use of the methods of the general tensor calculus, *cf.* Ehlers & Volk [34]. Thus,

$$\bar{\kappa}_S = \tfrac{1}{2}\,\overset{3}{\mathbf{E}}\left(\operatorname{Grad}_S \bar{\varepsilon}_S + \operatorname{Grad}_S^{\overset{13}{T}} \bar{\varepsilon}_S - \operatorname{Grad}_S^{\overset{23}{T}} \bar{\varepsilon}_S\right)\overset{2}{\,\cdot\,}\,. \tag{9.13}$$

9.2.2 Balance relations

Proceeding from the general geometrically non-linear formulation, the governing balance relations of the problem under consideration can be taken from the relevant literature on the theory of porous media, *cf.*, *e.g.*, Bowen [5, 6], de Boer & Ehlers [12] or Ehlers [20, 22]. However, under the present circumstances, the balance relations of the standard formulation of the classical TPM must be extended by the introduction of elements of multiphase micropolar theories as were discussed by Diebels & Ehlers [17] and Diebels [16]. In the framework of a purely mechanical theory of micropolar multiphasic materials, the following balance equations for mass, momentum and moment of momentum (m. o. m.) hold:

$$(\rho^\alpha)'_\alpha + \rho^\alpha \operatorname{div} \overset{\prime}{\mathbf{x}}_\alpha = \hat{\rho}^\alpha\,,$$

$$\rho^\alpha \overset{\prime\prime}{\mathbf{x}}_\alpha = \operatorname{div}\mathbf{T}^\alpha + \rho^\alpha\,\mathbf{b}^\alpha + \hat{\mathbf{p}}^\alpha\,, \tag{9.14}$$

$$\rho^\alpha (\bar{\boldsymbol{\Theta}}^\alpha \bar{\boldsymbol{\omega}}_\alpha)'_\alpha = \mathbf{I}\times\mathbf{T}^\alpha + \operatorname{div}\mathbf{M}^\alpha + \rho^\alpha\,\mathbf{c}^\alpha + \hat{\mathbf{m}}^\alpha\,.$$

In these relations, $\overset{\prime\prime}{\mathbf{x}}_\alpha$ and $\bar{\boldsymbol{\omega}}_\alpha = (\bar{\boldsymbol{\varphi}}_\alpha)'_\alpha$ are the translational acceleration and the total rotational velocity of φ^α, whereas $\operatorname{div}(\,\cdot\,)$ is the divergence operator corresponding to the spatial gradient $\operatorname{grad}(\,\cdot\,) = \frac{\partial(\,\cdot\,)}{\partial\mathbf{x}}$. Furthermore, $\mathbf{T}^\alpha$ is the non-symmetric *Cauchy* stress tensor of micropolar constituents, $\mathbf{b}^\alpha$ is the volume force per unit of φ^α-th mass, $\bar{\boldsymbol{\Theta}}^\alpha$ is the tensor of microinertia, $\mathbf{M}^\alpha$ is the couple stress tensor and $\mathbf{c}^\alpha$ is the body couple stress vector per unit of φ^α-th mass. In addition, the quantities $\hat{\rho}^\alpha$, $\hat{\mathbf{p}}^\alpha$ and $\hat{\mathbf{m}}^\alpha$ represent the density production and the direct parts of the momentum and the moment of momentum productions. For an intensive discussion of the definition of interaction or production terms in porous media theories, respectively, the reader is referred to [16, 22, 26].

The production terms included into (9.14) are restricted to the constraints

$$\sum_\alpha \quad \hat{\rho}^\alpha = 0\,,$$

$$\sum_\alpha \quad (\hat{\mathbf{p}}^\alpha + \hat{\rho}^\alpha \, \overset{'}{\mathbf{x}}_\alpha) = 0\,, \tag{9.15}$$

$$\sum_\alpha \quad [\,\hat{\mathbf{m}}^\alpha + \mathbf{x} \times (\hat{\mathbf{p}}^\alpha + \hat{\rho}^\alpha \, \overset{'}{\mathbf{x}}_\alpha) + \hat{\rho}^\alpha \, \bar{\Theta}^\alpha \, \bar{\omega}_\alpha\,] = 0\,.$$

Based on the fact that the micropolar approach proceeds from the assumption that the usual material point of the standard continuum exceeds to a microparticle, it is easily shown that

$$(\bar{\Theta}_\alpha)'_\alpha = 2\,(\bar{\Omega}_\alpha\,\bar{\Theta}^\alpha)_{\mathrm{sym}}\,, \qquad \text{where} \qquad \bar{\Omega}_\alpha = -\overset{3}{\mathbf{E}}\,\bar{\omega}_\alpha \tag{9.16}$$

defines the skew-symmetric micropolar gyration tensor.

Restricting the considerations to the case of quasi-static processes of fluid-saturated porous skeleton materials in the absence of mass exchanges ($\hat{\rho}^\alpha = 0$), the balance equations (9.14) read:

$$(\rho^\alpha)'_\alpha + \rho^\alpha \operatorname{div} \overset{'}{\mathbf{x}}_\alpha = 0\,,$$

$$0 = \operatorname{div}\mathbf{T}^\alpha + \rho^\alpha\,\mathbf{b} + \hat{\mathbf{p}}^\alpha\,, \tag{9.17}$$

$$0 = \mathbf{I} \times \mathbf{T}^\alpha + \operatorname{div}\mathbf{M}^\alpha + \rho^\alpha\,\mathbf{c}^\alpha + \hat{\mathbf{m}}^\alpha\,.$$

In $(9.17)_2$, it has been assumed that $\mathbf{b}^\alpha \equiv \mathbf{b}$ is the overall gravity. Furthermore, to obtain the above representation of $(9.17)_3$, the simplifying assumption $(\bar{\Omega}_\alpha\,\Theta^\alpha)_{\mathrm{sym}} = 0$ has been made. This assumption either incorporates the consideration of spherical microparticles or it represents the result of an averaging process over an assembly of randomly shaped microparticles.

Moreover, the consideration of a triphasic model consisting of a micropolar skeleton and non-polar pore-fluids (pore-liquid φ^L and pore-gas φ^G), where $\hat{\mathbf{m}}^\beta = 0$ [14], replaces $(9.17)_3$ by

$$0 = \mathbf{I} \times \mathbf{T}^S + \operatorname{div}\mathbf{M}^S + \rho^S\mathbf{c}^S\,,$$
$$\mathbf{T}^\beta = (\mathbf{T}^\beta)^T. \tag{9.18}$$

Furthermore, the saturation condition (9.2) yields

$$n^S + n^F = 1\,, \qquad \text{where} \qquad n^F = n^L + n^G\,. \tag{9.19}$$

Therein, the volume fraction n^F of both fluids can also be understood as the porosity. Following this, saturation functions s^β can be defined via

$$s^L + s^G = 1\,, \qquad \text{where} \qquad s^L = \frac{n^L}{n^F}\,, \qquad s^G = \frac{n^G}{n^F}\,. \tag{9.20}$$

In the framework of a quasi-static description of a triphasic medium consisting of a materially incompressible solid skeleton ($\rho^{SR} = \text{const.}$), a materially incompressible pore-liquid (ρ^{LR} = const.) and a materially compressible pore-gas ($\rho^{GR} \neq$ const.), the set of governing equations is given by the sum of the momentum balance equations, the momentum balance equations of the liquid and gas constituents, the skeleton moment of momentum balance equation and the liquid and gas mass balance equations:

$$
\begin{aligned}
\mathbf{0} &= \operatorname{div}\left(\mathbf{T}^S + \mathbf{T}^L + \mathbf{T}^G\right) + \left(\rho^S + \rho^L + \rho^G\right)\mathbf{b}\,, \\
\mathbf{0} &= \operatorname{div}\mathbf{T}^L + \rho^L\,\mathbf{b} + \hat{\mathbf{p}}^L\,, \\
\mathbf{0} &= \operatorname{div}\mathbf{T}^G + \rho^G\,\mathbf{b} + \hat{\mathbf{p}}^G\,, \\
\mathbf{0} &= \mathbf{I} \times \mathbf{T}^S + \operatorname{div}\mathbf{M}^S + \rho^S\mathbf{c}^S\,, \\
(n^L)'_S &+ \operatorname{div}\left(n^L\,\mathbf{w}_L\right) + n^L\operatorname{div}\left(\mathbf{u}_S\right)'_S = 0\,, \\
(\rho^G)'_S &+ \operatorname{div}\left(\rho^G\,\mathbf{w}_G\right) + \rho^G\operatorname{div}\left(\mathbf{u}_S\right)'_S = 0\,.
\end{aligned}
\tag{9.21}
$$

Note in passing that the mass balance equation of the materially incompressible liquid reduces to a volume balance. The above equations are governed by a set of primary variables given by the solid displacement $\mathbf{u}_S$, the liquid and gas seepage velocities $\mathbf{w}_L$ and $\mathbf{w}_G$, the total average grain rotation $\bar{\varphi}_S$ and the effective liquid and gas pressures p^{LR} and p^{GR}, where the liquid pressure p^{LR} can be replaced equivalently by the liquid saturation s^L. For a numerical treatment of the above equations, it is considered that the volume fractions n^L and n^G as well as the saturation s^G can be expressed by the solid volume fraction n^S and the liquid saturation s^L via

$$
n^L = (1 - n^S)s^L\,, \quad n^G = (1 - n^S)(1 - s^L)\,, \quad s^G = 1 - s^L\,.
\tag{9.22}
$$

However, the above model can easily be simplified to non-polar solids by dropping $(9.21)_4$ or to binary media by either dropping $(9.21)_2$ and $(9.21)_5$ or by dropping $(9.21)_3$ and $(9.21)_6$. In case of binary media, the mass balance of the pore-liquid or of the pore-gas is usually extended by adding the solid mass balance equation. Thus, one obtains

$$
\operatorname{div}\left[n^F\,\mathbf{w}_F + (\mathbf{u}_S)'_S\right] = 0
\tag{9.23}
$$

if the skeleton is fully liquid-saturated ($\varphi^F = \varphi^L$) or

$$
n^F (\rho^{FR})'_S + \operatorname{div}\left(n^F \rho^{FR}\,\mathbf{w}_F\right) + \rho^{FR}\operatorname{div}\left(\mathbf{u}_S\right)'_S = 0
\tag{9.24}
$$

if the skeleton is fully gas-saturated ($\varphi^F = \varphi^G$). Finally, it may be noted that, in case of liquid-saturated media, (9.23) represents a volume balance relation or an incompressibility constraint of this model.

9.3   Constitutive setting

9.3.1   General setting

To close the triphasic model under consideration, the set of equations (9.21) must be completed by constitutive equations for the solid and the fluid stresses $\mathbf{T}^S$ and $\mathbf{T}^\beta$, the solid couple stress $\mathbf{M}^S$ and the momentum production terms (volumetrical interaction forces) $\hat{\mathbf{p}}^S = \hat{\mathbf{p}}^L + \hat{\mathbf{p}}^G$, whereas the gravity $\mathbf{b}$ and the body couple $\mathbf{c}^S$ are understood as prescribed quantities. Note in passing that the above model is downward compatible to simpler situations. In particular, by setting $\mathbf{M}^S = \mathbf{0}$ and $\mathbf{c}^S = \mathbf{0}$ to result in $\mathbf{T}^S = (\mathbf{T}^S)^T$ and by setting $\overset{*}{\boldsymbol{\varphi}}_S = \mathbf{0}$ to result in $\bar{\boldsymbol{\varepsilon}}_{S\,\mathrm{skw}} = \mathbf{0}$, the description of a standard non-polar skeleton material is naturally included. In coupled solid-fluid models, the stress state of the solid and the fluid constituents is usually separated into two parts (Ehlers [20]), where the first part is governed by the pore-fluid pressures, while the second part, the extra term $(\,\cdot\,)_E$, results from the solid deformation (effective stress) and the pore-fluid flow (frictional stress). In the classical literature (*cf.*, *e.g.*, Bishop [3]), this separation is known as the *effective stress principle*. Extending this principle to the interaction force in fully coupled situations, the following relations hold:

$$
\begin{aligned}
\mathbf{T}^S &= -n^S\,p\,\mathbf{I} + \mathbf{T}^S_E\,, \\
\mathbf{T}^\beta &= -n^\beta\,p^{\beta R}\,\mathbf{I} + \mathbf{T}^\beta_E\,, \\
\hat{\mathbf{p}}^\beta &= p^{\beta R}\,\mathrm{grad}\,n^\beta + \hat{\mathbf{p}}^\beta_E\,.
\end{aligned}
\tag{9.25}
$$

Therein, the pore pressure p is given through (*Dalton*'s law)

$$
p = s^L p^{LR} + s^G p^{GR}.
\tag{9.26}
$$

Note in passing that in case of a binary model of a solid skeleton and a single pore-fluid, p either changes to a *Lagrange*an multiplier (materially incompressible solid and materially incompressible pore-liquid) or to the gas-pressure (materially incompressible solid and materially compressible pore-gas).

9.3.2   The fluid constituents

In soil mechanics as well as in several further applications (*e.g.*, hydraulics), the fluid friction forces $\mathrm{div}\,\mathbf{T}^\beta_E$ are usually assumed to be negligible in comparison to the viscous interaction terms $\hat{\mathbf{p}}^\beta_E$, thus leading to

$$
\mathbf{T}^\beta_E \approx \mathbf{0}\,.
\tag{9.27}
$$

This assumption can formally be proved by the methods of dimensional analysis [28]. Furthermore,

$$\hat{\mathbf{p}}_E^\beta = - \frac{(n^\beta)^2 \, \gamma^{\beta R}}{k^\beta} \, \mathbf{w}_\beta \,, \tag{9.28}$$

where $\gamma^{FR} = \rho^{FR} |\mathbf{b}|$ is the effective (true) specific weight of φ^β and k^β is the *Darcy* permeability coefficient. Inserting (9.28) into the quasi-static fluid momentum balance relations $(9.21)_2$ or $(9.21)_3$ leads to the *Darcy* equations

$$n^\beta \mathbf{w}_\beta = - \frac{k^\beta}{\gamma^{\beta R}} \, (\operatorname{grad} p^{\beta R} - \rho^{\beta R} \mathbf{b}) \,. \tag{9.29}$$

In the present considerations, k^β is assumed to depend on the degree of saturation s^β and on the solid deformation [24] via

$$k^\beta = k_r^\beta \, k_0^L \,, \quad \text{where} \quad k_r^\beta = (s^\beta)^\lambda \left(\frac{1 - n^S}{1 - n_{0S}^S} \right)^\kappa . \tag{9.30}$$

In the above representation, λ and κ are material parameters, whereas n_{0S}^S represents the solid volume fraction in the reference configuration of φ^S. Furthermore, k_r^β can be understood as a weighting function (relative permeability), whereas k_0^L is the initial *Darcy* permeability under fully saturated conditions ($s^L = 1$), which is related to the intrinsic (or physical) permeability K_0^S through

$$k_0^L = \frac{\gamma^{LR}}{\mu^{LR}} \, K_0^S \,, \tag{9.31}$$

where μ^{LR} defines the effective shear viscosity of the pore-liquid. Assuming geometrically linear conditions for the solid constituent, the solid deformation enters (9.30) through

$$n^S = n_{0S}^S \, (1 - \operatorname{div} \mathbf{u}_S) \,. \tag{9.32}$$

The effective density functions of the materially incompressible pore-liquid and the materially compressible pore-gas are given by

$$\rho^{LR} = \text{const.} \,, \quad \rho^{GR} = \frac{p^{GR}}{\bar{R} \, \theta} \,. \tag{9.33}$$

Recall that $(9.33)_1$ stems from the property of liquid incompressibility, whereas $(9.33)_2$ is known as the ideal gas law (*Boyle–Mariotte*'s law). Therein, $\bar{R}$ denotes the specific gas constant of the pore-gas and θ the absolute *Kelvin*'s temperature. However, in the present investigations, it is assumed that the overall model can be described under isothermal conditions ($\theta = \text{const.}$).

The effective liquid and gas pressures p^{LR} and p^{GR} are coupled by the capillary pressure $p^C > 0$ through

$$p^{LR} = p^{GR} - p^C, \tag{9.34}$$

where p^C depends on the liquid saturation s^L or, vice versa, s^L depends on p^C. In particular, use is made of the *van Genuchten* model [56] given by

$$s^L = [\,1 + (\alpha\, p^C)^n\,]^{-m}. \tag{9.35}$$

Therein, α, n and m are material constants, where m and n are often used as coupled variables through $m = 1 - \frac{1}{n}$.

9.3.3 The solid constituent

Following the geometrically linear approach to elasto-plasticity, both the *Cosserat* strain $\bar{\varepsilon}_S$ and the curvature tensor $\bar{\kappa}_S$ are additively decomposed into elastic and plastic parts:

$$\begin{aligned}
\bar{\varepsilon}_S &= \bar{\varepsilon}_{Se} + \bar{\varepsilon}_{Sp}, \\
\bar{\kappa}_S &= \bar{\kappa}_{Se} + \bar{\kappa}_{Sp}.
\end{aligned} \tag{9.36}$$

Note that, once $(9.36)_1$ is given, the decomposition $(9.36)_2$ is a natural consequence of the micropolar compatibility condition (9.13). This has been pointed out in more detail by Ehlers & Volk [34].

As was discussed in [34], the non-symmetric solid extra stress and the couple stress are given by

$$\begin{aligned}
\mathbf{T}^S_E &= 2\,\mu^S\,\bar{\varepsilon}_{Se\,\mathrm{sym}} + 2\,\mu^S_c\,\bar{\varepsilon}_{Se\,\mathrm{skw}} + \lambda^S\,(\,\bar{\varepsilon}_{Se} \cdot \mathbf{I}\,)\,\mathbf{I}, \\
\mathbf{M}^S &= 2\,\mu^S_c\,(l^S_c)^2\,\bar{\kappa}_{Se}.
\end{aligned} \tag{9.37}$$

In the above equations, μ^S and λ^S are the *Lamé* constants of the porous skeleton material, whereas μ^S_c is an additional parameter governing the influence of the skew-symmetric part of the elastic *Cosserat* strain on the effective stress of the skeleton material. Furthermore, the symmetric part of $\mathbf{T}^S_E$ is equivalent to the stress tensor of non-polar skeleton materials, whereas the skew-symmetric part is directly related to the independent micropolar rotation $\overset{*}{\varphi}_S$ through $(9.11)_2$. Finally, as was pointed out, *e.g.*, by de Borst [15], l^S_c represents an intrinsic length scale parameter relating the couple stress to the elastic curvature tensor.

In order to describe the plastic or the viscoplastic material properties of both non-polar and micropolar skeleton materials, one has to consider a convenient yield function to bound the elastic domain. In extension of the yield criterion by Ehlers [20] towards micropolar cohesive-frictional materials, it is assumed that

$$\bar{F}(\mathbf{T}^S_E;\, p,\, p^c) = \bar{\Phi}^{\frac{1}{2}} + \beta\,\mathrm{I} + \epsilon\,\mathrm{I}^2 + \tfrac{1}{2}\,k_M\,(\mathbf{M}^S \cdot \mathbf{M}^S)^{\frac{1}{2}} - \kappa = 0,$$

with

$$\bar{\Phi} = \mathrm{II}^{D}_{\mathrm{sym}} (1 + \gamma\,\vartheta)^{m} + k_{T}\,\mathrm{II}_{\mathrm{skw}} + \tfrac{1}{2}\alpha\,\mathrm{I}^{2} + \delta^{2}\,\mathrm{I}^{4}, \qquad (9.38)$$

$$\vartheta = \frac{\mathrm{III}^{D}_{\mathrm{sym}}}{(\mathrm{II}^{D}_{\mathrm{sym}})^{\frac{3}{2}}},$$

holds [34]. Therein, I, $\mathrm{II}^{D}_{\mathrm{sym}}$ and $\mathrm{III}^{D}_{\mathrm{sym}}$ are the principal invariants of $\mathbf{T}^{S}_{E}$ and of the symmetric part of the effective stress deviator $(\mathbf{T}^{S}_{E})^{D}$, whereas $\mathrm{II}_{\mathrm{skw}}$ defines the second principal invariant of the skew-symmetric part of $\mathbf{T}^{S}_{E}$. Next, the vectors $\boldsymbol{p} = (\alpha,\ \beta,\ \gamma,\ \delta,\ \epsilon,\ m,\ \kappa)^{T}$ and $\boldsymbol{p}^{c} = (k_{M},\ k_{T})^{T}$ contain two sets of material parameters, where the parameters of $\boldsymbol{p}$ govern the non-polar part and the parameters of $\boldsymbol{p}^{c}$ the micropolar part of the yield function. In case of non-polar materials ($\boldsymbol{p}^{c} \equiv \mathbf{0}$), the yield criterion exhibits a closed single-surface yield function in the principal stress space, *cf.* Figure 9.2.

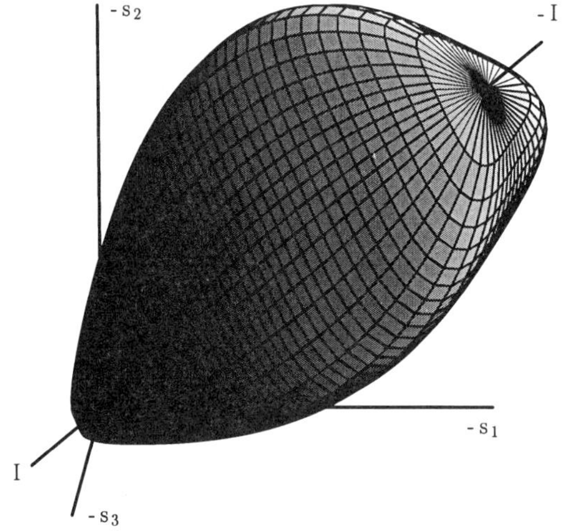

FIGURE 9.2. Single-surface yield criterion for non-polar cohesive-frictional materials; s_1, s_2, s_3: principal stresses of $\mathbf{T}^{S}_{E}$ (tension positive).

Proceeding either from the viscoplastic approach or from the ideal plasticity concept, $\boldsymbol{p}$ as well as $\boldsymbol{p}^{c}$ are constant during the deformation process. However, while $\boldsymbol{p}$ can be computed from standard experimental data by use of an optimization procedure [32], k_{M} and k_{T} have not been satisfactorily determined yet. This is due to the fact that these parameters as well as the internal length scale l^{S}_{c} and the micropolar shear modulus μ^{S}_{c} must be computed by a back analysis of a typical non-homogeneous boundary-value problem including a shear band localization, since these parameters strongly depend on the micropolar rotation which, however, is only active in the localization zones.

Proceeding from the fact that the associated plasticity concept cannot be applied to frictional materials (*cf.*, *e.g.*, Ehlers & Volk [34]), the plastic potential

$$\bar{G} = \bar{\Gamma}^{\frac{1}{2}} + \beta\,\mathrm{I} + \epsilon\,\mathrm{I}^2 - g\,(\mathrm{I}) = 0\,,$$

$$\bar{\Gamma} = \mathrm{II}^D_{\mathrm{sym}} + k_T\,\mathrm{II}_{\mathrm{skw}}\tfrac{1}{2} + \alpha\,\mathrm{I}^2 + \delta^2\,\mathrm{I}^4 \tag{9.39}$$

is considered, where $g(\mathrm{I})$ serves to relate the dilatation angle to experimental data. From the concept of a plastic potential, it is straight forward to obtain the evolution equation (flow rule) for the plastic *Cosserat* strain $\bar{\varepsilon}_{Sp}$ via

$$(\bar{\varepsilon}_{Sp})'_S = \Lambda\,\frac{\partial\bar{G}}{\partial\mathbf{T}^S_E}\,, \tag{9.40}$$

where Λ is the plastic multiplier. As was pointed out by Ehlers & Volk [34], there exists no evolution equation for the plastic rate of curvature tensor independent from both the evolution equation (9.40) and the micropolar compatibility condition (9.13). Thus, once (9.40) is given, the most convenient possibility to obtain an evolution equation for $\bar{\kappa}_{Sp}$ directly results from (9.40) and (9.13). Thus,

$$(\bar{\kappa}_{Sp})'_S = \tfrac{1}{2}\,\overset{3}{\mathbf{E}}\left(\mathrm{Grad}_S\,(\bar{\varepsilon}_{Sp})'_S + \mathrm{Grad}_S^{\overset{13}{T}}\,(\bar{\varepsilon}_{Sp})'_S - \mathrm{Grad}_S^{\overset{23}{T}}\,(\bar{\varepsilon}_{Sp})'_S\right)\overset{2}{}. \tag{9.41}$$

In contrast to the investigation of elasto-plastic problems of non-polar materials, where the standard compatibility condition is generally fulfilled, one has to take into account the micropolar compatibility condition explicitly, if one considers elasto-plastic problems of micropolar materials [34]. Following this, one has two principally different choices to assume an evolution equation for the plastic rate of curvature tensor. The first and simplest one is given by the consideration of (9.41), where the micropolar compatibility condition is automatically fulfilled. The second one could proceed, on the one hand, from an evolution equation for $\bar{\kappa}_{Sp}$ independent from (9.41) but must then take into account, on the other hand, the micropolar compatibility condition as a constraint to the computation of the elasto-plastic process under study. Using the second choice without taking into account the micropolar compatibility constraint is not only insufficient but also leads to an extremely unsatisfactory convergence of the numerical computations, whenever the micropolar rotation is active.

In the framework of viscoplasticity using the overstress concept of *Perzyna* type [49], the plastic multiplier included in (9.40) is given by

$$\Lambda = \frac{1}{\eta}\left\langle\frac{\bar{F}(\mathbf{T}^S_E)}{\sigma_0}\right\rangle^r\,. \tag{9.42}$$

Therein, $\langle \cdot \rangle$ are the *Macauley* brackets, η is the relaxation time, σ_0 is the reference stress and r is the viscoplastic exponent. However, in the framework of elasto-plasticity, where the plastic strains are rate-independent, Λ has to be computed from the *Kuhn–Tucker* conditions

$$\bar{F} \leq 0, \quad \Lambda \geq 0, \quad \Lambda \bar{F} = 0 \tag{9.43}$$

rather than from (9.42).

9.4 Discretization in space and time

9.4.1 Weak formulation of the governing field equations

Based on consideration of six independent fields, the solid displacement $\mathbf{u}_S$, the seepage velocities $\mathbf{w}_L$ and $\mathbf{w}_G$, the effective liquid and gas pressures p^{LR} and p^{GR} and the total average grain rotation $\bar{\varphi}_S$, the corresponding six equations of the weak formulation can be obtained from the kinematics, the balance relations (9.21) and the constitutive equations of Section 9.3. Concerning the quasi-static problem under study, the seepage velocities can be eliminated by use of the momentum balance equations of the pore-fluids φ^β, compare (9.29). Thus, $\mathbf{w}_\beta$ looses the status of an independent field variable and the number of equations of the weak formulation reduces to four. Following the work by Lewis & Schrefler [45] and Ehlers and coworkers [24, 26], these equations are given, in the framework of the standard *Galerkin* procedure (*Bubnov–Galerkin*), firstly by the sum of the solid and fluid momentum balance equations $(9.21)_1$ or the mixture momentum balance, respectively, multiplied by the test function $\delta\mathbf{u}_S$, secondly by the solid moment of momentum balance $(9.21)_4$ multiplied by the test function $\delta\bar{\varphi}_S$, thirdly by the volume balance relation $(9.21)_5$ of the pore-liquid multiplied by the test function δp^{LR} and finally by the mass balance equation $(9.21)_6$ of the pore-gas multiplied by δp^{GR}. Thus, the weak formulation of the triphasic model reads

$$\int_\Omega [\, \mathbf{T}_E^S - (s^L p^{LR} + s^G p^{GR})\, \mathbf{I} \,] \cdot \operatorname{grad}\delta\mathbf{u}_S \, \mathrm{d}v$$

$$= \int_\Omega (n^S \rho^{SR} + n^L \rho^{LR} + n^L \rho^{LR})\, \mathbf{b} \, \cdot \, \delta\mathbf{u}_S \, \mathrm{d}v + \int_{\Gamma_t} \bar{\mathbf{t}} \cdot \delta\mathbf{u}_S \, \mathrm{d}a \,,$$

$$\int_\Omega \mathbf{M}^S \cdot \operatorname{grad}\delta\bar{\varphi}_S \, \mathrm{d}v - \int_\Omega (\mathbf{I} \times \mathbf{T}^S) \cdot \delta\bar{\varphi}_S \, \mathrm{d}v = 0 \,, \tag{9.44}$$

$$\int_\Omega \left(\frac{k^L}{\gamma^{LR}}\,\mathrm{grad}\,p^{LR} \cdot \mathrm{grad}\,\delta p^{LR} + [\,(n^L)'_S + n^L\mathrm{div}\,(\mathbf{u}_S)'_S\,]\,\delta p^{LR} \right) \mathrm{d}v$$

$$= \int_\Omega k^G\,\frac{\mathbf{b}}{g} \cdot \mathrm{grad}\,\delta p^{LR}\mathrm{d}v - \int_{\Gamma_v} \bar{v}^G\,\delta p^{LR}\,\mathrm{d}a\,,$$

$$\int_\Omega \left(\frac{k^G}{g}\,\mathrm{grad}\,p^{GR} \cdot \mathrm{grad}\,\delta p^{GR} + [\,(\rho^G)'_S + \rho^G\mathrm{div}\,(\mathbf{u}_S)'_S\,]\,\delta p^{GR} \right) \mathrm{d}v$$

$$= \int_\Omega k^G\,\rho^{GR}\,\frac{\mathbf{b}}{g} \cdot \mathrm{grad}\,\delta p^{GR}\mathrm{d}v - \int_{\Gamma_q} \bar{q}^G\,\delta p^{GR}\,\mathrm{d}a\,.$$

However, if the triphasic model reduces to a binary one, the equations $(9.44)_3$ and $(9.44)_4$ are replaced by

$$\int_\Omega \left(\frac{k^F}{\gamma^{FR}}\,\mathrm{grad}\,p \cdot \mathrm{grad}\,\delta p + \mathrm{div}\,(\mathbf{u}_S)'_S\,\delta p \right) \mathrm{d}v$$

$$= \int_\Omega k^F\,\frac{\mathbf{b}}{g} \cdot \mathrm{grad}\,\delta p\,\mathrm{d}v - \int_{\Gamma_v} \bar{v}^F\,\delta p\,\mathrm{d}a, \tag{9.45}$$

if the skeleton is fully liquid-saturated ($\varphi^F = \varphi^L$), or by

$$\int_\Omega \left(\frac{k^F}{g}\,\mathrm{grad}\,p \cdot \mathrm{grad}\,\delta p + [\,n^F(\rho^{FR})'_S + \rho^{FR}\mathrm{div}\,(\mathbf{u}_S)'_S\,]\,\delta p \right) \mathrm{d}v$$

$$= \int_\Omega k^F\,\rho^{FR}\,\frac{\mathbf{b}}{g} \cdot \mathrm{grad}\,\delta p\,\mathrm{d}v - \int_{\Gamma_q} \bar{q}^G\,\delta p\,\mathrm{d}a, \tag{9.46}$$

if the skeleton is fully gas-saturated ($\varphi^F = \varphi^G$), *cf.* (9.23) and (9.24). In the weak formulation of the problem under study given by the above equations (9.44)–(9.46), $\bar{\mathbf{t}}$ is the external load vector acting on the *Neumann* boundary Γ_t of the overall model. Furthermore, $\bar{v} = n^L\,\mathbf{w}_L \cdot \mathbf{n}$ is the efflux of liquid volume through the *Neumann* boundary Γ_v, whereas $\bar{q} = \rho^G\,\mathbf{w}_G \cdot \mathbf{n}$ characterizes the efflux of gaseous mass through the *Neumann* boundary Γ_q; $\mathbf{n}$ is the outward oriented unit surface normal. To obtain the moment of momentum equation $(9.44)_2$, it has been assumed that there is no external loading by volume couples $\mathbf{c}^S$ and by surface couples $\mathbf{m}^S = \mathbf{M}^S\,\mathbf{n}$.

The equations (9.44) represent the weak form of the so-called displacement-rotation-pressures formulation of the strongly coupled solid-fluid problem of triphasic media. In case of non-polar skeleton materials, these equations reduce to the well-known displacement-pressures formulation, where $(9.44)_2$ is dropped. Furthermore, if binary media are concerned, $(9.44)_3$ and $(9.44)_4$ are replaced by (9.45) or by (9.46), respectively, thus resulting in either the displacement-rotation-pressure formulation or in the displacement-pressure formulation. In addition, it may be noted that the possibility to deal with an empty skeleton material is always included by simply disregarding the liquid and gas equations $(9.44)_3$ and $(9.44)_4$ together with $p^{LR} = p^{GR} \equiv 0$. Finally, in the framework of the standard *Galerkin* procedure, the included test functions $\delta \mathbf{u}_S$, $\delta \bar{\varphi}_S$ δp^{LR} and δp^{GR} correspond to the respective field quantities and, as a result, vanish at the *Dirichlet* boundaries with prescribed displacements, rotations and pressure values.

9.4.2 Spatial discretization

In the framework of the finite element method (FEM), the spatial discretization (semi-discretization with respect to the space variable $\mathbf{x}$) of the field equations (9.44) is based on quadratic shape functions for the solid displacement $\mathbf{u}_S$ and linear shape functions for the total average grain rotation $\bar{\varphi}^S$ and the fluid pressures p^{LR} and p^{GR} (external variables). Furthermore, both the evolution equation (9.40) for the plastic strain tensor $\bar{\varepsilon}_{Sp}$ (internal variable) and the plastic multiplier Λ are computed, in the sense of the collocation method, at the integration points of the numerical quadrature. Note again that there is no independent evolution equation for the plastic curvature tensor, since $(\bar{\kappa}_{Sp})'_S$ is obtained from $(\bar{\varepsilon}_{Sp})'_S$ by (9.41). For a mesh of N_u nodes and N_q integration points, the space-discrete variables of the semi-discrete problem are collected in the vectors

$$
\begin{aligned}
u &= ((\mathbf{u}_S^1, \bar{\varphi}_S^1, p_1^{LR}, p_1^{GR}), \quad \ldots, \quad (\mathbf{u}_S^{N_u}, \bar{\varphi}_S^{N_u}, p_{N_u}^{LR}, p_{N_u}^{GR}))^T, \\
q &= ((\bar{\varepsilon}_{Sp}^1, \Lambda^1), \qquad\qquad \ldots, \quad (\bar{\varepsilon}_{Sp}^{Nq}, \Lambda^{Nq}))^T.
\end{aligned} \tag{9.47}
$$

Using the abbreviation $(\cdot)' := (\cdot)'_S$ and the vector $y := (u^T, q^T)^T$, one obtains the semi-discrete initial-value problem

$$
\mathbf{F}(t,\, y,\, y') \equiv \begin{bmatrix} \mathbf{F}_1(t,\, u,\, u',\, q) \\ \mathbf{F}_2(t,\, q,\, q',\, u) \end{bmatrix} \equiv \begin{bmatrix} M u' + k(u,\, q) - f \\ A\, q' - g(q,\, u) \end{bmatrix} \overset{!}{=} 0 \tag{9.48}
$$

of first order in the time variable t, where $t \geq t_0$ and $y(t_0) = y_0$ are the corresponding initial conditions [18, 35]. In (9.48), the first equation $(\mathbf{F}_1 = 0)$ represents the discretization of the governing field equations, where M is the generalized mass matrix, k is the generalized stiffness vector and f is the vector of the external forces. The second equation $(\mathbf{F}_2 = 0)$ exhibits the plastic or the viscoplastic evolution equations together with the constraints resulting from the *Kuhn–Tucker* conditions of the elastoplastic formulation. The introduction of the matrix A formally allows for a joint formulation of elasto-viscoplastic and elasto-plastic problems. Finally,

g represents the right-hand side of the evolution equations and constraints, which are element-wise decoupled as a result of their evaluation at the integration points of the finite elements [35].

As a result of the quasi-static problem under consideration, it may occur that the generalized mass matrix M is not regular. Then, the system (9.48) turns out to be a system of differential-algebraic equations (DAE) of index one in the time variable. Details on the solution of DAE systems can be taken from the literature [38, 39].

9.4.3 Time adaptivity

The time integration as well as the following time-adaptive strategy are based on one-step methods with an embedded time-step control, where the solution at time t_{n+1} only depends on the solution at time t_n. This choice is of essential importance with respect to space-adaptive methods (refinements as well as coarsenings), since the transfer of the numerical solution thus only includes two meshes, *cf.* Diebels *et al.* [18], Ellsiepen [35] Ehlers & Ellsiepen [26] and Ehlers *et al.* [27]. Based on the fact that it may occur that the system (9.48) is a DAE system of index one, it is convenient to apply diagonally implicit *Runge–Kutta* methods (DIRK) with suitable stability properties. With respect to both the size of the system and the treatment of elasto-plastic problems, DIRK methods yield the advantage of being able to solve the non-linear equation systems in a decoupled way. In addition, embedded methods allow for an efficient estimation of the time error [18, 26, 35]. In particular, one obtains two numerical solutions of (9.48) at time t_{n+1}, namely y_{n+1} with the order r and $\hat{y}_{n+1}$ with the order $\hat{r} \leq r$. As a result, an embedded error estimation is given by the difference of these solutions through

$$ERR \approx \|y_{n+1} - \hat{y}_{n+1}\|. \tag{9.49}$$

As was pointed out by Diebels *et al.* [18], this type of an error estimation is "cheap", since it does not require the additional solution of non-linear systems but only a weighted sum of already computed quantities. Following this, *Runge–Kutta* methods with embedded error estimators are well suited for large equation systems. In the present contribution, the numerical examples are carried out by use of a 2-stage singly diagonally implicit *Runge–Kutta* method (SDIRK) with order $r = 2$ and embedded order $\hat{r} = 1$ [35].

Using the relative and absolute tolerances ϵ_r and ϵ_a together with the weighted error measures

$$e_u := \left(\frac{1}{N} \sum_{k=1}^{N} \left[\frac{u_{n+1}^k - \hat{u}_{n+1}^k}{\epsilon_r \, |u_n^k| + \epsilon_a} \right]^2 \right)^{\frac{1}{2}}, \; e_q := \max_k \left| \frac{q_{n+1}^k - \hat{q}_{n+1}^k}{\epsilon_r \, |q_n^k| + \epsilon_a} \right|, \tag{9.50}$$

where $N = \dim \boldsymbol{u}$, the time-step is accepted if $e_y := \max\{e_u, e_q\} \leq 1$ and rejected otherwise. In both cases, a new step size is predicted from the above error measures together with the order $\hat{r}$ of the embedded method by

$$\Delta t_{\text{new}} := \Delta t_{\text{old}} \; \min\left\{f_{\max}, \, \max\left\{f_{\min}, \, f_{\text{safety}} \; e_y^{-\frac{1}{\hat{p}+1}}\right\}\right\}. \tag{9.51}$$

Therein, $f_{\text{safety}} < 1$ is a safety factor, which prevents an oscillation of the time-step size, whereas $f_{\max} > 1$ and $f_{\min} < 1$ are used to limit the step size variation. Concerning further details of this procedure, the reader is referred to [35].

9.4.4 Space adaptivity

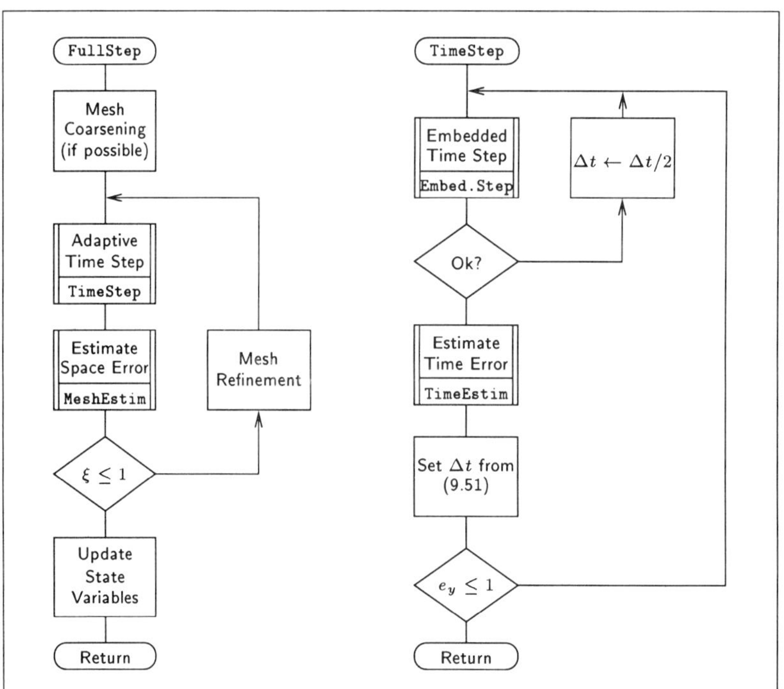

FIGURE 9.3. Algorithm for a time- and space-adaptive step.

Adaptive mesh size control. Concerning the model under consideration, no mathematically founded methods are known so far to estimate the spatial error [23]. Thus, the following procedure is applied, *cf.* Figure 9.3. A time-step of the non-stationary problem is treated as a stationary problem, where the initial conditions are taken from the solution of the previous step. In order to estimate the spatial error of the discretized problem, the gradient-based error indicator of *Zienkiewicz–Zhu* type [58] is extended in such a way that all the driving quantities of saturated and unsaturated non-polar and micropolar elasto-plastic and elasto-viscoplastic materials are included.

Apart from the standard consideration of the effective solid stresses representing the elastic part of the problem, the error indicator is extended towards the plastic part of the strain state representing the accumulated plasticity and towards the seepage velocities representing the viscosities of the pore-fluids. As was pointed out in the introduction to this article, there is no need to either change or extend this set of mechanical quantities incorporated into the error indicator, even in case of micropolar problems. This essential statement can be obtained as a result of the moment of momentum balance $(9.21)_4$ together with the micropolar compatibility condition (9.13). From $(9.21)_4$, the couple stress tensor $\mathbf{M}^S$ representing the elastic curvature is included through the skew-symmetric part of the effective stress $\mathbf{T}_E^S$, whereas the plastic curvature $\bar{\boldsymbol{\kappa}}_{Sp}$ is included through the plastic strain $\bar{\varepsilon}_{Sp}$ by (9.13).

Proceeding from the L_2-norm $\|\cdot\|_2$ and the corresponding element-wise norm $\|\cdot\|_{2,e}$ (per element e), smoothed values $(\cdot)^*$ are computed on the basis of the FEM quantities $(\cdot)^h$. Following this, the error indicators

$$
\begin{aligned}
\eta_1^{(e)} &:= \|\mathbf{T}_E^{S*} - \mathbf{T}_E^{Sh}\|_{2,e}, & \eta_3^{(e)} &:= \|\mathbf{w}_L^* - \mathbf{w}_L^h\|_{2,e}, \\
\eta_2^{(e)} &:= \|\bar{\varepsilon}_{Sp}^* - \bar{\varepsilon}_{Sp}^h\|_{2,e}, & \eta_4^{(e)} &:= \|\mathbf{w}_G^* - \mathbf{w}_G^h\|_{2,e}
\end{aligned}
\tag{9.52}
$$

are applied, where η_1 considers the solid elasticity through the stresses and (indirectly) through the couple stresses. In case of inelastic deformations, η_2 considers the accumulated solid plasticity or viscoplasticity through the plastic strains and (indirectly) through the plastic curvatures. Finally, η_3 and η_4 consider the pore-liquid and the pore-gas flow processes through their seepage velocities. The domain integrals

$$
\begin{aligned}
W_1 &:= \|\mathbf{T}_E^{Sh}\|_2, & W_3 &:= \|\mathbf{w}_L^h\|_2, \\
W_2 &:= \|\bar{\varepsilon}_{Sp}^h\|_2, & W_4 &:= \|\mathbf{w}_G^h\|_2
\end{aligned}
\tag{9.53}
$$

serve as reference quantities of the respective error indicators. For practical reasons, the absolute errors $\eta_i^{(e)}$ are transferred into dimensionless (relative) errors by dividing through W_i. Consequently, tolerance-weighted error measures $\xi_{e,i}$ can be defined on the basis of user-specified relative and absolute tolerances, ϵ_r and $\epsilon_{a,i}$:

$$
\xi_{e,i} = \frac{\eta_i^{(e)}}{\epsilon_r\, W_i + \epsilon_{a,i}}, \qquad i = 1, 2, 3, 4.
\tag{9.54}
$$

In contrast to the usual considerations on spatial error measures, where a user-specified combination of the absolute element-wise errors $\xi_{e,i}$ is taken to contribute to the global error measure ξ [27], it has been shown by Ehlers *et al.* [23] that the maximum error indicator ξ_e of each element is a very convenient measure to contribute to ξ. Thus,

$$\xi_e = \max_{i=1,2,3,4} \left(\frac{\eta_i^{(e)}}{\epsilon_r\, W_i + \epsilon_{a,i}} \right), \qquad \xi = \left(\sum_{e=1}^{E} \xi_e^2 \right)^{\frac{1}{2}}. \qquad (9.55)$$

Following this, the solution on the actual mesh is accepted if $\xi \leq 1$ and not accepted else. In order to refine ($\xi > 1$) or to coarsen the mesh ($\xi \leq 1$), a new element radius h_{new} must be computed on the basis of a given density function. Concerning the choice of a convenient density function, it has been found by Ehlers *et al.* [23] that the function proposed by Ladevèse *et al.* [44] represents an optimal tool for mesh refinements and mesh coarsenings both in the framework of remeshing and hierarchical strategies. Thus, a new element radius h_{new} can be computed via

$$h_{\mathrm{new}} := h_{\mathrm{old}}\; \xi_e^{-\frac{1}{r+1}} \left[\sum_{e=1}^{N_e} \xi_e^{\frac{2}{r+1}} \right]^{-\frac{1}{2r}}, \qquad (9.56)$$

where r is again the convergence order of the FEM discretization. Given (9.56), h_{new} represents the new element radius optimized per element with respect to the number of elements, *cf.* Gallimard *et al.* [37], Ellsiepen [35] and Ehlers & Ellsiepen [26]. Concerning the following numerical examples, the present space-adaptive strategy proceeds both remeshing and hierarchical h-adaptive schemes.

Remeshing strategy. Proceeding from a remeshing strategy means that a completely new mesh must be created, whenever a modification of the mesh is necessary. Therefore, after having evaluated the density function, the new element sizes h_{new} are written into a file, thus delivering the basic information for the mesh generator during the creation of the new mesh. In the present case, this procedure is based on a modified version of the triangular mesh generator `Triangle` presented by Shewchuk [51]. Computing time-dependent problems, the complete data of the current mesh has to be transferred to the new mesh in order to avoid a restart of the computation. When transferring data between FE meshes, two different data types have to be considered: data at nodal points and data at integration points. Concerning the transfer of nodal data, the first task is to find the specific element in the old mesh, wherein a given nodal point $\hat{P}$ of the new mesh is located, *cf.* Figure 9.4.

The location of an element in an FE mesh plays a crucial role in the data transfer of remeshing h-adaptive methods. Following this, an efficient data transfer can only be realized if an efficient algorithm for the element location is available. The algorithm used in this presentation consists of a combination of two methods: (1) a quadtree search [43] and (2) an inversion of the shape functions. In order to locate an element, one firstly uses the

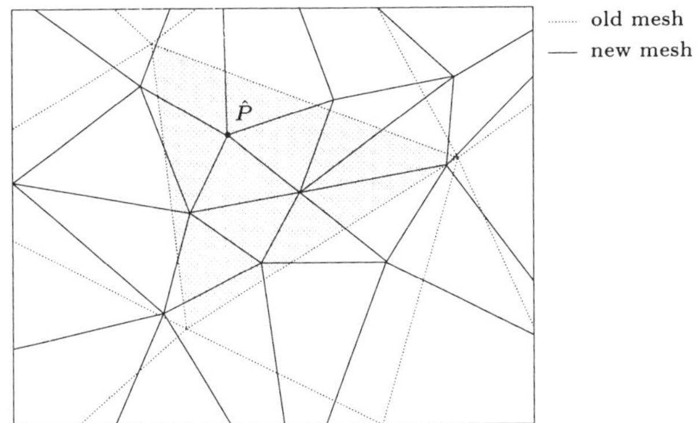

FIGURE 9.4. Data transfer of nodal points.

quadtree search to reduce the amount of the possible elements in the whole mesh. Secondly, the inversion of the shape functions is used to exactly locate the requested element. For example, proceeding from biquadratic shape functions as are widely used in the framework of multi-phasic problems, an efficient method for inverting shape functions was shown by Crawford *et al.* [9]. Finally, the correct element is found, if certain conditions for the local coordinates ξ and η hold. For triangular elements, these conditions are

$$
\begin{aligned}
0 \;\leq\; \xi \;&\leq\; 1, \\
0 \;\leq\; \eta \;&\leq\; 1, \\
\xi + \eta \;&\leq\; 1.
\end{aligned}
\tag{9.57}
$$

After having found the requested element, the local coordinates $(\xi_{\hat{P}}, \eta_{\hat{P}})$ are computed and the FE shape functions are evaluated to yield the transferred data at the nodal point $\hat{P}$ of the new mesh, *e.g.*, the horizontal displacement $u_{\hat{P}}$ in a triangular element with quadratic shape functions (6 nodes):

$$
u_{\hat{P}} = \sum_{i=1}^{6} N_i(\xi_{\hat{P}}, \eta_{\hat{P}})\, u_i.
\tag{9.58}
$$

When transferring data at integration points, the element-wise data, in a first step, must be projected onto the nodal points. Therefore, a function $f(x_1, x_2)$ defined by

$$
f(x_1, x_2) = \sum_i a_i\, \phi_i(x_1, x_2)
\tag{9.59}
$$

has to be created. In the above equation, a_i are the coefficients and $\phi_i(x_1, x_2)$ are the corresponding bases of a chosen function. For a quadratic function, these terms yield

$$
\begin{aligned}
a &= (\; a_1, \;\; a_2, \;\; a_3, \;\; a_4, \;\; a_5, \;\; a_6 \;)^T, \\
\phi &= (\; 1, \;\; x_1, \;\; x_2, \;\; x_1 x_2, \;\; x_1^2, \;\; x_2^2 \;)^T.
\end{aligned}
\tag{9.60}
$$

By minimizing the sum of the quadratic difference between the value f_k of the data and the function $f(x_1, x_2)$ over all integration points K,

$$\mathcal{S}(a_1, \dots, a_n) := \sum_{k=1}^{K} \left(f_k - \sum_i a_i \, \phi_i(x_{1k}, x_{2k}) \right)^2 \overset{!}{\longrightarrow} \text{ min.,} \qquad (9.61)$$

the coefficients a_i of the function $f(x_1, x_2)$ can be computed. After having evaluated this function at the nodal points, the same strategy as for the transfer of nodal data can be applied.

Hierarchical strategy based on bisection. In a hierarchical strategy, refinement or de-refinement of meshes can be carried out by adding or removing FE edges. In addition, it is very important for the stability of the adaptive process that degenerated elements are avoided. This, however, strongly depends on how the refinement or the de-refinement process is carried out. Using triangular elements, the *Newest Vertex Bisection* by Mitchell [46] in combination with the recursive algorithm by Kossaczký [42] was found very stable in a lot of adaptive computations [26, 35].

Using the *Newest Vertex Bisection* strategy, one firstly marks the edges of the initial FE mesh which have to be bisected during a first modification of the mesh. Basically, any edge can be chosen but it is obviously reasonable to mark the longest edge of each element. Subsequently, that edge of a triangle is marked for bisection which faces the most recently generated vertex: the *Newest Vertex*. This condition has to be accomplished by both neighbouring elements of the dividing edge. An example for this strategy is shown in Figure 9.5. Therein, the shaded triangle is the element to be refined. The marked edges are shown by the small arrows in each element.

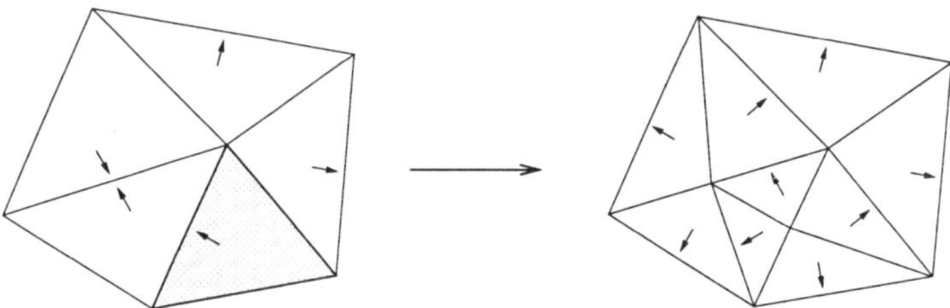

FIGURE 9.5. Newest vertex bisection.

In case of $S \geq 1$, the element remains unchanged; for $\frac{1}{2} \leq S < 1$, the element is bisected; for $\frac{1}{3} \leq S < \frac{1}{2}$, the element is divided into three new elements, *etc.* De-refinement is only possible, if all neighbouring elements having been refined by bisection of the same edge satisfy the condition $S \geq 2$.

The data transfer in the hierarchical strategy is totally different from the data transfer of the previous method. In the present case, only a local data transfer of the modified elements has to be carried out. Furthermore, an element location algorithm is not necessary, because, while refining or de-refining an element by adding or removing FE edges, the location of the element is obviously already known. The transfer of the nodal data even drops out in the case of de-refinement. The actual transfer of the data, however, is handled as was shown above in the remeshing strategy by evaluation of the shape functions.

9.5   Numerical examples

parameter	symb.	value	symb.	value
Lamé constants	μ^S	$5\,583\,\mathrm{kN/m^2}$	λ^S	$8\,375\,\mathrm{kN/m^2}$
effective densities	ρ^{SR}	$2\,600\,\mathrm{kg/m^3}$	ρ^{FR}	$1\,000\,\mathrm{kg/m^3}$
volume fractions	n_0^S	0.67	n_0^F	0.33
liquid weight	γ^{FR}	$10\,\mathrm{kN/m^3}$		
permeability	k_0^L	$1.2 \cdot 10^{-7}\,\mathrm{m/s}$		
parameters	λ	3.0	κ	1.0
parameters of the	α	$1.074\,0 \cdot 10^{-2}$	β	$0.119\,6$
single-surface	γ	1.555	δ	$1.377 \cdot 10^{-4}\,\mathrm{m^2/kN}$
yield criterion	ϵ	$4.330 \cdot 10^{-6}\,\mathrm{m^2/kN}$	κ	$10.27\,\mathrm{kN/m^2}$
	m	$0.593\,5$		
viscoplasticity	η	$2 \cdot 10^3\,\mathrm{s}$	σ_0	$10.27\,\mathrm{kN/m^2}$
	r	1		
Cosserat	l_c^S	$1 \cdot 10^{-3}\,\mathrm{m}$	μ_c^S	$4 \cdot 10^3\,\mathrm{kN/m^2}$
parameters	k_M	12	k_T	0

TABLE 9.1. Material parameters.

The numerical examples presented here concern the wide range of applications of the TPM approach to geomechanical problems like the leaking and wetting of a porous column, the saturated and the unsaturated consolidation and two principally different localization phenomena like the well-known biaxial experiment and the base failure problem. The examples generally proceed from the material parameters included into Table 9.1 and are computed by use of the FE package PANDAS, where the the time- and space-adaptive methods introduced in Section 9.4 are widely applied.

Proceeding from the fact that the regularization of localization problems on the basis of the inclusion of micropolarity, viscoplasticity and fluid viscosity has been demonstrated in a series of articles [18, 23, 26, 27, 33, 34], it is also the goal of the present paper to show how these regularization methods combine with time- and space-adaptive strategies. In particular, the first localization example (the biaxial experiment) concerns an empty elasto-plastic skeleton, where the shear band computation is regularized by the inclusion of micropolarity, the second localization example exhibits the same basic situation, however applied to a fluid-saturated elasto-viscoplastic material, whereas the third localization example (the base failure problem) includes the combination of micropolarity and fluid viscosity.

9.5.1 Leaking and wetting of a porous column

(a) The leaking problem. The present example exhibits a rigid soil

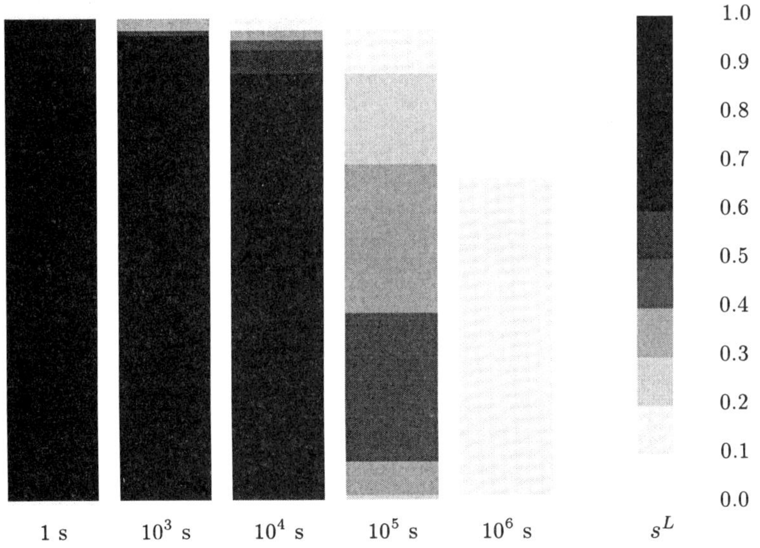

FIGURE 9.6. Progression of liquid saturation s^L.

column of 1 m height which is fully liquid-saturated ($s^L = 1$) at time $t_0 = 0$. In order to correctly describe the leakage process, a triphasic medium is considered, where at times $t \geq t_0$ the values of the liquid saturation s^L characterize the distribution of the pore-fluids water and air throughout the soil column. In particular, the initial conditions are prescribed through $s^L = 0$ at the top and at the bottom of the sample. Figure 9.6 shows the progression of the leaking process driven by gravitation, whereas Figure 9.7 represents the efflux of liquid volume through the bottom of the soil column.

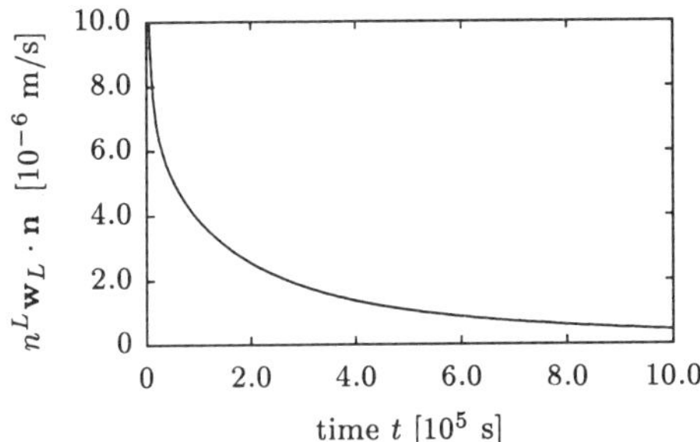

FIGURE 9.7. Leakage vs. time at the bottom of the column.

(b) The wetting problem. Basically, the wetting problem concerns the same soil column as before. However, the present sample is assumed to be fully gas-saturated ($s^L = 0$) at time $t_0 = 0$. Furthermore, the prescribed initial conditions are given by $s^L = 0$ at the top and $s^L = 1$ at the bottom of the column. Figure 9.8 shows the progression of the wetting process driven by capillary suction *vs.* gravitation, whereas Figure 9.9 represents the final distribution of the liquid-saturation *vs.* the column height.

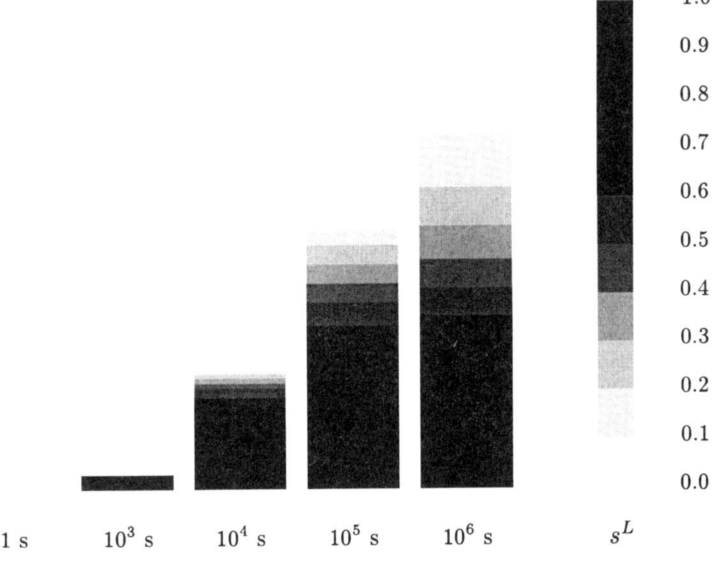

FIGURE 9.8. Progression of liquid saturation s^L.

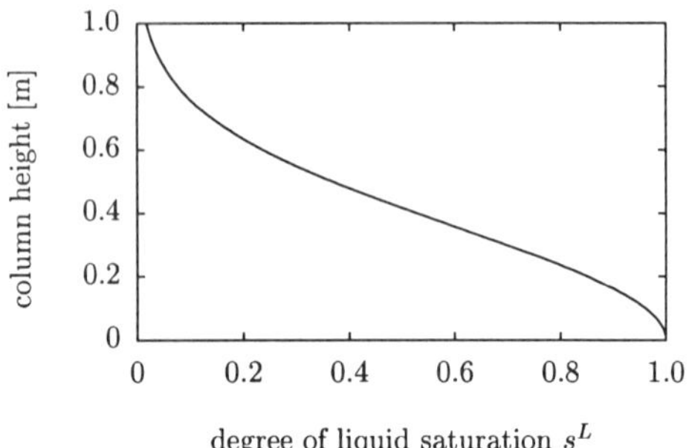

FIGURE 9.9. Final distribution of the liquid saturation vs. column height.

9.5.2 Saturated and unsaturated consolidation

(a) The classical consolidation problem. In the classical literature, *cf.*, *e.g.*, Terzaghi [53], the consolidation problem is defined by the onset of an additional external load (*e.g.*, a building) onto a fluid-saturated porous elastic solid (*e.g.*, a soil), *cf.* Figure 9.10. As a result, a time-depending settlement process occurs which is accompanied by a drainage of the viscous pore-content. The present example concerns the well-known problem

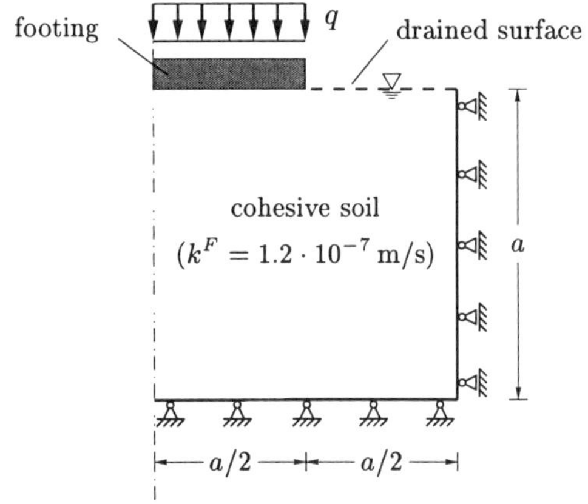

FIGURE 9.10. Rigid strip footing on a water-saturated half-space.

of a rigid strip footing on a soil half-space. Furthermore, the computations are based on the standard quasi-static formulation of a materially incompressible linearly elastic (non-polar) skeleton which is fully saturated by a viscous and materially incompressible pore-water. In order to model the half-space, the dimension a is chosen large enough so that the impermeable boundaries at both sides and at the bottom of the sample do not influence

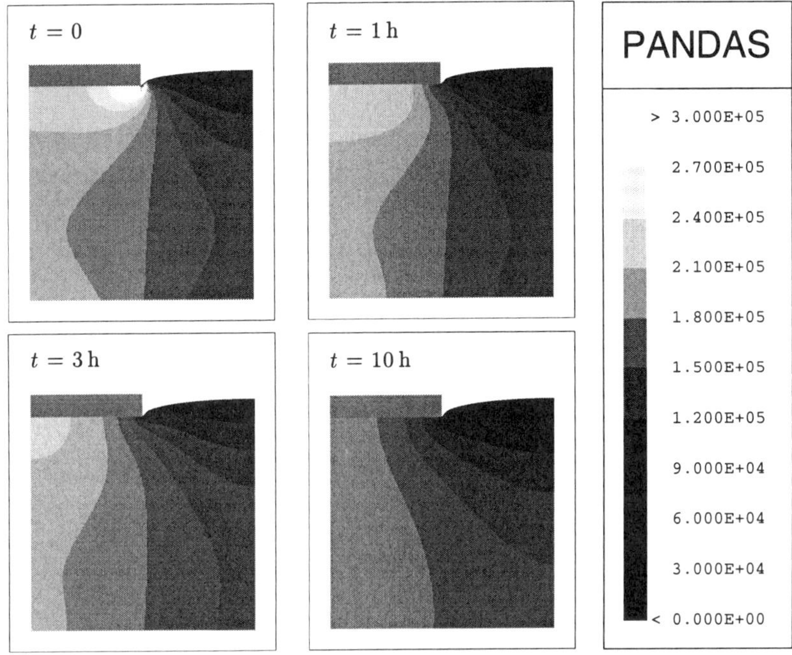

FIGURE 9.11. Decrease of the pore-water pressure p $[\mathrm{N/m}^2]$ at the deformed soil skeleton (scaling factor 10).

the numerical solution.

In particular, the external load q is linearly increased from zero to $q_{max} = 300\,\mathrm{kN/m}^2$. As a reaction on the external load, there is not an instantaneous but a time-dependent deformation of the elastic skeleton due to the drainage process of the pore-water. Figure 9.11 exhibits the decrease of the pore-water pressure p (excess over the atmospheric pressure) throughout the consolidation process, where $t = 0$ indicates the time when q_{max} is reached and kept constant for $t > 0$. The development of the pressure isolines between $t = 0$ and $t = 10\,\mathrm{h}$ shows that, firstly, the maximum pressure is directly under the footing, while $p = 0$ is prescribed at the drained surface. Secondly, the pore-water pressure decreases with time up to an overall value of $p = 0$ at $t \to \infty$.

(b) Soil subsidence by loss of ground water. Apart from the standard consolidation problem described above, it may occur under more realistic circumstances that the soil does not only behave purely elastically. Instead, it generally behaves elasto-plastically in a wide range of deformations. Moreover, in addition to an external loading process, there may be further reasons for a soil subsidence. Therefore, the present example concerns, firstly, the standard strip footing situation on a liquid-saturated half-

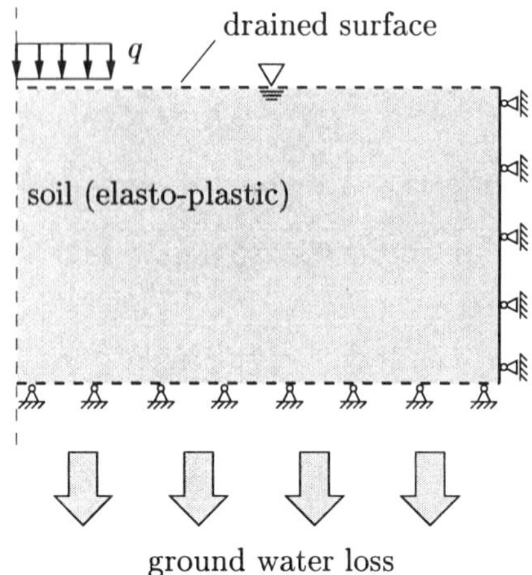

FIGURE 9.12. Strip footing and loss of ground water.

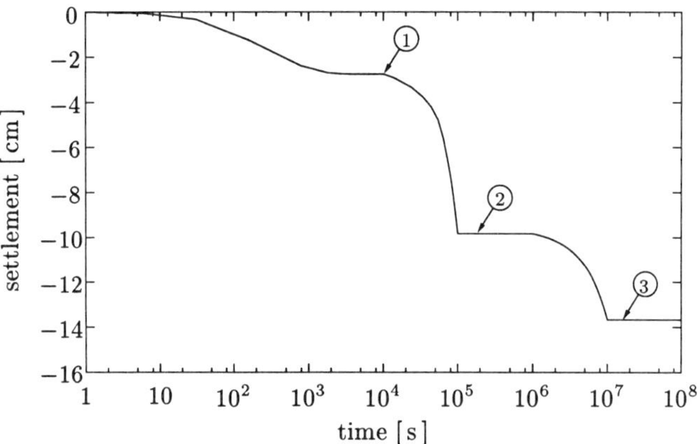

FIGURE 9.13. Time-settlement curve.

space, where the soil is considered as a materially incompressible elasto-plastic (non-polar) skeleton. Secondly, when the maximum settlement of the elasto-plastic consolidation process is reached, it is assumed that there is a loss of ground-water through the bottom of the domain under consideration (see Figure 9.12). This is modeled by switching the boundary condition at the bottom from impermeable to permeable, where, in addition, a fluid suction $p < 0$ is prescribed. Furthermore, $s^L = 0$ is prescribed at the top boundary, so that an unsaturated domain is generated in the sense of the general triphasic medium. In reality, these boundary conditions correspond, *e.g.*, to the case where the liquid pore-content in the area under the considered domain is pumped out. Problems like this occur, for example, in the surrounding of open coal mines or near oil and gas reservoirs under

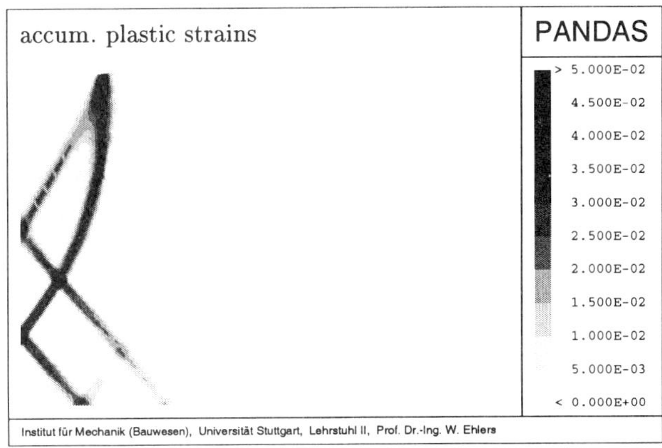

FIGURE 9.14. Surface subsidence after elasto-plastic consolidation.

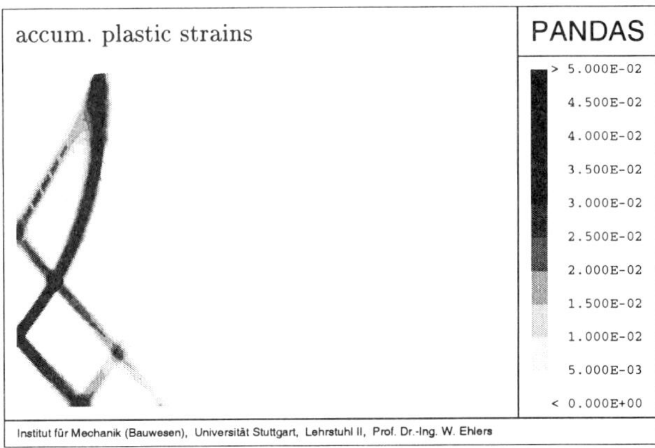

FIGURE 9.15. Surface subsidence after loss of ground water.

exploitation.

Figure 9.13 shows the time-settlement curve of the whole process including the pre-consolidation domain. In particular, point (1) indicates the end of the pre-consolidation process driven by gravitation, where, in addition, the external load q is applied. Point (2) characterizes the end of the standard consolidation process under fully liquid-saturated conditions, where, at the same time, the loss of ground water is initiated. Finally, point (3) exhibits the final situation, when the whole process has come to an end. In addition to Figure 9.13, Figures 9.14 and 9.15 present the accumulated plastic strains after having reached the consolidation points (2) and (3). Note in passing that, obviously, the plastic strains localize in small bands, the so-called shear bands, thus giving a hint on a possible base failure problem.

9.5.3   The biaxial experiment

(a) The empty elasto-plastic micropolar solid (hierarchical h-adaptive scheme). The present example exhibits a simulation of a biaxial experiment, where plane strain conditions are prescribed, *cf.* Figure 9.16. In particular, the computations are carried out on the basis of time-

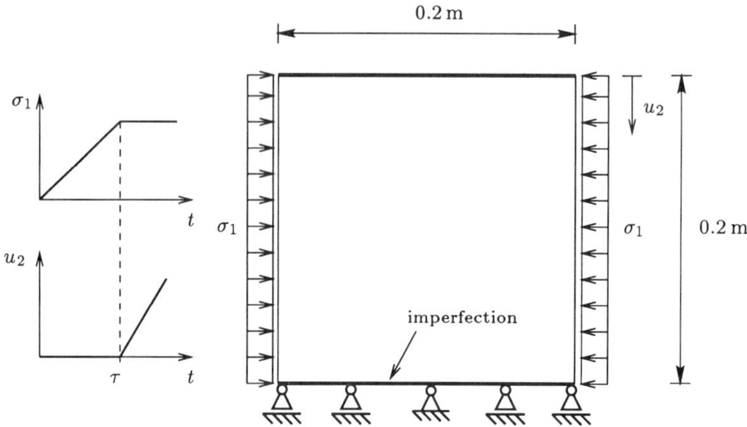

FIGURE 9.16. Biaxial experiment: initial boundary-value problem.

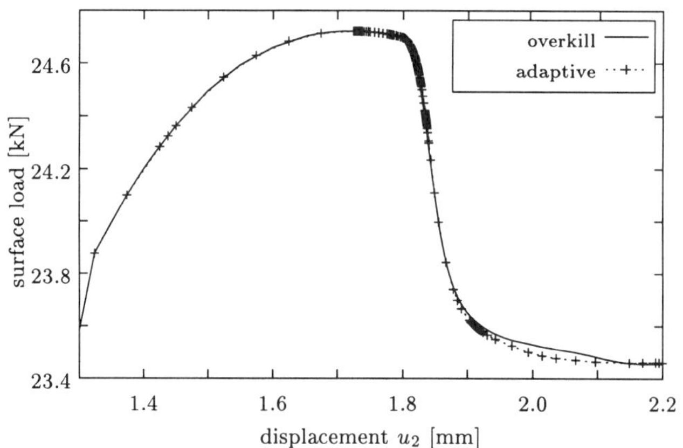

FIGURE 9.17. Biaxial experiment: load-deflection curves.

and space-adaptive methods, where use is made of the hierarchical refinement and de-refinement scheme. Furthermore, an empty granular micropolar elasto-plastic solid skeleton is considered, where at time $0 \leq t \leq \tau$ a linearly increasing horizontal stress σ_1 is applied, which is kept constant when $t > \tau$. Simultaneously, the specimen is loaded with a displacement-driven vertical stress such that $u_2 = (u_2)'_S (t - \tau)$. From the numerical

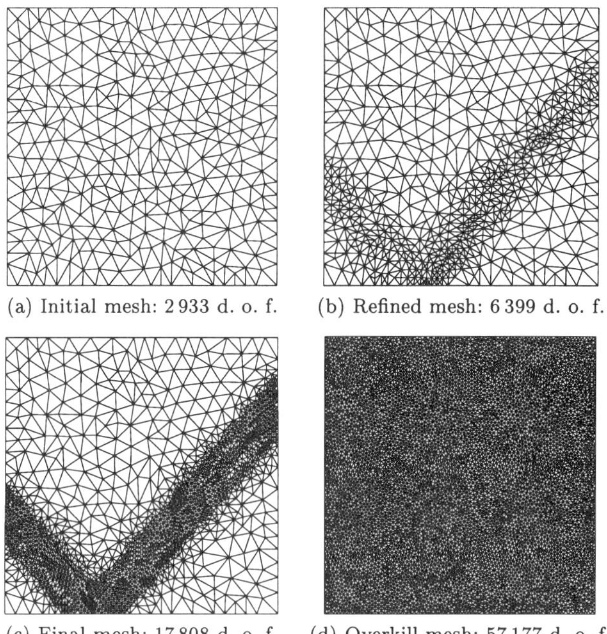

(a) Initial mesh: 2 933 d. o. f.   (b) Refined mesh: 6 399 d. o. f.

(c) Final mesh: 17 808 d. o. f.   (d) Overkill mesh: 57 177 d. o. f.

FIGURE 9.18. Hierarchical adaptive mesh refinement (micropolar formulation).

point of view, Figure 9.16 characterizes a homogeneous problem. In order to obtain a shear band localization, an imperfection at the bottom of the material is included, thus initiating the shear band development.

Since there is no possibility to obtain an analytical solution of the problem, the adaptive computations are compared to a so-called overkill solution carried out on an extremely fine mesh which is kept constant during the numerical solution process. Figure 9.17 shows the load-deflection curves of the adaptive and the overkill solutions, where, as a result of the time- and space-adaptive strategy, both solutions are equivalent up to a great extent. It is furthermore seen by the crosses marking the varying time step sizes of the adaptive solution that very small time steps occur when the localization initiates and the load-defection curve decreases nearly vertically.

The computations start with an initial mesh of triangular elements with 2933 d. o. f., compare Figure 9.18. During the computational process, a mesh refinement takes place initiated by the onset of plastic yielding at the imperfection point. Based on prescribed values of the minimum element size, the final mesh results in 17808 d. o. f. To compare the solution of the adaptive computation with an overkill solution, which is assumed to be close to the "true" solution of the problem, a considerably finer mesh of 57177 d. o. f. is used. The regularization of the problem is seen from the

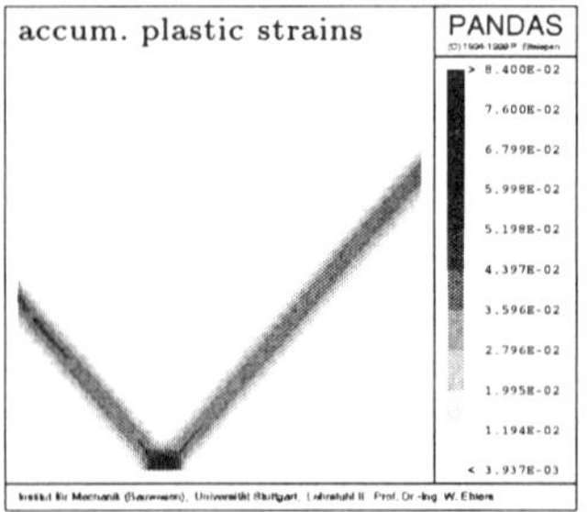
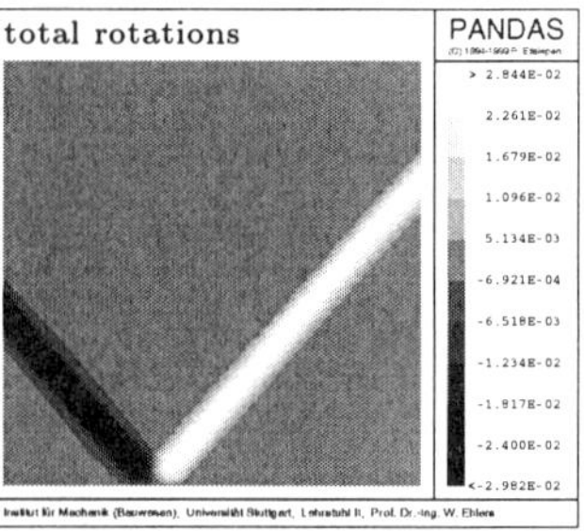

FIGURE 9.19. **Accumulated plastic strains and total rotations (micropolar formulation).**

fact that the distribution of the plastic strains, compare Figure 9.19 (left), exhibits the shear band development which clearly spans several element layers.

Finally, Figure 9.19 (right) shows the distribution of the total rotation produced by micropolarity. Following this, it is clearly seen that there is approximately no rotation outside the shear bands. Inside the shear bands, the rotation is either positive or negative. As a result, the rotation is zero at the crossing point of positively and negatively rotating shear bands. In the present example, such a point is given at the imperfection which causes the shear band initiation.

(b) The liquid-saturated elasto-viscoplastic solid (hierarchical and remeshing h-adaptive schemes). The following considerations concern the same initial boundary-value problem as before, however, in contrast to the previous section, the present example makes use of a binary material, namely a liquid-saturated solid skeleton consisting of an incompressible pore-fluid (the pore-water) and a materially incompressible elasto-viscoplastic skeleton in the framework of the standard non-polar formulation. Following this, the regularization of shear band phenomena is based on the included solid viscoplasticity and on the fluid viscosity. Furthermore, the boundary-value problem is defined as was described in the preceding example, except for the fact that here, after having applied the lateral load (phase 1), the rigid load platen is only driven into the specimen (phase 2) when the consolidation process resulting from σ_1 is finished. Furthermore, to compare the effort of hierarchical and remeshing strategies, the computations are carried out by use of both hierarchical and the remeshing methods. In particular, the specimen is loaded displacement driven by a rigid load platen with $\dot{u}_2 = 2 \cdot 10^{-7}$ m/s. While the top and bottom boundaries of the specimen are impermeable, the side boundaries are assumed to be ideally permeable. Furthermore, the side boundaries are stabilized by a linearly increasing stress of $\sigma_1 \le 50$ kN/m^2 (Figure 9.16).

To compare the hierarchical (H) and the remeshing strategies (R), both the computing times and the accuracy of the solutions are considered. In particular, it turns out that there is a substantial difference between the computations based on the hierarchical and on the remeshing strategy. Proceeding from the same user-defined tolerances, the remeshing strategy, as far as this example is concerned, is more efficient than the hierarchical procedure. This is seen from the fact that the hierarchical scheme, on an SGI Power Challenge R 10000/195, needs a computing time of 6 h 56 min to result in 32 046 elements, whereas the remeshing strategy comes along with only 5 h 21 min and 26 947 elements. The mesh refinement of the remeshing process together with an overkill mesh can be taken from Figure 9.20.

However, although there were considerable differences in the performance of both strategies, there is no significant difference in the corresponding load-deflection curves, *cf.* Figure 9.21. Furthermore, it is seen from Figure 9.21 by the crosses marking the varying time step sizes of the adaptive solution obtained by use of the remeshing strategy that very small time steps occur when the localization is initiated.

(c) The liquid-saturated elasto-viscoplastic solid (compressive and dilatant shear bands). The following considerations concern the same example as before, namely the liquid-saturated solid skeleton consisting of an incompressible pore-fluid (the pore-water) and a materially incompressible elasto-viscoplastic skeleton. Following this, the regularization of shear band phenomena is again based on the included solid viscoplasticity and on the fluid viscosity. In particular, two different lateral loads are applied, thus leading to both contractant and dilatant shear bands, compare Figures 9.22 and 9.23. The reason for the onset of either contractant or dilatant shear bands stems from the fact that, when viscoplastic yielding occurs, a lateral load of $\sigma_1 = 140\,\mathrm{kN/m^2}$ leads to a yield point in the ductile regime, whereas a lateral load of $\sigma_1 = 110\,\mathrm{kN/m^2}$ produces a yield point in the brittle regime.

It is furthermore seen from Figure 9.22 that the final vertical displacement of $u_2 = 5\,\mathrm{mm}$ yields comparable shear band developments in both cases the contractant and the dilatant one. However, as a result of dilatancy, the maximum value of the accumulated plastic strains is approximately 50 % higher in the dilatant case than in the contractant one. The reason for this fact mainly results from the incompressibility constraint of the pore-fluid. To explain this behaviour in more detail, the reader is referred to the pore-fluid pressure and the seepage velocity developments of both computations. In the first case of Figure 9.22, one obtains a high pore-fluid pressure and, as a result of contractancy, a pore-fluid that is pressed out of the shear band, whereas, in the second case of Figure 9.23, one observes a considerable fluid suction and, as a result, a seepage flow into the shear band. Furthermore, as a result of contractancy, the skeleton parts separated by the shear band glide upon each other very easily,

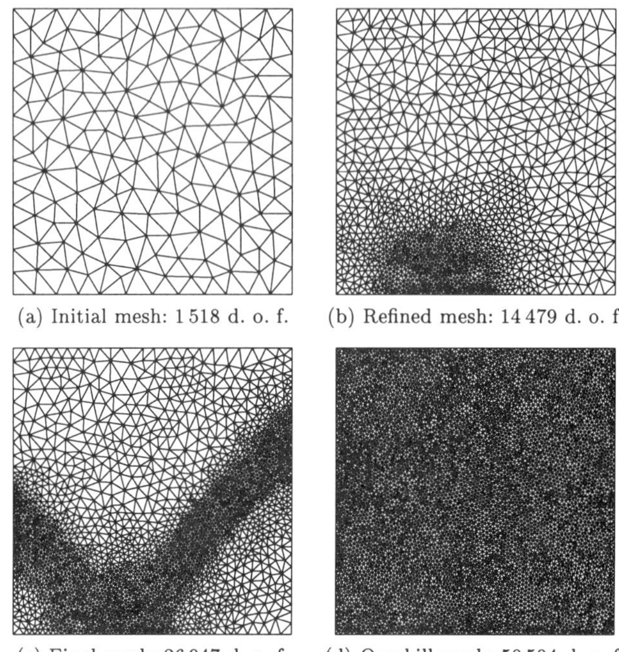

(a) Initial mesh: 1 518 d. o. f.   (b) Refined mesh: 14 479 d. o. f.

(c) Final mesh: 26 947 d. o. f.   (d) Overkill mesh: 58 504 d. o. f.

FIGURE 9.20. Remeshing scheme (standard formulation).

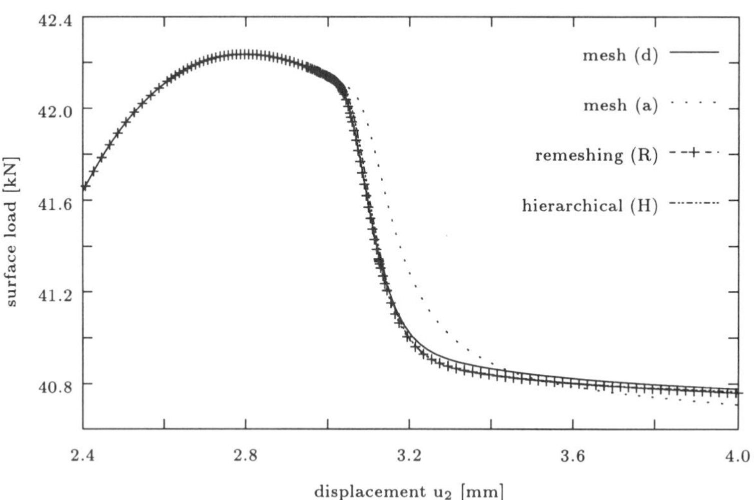

FIGURE 9.21. Biaxial experiment: load-deflection curves.

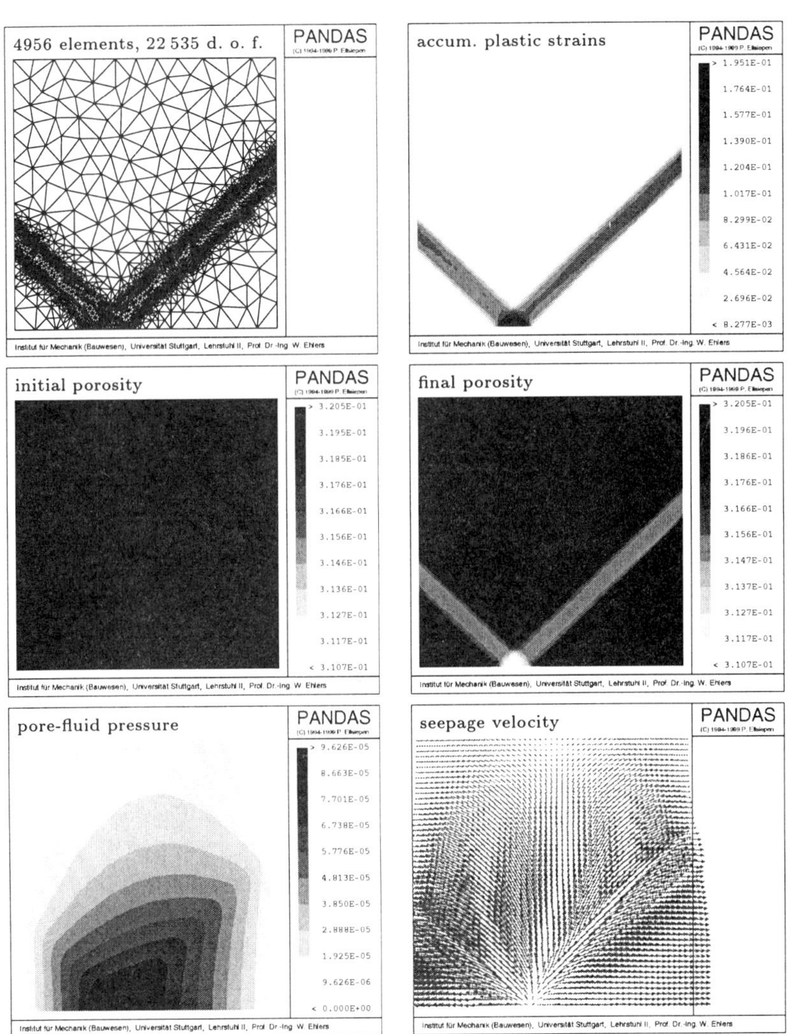

FIGURE 9.22. Contractant shear band development ($\sigma_1 = 140\,\mathrm{kN/m^2}$).

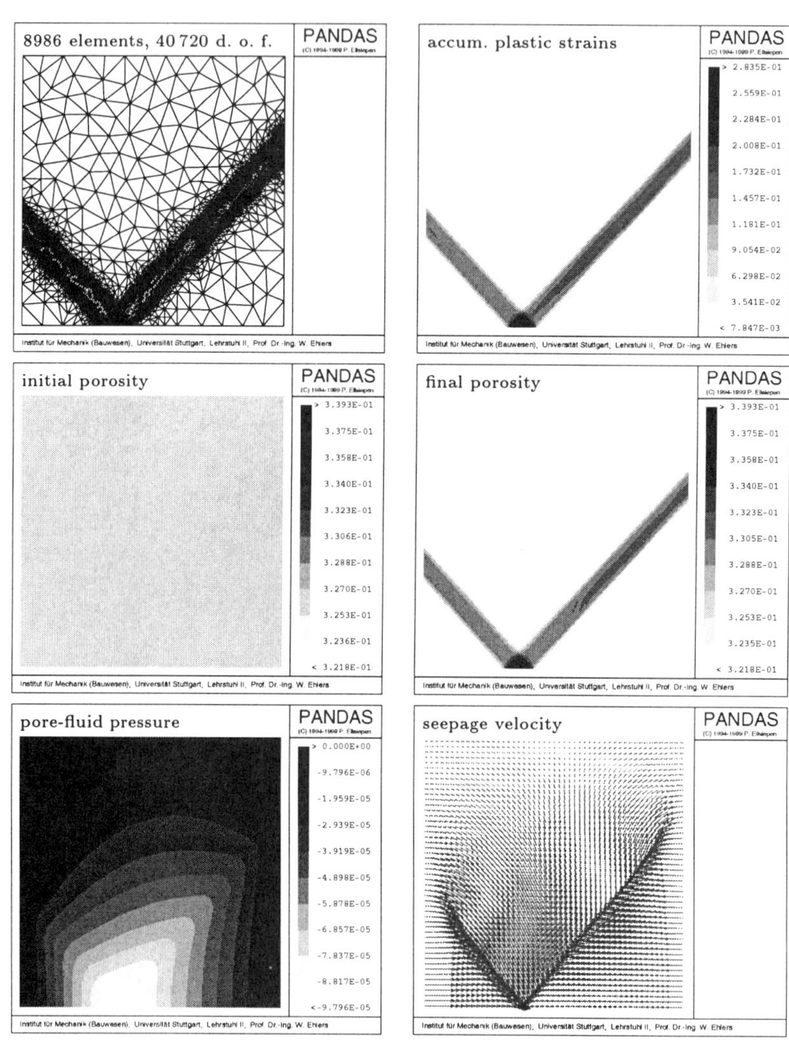

FIGURE 9.23. Dilatant shear band development ($\sigma_1 = 110\,\mathrm{kN/m^2}$).

whereas, as a result of dilatancy, they suck onto each other, thus gluing the skeleton parts together.

Proceeding from the porosity of the specimen after phase 1 of the loading process is passed, the final porosity development in the shearing zones either exhibits an increase or a decrease of only 3 %. Note in passing that the final mesh obtained by use of the hierarchical procedure ranges from 22 535 d. o. f. in the contractant case to 40 720 in the dilatant one. Obviously, this is again a result of either contractancy or dilatancy, since, in the dilatant case, gluing together the skeleton parts as a result of fluid suction needs a much larger loading force to obtain the final vertical displacement. In comparison of the meshes of Figures 9.22 and 9.23 with that of Figure 9.18 (c), one observes that the mesh refinement of the present viscoplastic computation mainly concentrates at the boundaries of the shear bands, whereas it is approximately constant in case of the micropolar elasto-plastic material. Following this, it is finally concluded that the gradients of the plastic strains are much steeper in the elasto-viscoplastic case than in the micropolar elasto-plastic one.

9.5.4   The base failure problem

The final example of this article exhibits the well-known base failure problem, compare Figure 9.24 and the book by Terzaghi & Jelinek [54]. It furthermore corresponds to the elasto-viscoplastic consolidation problem described in Subsection 9.5.2 (b). Concerning the problem under study, an external load q of a rigid strip footing is applied onto the surface of a liquid-saturated half-space. As a result, a more or less undeformed wedge (the active state) is driven into the half-space, whereas the remainder of the shaded zone is either interspersed with plastic deformations and a field of shear bands (the radial slip domain) or it glides upwards on the failure line (the passive state) which separates the plastic from the elastic domains of the overall problem. It is furthermore assumed that the binary model consists of a materially incompressible elasto-viscoplastic skeleton (standard formulation) saturated by an incompressible viscous pore-liquid. Assuming a rather high permeability through ($k_0^F = 10^{-2}\,\mathrm{m/s}$), the mathematical solution is only regularized by the included viscoplastic properties of the solid skeleton.

To get a high resolution result of this complex problem, the time- and space-adaptive strategies must be applied. The computations start with the initial mesh shown in Figure 9.25. Initiated by the singularity between the loaded and the non-loaded parts of the upper boundary, shear bands are initiated covering the whole area of the radial slip domain. As a result of the symmetry condition included in the solution of the boundary value problem, the shear bands are reflected at the symmetry line, thus delivering those shear band developments which include the final failure line, compare Figures 9.26 and 9.27. In particular, Figure 9.26 exhibits the final mesh

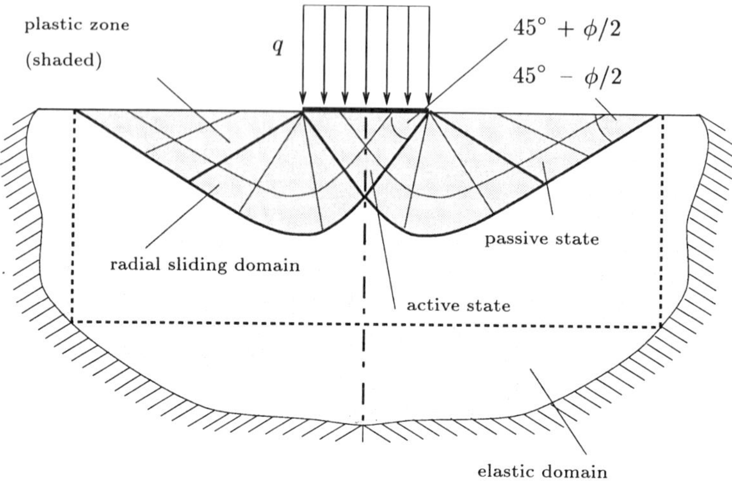

FIGURE 9.24. The classical base failure problem.

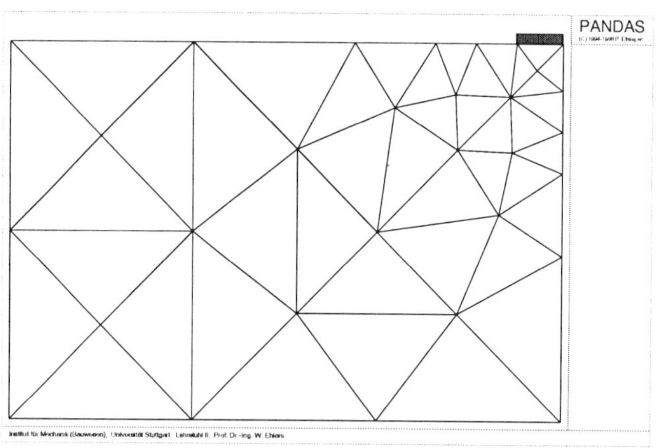

FIGURE 9.25. Initial mesh (symmetric problem).

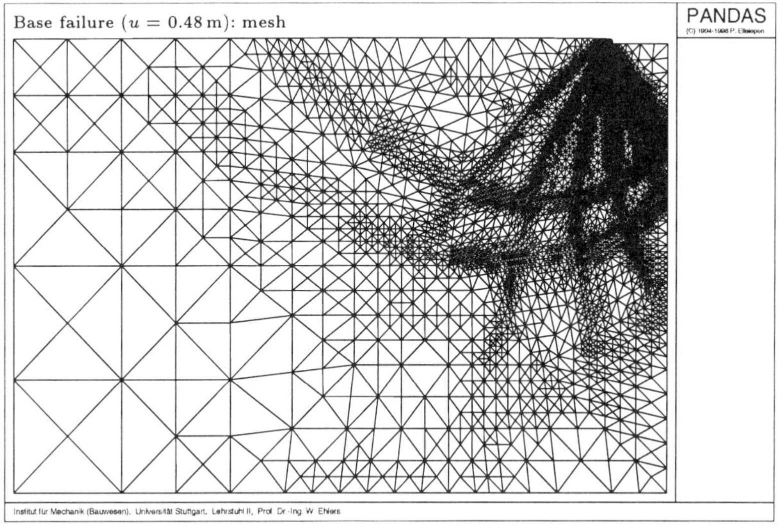

FIGURE 9.26. Base failure problem: Final mesh (30 643 nodes, 15 254 elements, 68 981 d. o. f.).

development at a vertical displacement of the rigid strip footing of 48 cm.

From this figure, it is furthermore not only seen that the mesh development excellently catches the shear band developments, compare Figure 9.27, but also that the problem is well regularized, since the occurring shear bands are mesh-independent and the shear band widths exceed the mesh size. Concerning the quantification of the FE problem in the final state, it should be noted that the problem is characterized by 30 643 nodes, 15 254 elements, 68 981 d. o. f. and 228 810 internal variables.

From the mathematical point of view, the base failure problem represents a very difficult initial boundary value problem including very complex phenomena such as shear bands initiated by a singularity. Following this, there are generally no mathematically ensured statements available on the existence and the uniqueness of the solutions. However, based on the excellent results of the verification examples included in Chapter 5, the theoretical methods described in the present article nevertheless allow for a numerical computation of these problems without any theoretical *a priori* statement. Following this, the foregoing applications show that the continuum mechanical basis together with the time- and space-adaptive methods are not only convenient but also appropriate to solve a variety of practical problems occurring in the framework of geomechanical engineering.

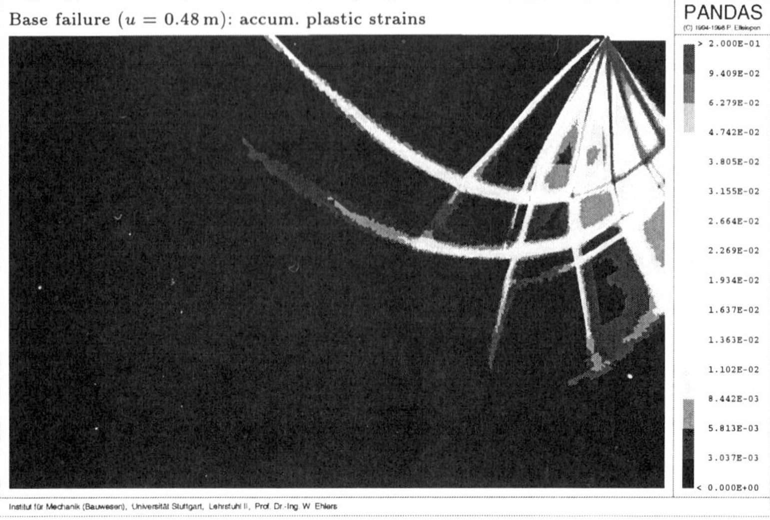

FIGURE 9.27. Base failure problem: Accumulated plastic strains.

9.6 Concluding remarks

In the present article, geomechanical problems have been discussed on the basis of the macromechanical approach of the theory of porous media. In particular, biphasic and triphasic fluid-saturated and unsaturated porous skeleton materials have been investigated in the framework of both the standard non-polar and the extended micropolar formulations including elastic, viscous and plastic properties of the constituents. Furthermore, a general strategy of treating volumetrically coupled solid-fluid problems within the finite element method was discussed and extended towards time- and space-adaptive strategies.

Based on this approach, a lot of examples have been computed by use of the FE tool PANDAS. In particular, biphasic and triphasic coupled solid-fluid problems have been treated including the mathematically difficult problems of shear band phenomena, where the regularization of a basically ill-posed problem, as for instance an empty skeleton material described within the standard elasto-plastic formulation, could be carried out by the inclusion of additional degrees of freedom in the sense of the *Cosserat* brothers (extended micropolar formulation) and/or by the inclusion of elasto-viscoplastic skeleton behaviour and/or by viscous pore-fluid properties (standard non-polar formulation).

The numerical examples demonstrate the efficiency of the proposed method. In particular, proceeding from the extended micropolar formulation opens the possibility to implicitly include the shear band width into the problem by proposing convenient numbers for the internal length scale parameter. Furthermore, since the micropolar rotations are only active in the localization zones, micropolarity is a very convenient regularization tool that does not affect the solution in the non-localizing zones. In contrast, although viscoplasticity is also very convenient to be applied as a regularization tool, it implies a principally different material behaviour and can thus only be used if the material also behaves viscoplastically in reality. On the other hand, since the inclusion of a viscous pore-fluid concerns a different mechanical problem than the description of an empty porous material, regularization by fluid viscosity is clearly restricted to volumetrically coupled saturated problems.

The work presented here and the methods to describe coupled solid-fluid problems can of course not only be applied to geomechanical problems or general environmental mechanics, respectively, but also to mechanical engineering problems, as for example to porous polymeric and aluminium foams as very promising new materials for the transport industry (*cf.*, *e.g.*, the work by Ehlers & Droste [25]), or to biomechanical problems, as for instance soft tissues or articular cartilage (*cf.*, *e.g.*, Mow *et al.* [47] or Ehlers & Markert [30, 31]).

Acknowledgments: The work presented in this article is a result of several years of investigation at my institute. In particular, the numerical computations have been carried out by my coworkers or former coworkers Martin Ammann, Peter Blome, Peter Ellsiepen and Wolfram Volk. Their work was essential for the preparation of this article and is herewith gratefully acknowledged.

References

[1] M.A. BIOT: General theory of three-dimensional consolidation, *J. Appl. Phys.* **12**, pp. 155–164, 1941.

[2] M.A. BIOT: Theory of propagation of elastic waves in a fluid-saturated porous solid, I: low frequency range, *J. Acoust. Soc. Am.* **28**, pp. 168–178, 1956.

[3] A.W. BISHOP: The effective stress principle, *Teknisk Ukeblad* **39**, pp. 859–863, 1959.

[4] R.M. BOWEN: Theory of mixtures, in: A.C. ERINGEN (ed): *Continuum Physics; Volume III: Mixtures and EM Field Theories*, Academic Press, New York, 1976, pp. 1–127.

[5] R.M. BOWEN: Incompressible porous media models by use of the theory of mixtures, *Int. J. Engng. Sci.* **18**, pp. 1129–1148, 1980.

[6] R.M. BOWEN: Compressible porous media models by use of the theory of mixtures, *Int. J. Engng. Sci.* **20**, pp. 697–735, 1982.

[7] R.B.J. BRINKGREVE: *Geomaterial Models and Numerical Analysis of Softening*, Ph.D. thesis, CIP-Gegevens Koniglijke Bibliotheek, Den Haag, 1994.

[8] O. COUSSY: *Mechanics of Porous Continua*, Wiley, Chichester, 1995.

[9] R.H. CRAWFORD, D.C. ANDERSON, W.N. WAGGENSPACK: Mesh rezoning of 2-d isoparametric elements by inversion, *Int. J. Num. Meth. Engng.* **28**, pp. 523–531, 1989.

[10] R. DE BOER: Highlights in the historical development of porous media theory: toward a consistent macroscopic theory, *Applied Mechanics Review* **49**, pp. 201–262, 1996.

[11] R. DE BOER: *Theory of Porous Media*, Springer-Verlag, Berlin, 2000.

[12] R. DE BOER, W. EHLERS: *Theorie der Mehrkomponentenkontinua mit Anwendung auf Bodenmechanische Probleme*, Forschungsberichte aus dem Fachbereich Bauwesen **40**, Universität-GH-Essen, 1986.

[13] R. DE BOER, W. EHLERS: Uplift, friction and capillarity – three fundamental effects for liquid-saturated porous media, *Int. J. Solids Struct.* **26**, pp. 43–57, 1990.

[14] R. DE BOER, W. EHLERS, S. KOWALSKI, J. PLISCHKA: *Porous Media – A Survey of Different Approaches*, Forschungsberichte aus dem Fachbereich Bauwesen **54**, Universität-GH-Essen, 1991.

[15] R. DE BORST: Simulation of strain localization: A reappraisal of the Cosserat continuum, *Engng. Comput.* **8**, pp. 317–332, 1991.

[16] S. DIEBELS: *Mikropolare Zweiphasenmodelle: Formulierung auf der Basis der Theorie Poröser Medien*, Habilitation, Bericht Nr. II-4, Institut für Mechanik (Bauwesen), Universität Stuttgart, 2000.

[17] S. DIEBELS, W. EHLERS: On fundamental concepts of multiphase micropolar materials, *Technische Mechanik* **16**, pp. 77–88, 1996.

[18] S. DIEBELS, P. ELLSIEPEN, W. EHLERS: Error-controlled *Runge-Kutta* time integration of a viscoplastic hybrid two-phase model, *Technische Mechanik* **19**, pp. 19–27, 1999.

[19] W. EHLERS: *Poröse Medien – ein Kontinuumsmechanisches Modell auf der Basis der Mischungstheorie*, Forschungsberichte aus dem Fachbereich Bauwesen **47**, Universität-GH-Essen, 1989.

[20] W. EHLERS: Constitutive equations for granular materials in geomechanical context, in: K. HUTTER (ed.): Continuum mechanics in environmental sciences and geophysics, *CISM Courses and Lectures* **337**,
Springer-Verlag, Wien, 1993, pp. 313–402.

[21] W. EHLERS: A single-surface yield function for geomaterials, *Arch.
Appl. Mech.* **65**, pp. 246–259, 1995.

[22] W. EHLERS: Grundlegende konzepte in der theorie poröser medien,
Technische Mechanik **16**, pp. 63–76, 1996.

[23] W. EHLERS, M. AMMANN, S. DIEBELS: *h*-Adaptive FE methods applied to single- and multiphase problems, *Int. J. Numer. Meth. Engng.*
54, pp. 219–239, 2002.

[24] W. EHLERS, P. BLOME: A triphasic model for unsaturated soils based
on the theory of porous media, in: G. CAPRIZ, V.N. GHIONNA, P.
GIOVINE, N. MORACI (eds.): Mathematical Models in Soil Mechanics,
Mathematical and Computer Modelling, Elsevier Science Publishers,
Amsterdam, 2002, in press.

[25] W. EHLERS, A. DROSTE: A continuum model for highly porous aluminium foam, *Technische Mechanik* **19**, pp. 341–350, 1999.

[26] W. EHLERS, P. ELLSIEPEN: Theoretical and numerical methods in environmental continuum mechanics based on the theory of porous media, in: B. A. SCHREFLER (ed.): Environmental geomechanics, *CISM
Courses and Lectures* **417**, Springer-Verlag, Wien, 2001, pp. 1–81.

[27] W. EHLERS, P. ELLSIEPEN, M. AMMANN: Time- and space-adaptive
methods applied to localization phenomena in empty and saturated
micropolar and standard porous materials, *Int. J. Numer. Meth. Engng.* **52**, pp. 503-526, 2001.

[28] W. EHLERS, P. ELLSIEPEN, P. BLOME, D. MAHNKOPF, B. MAR-
KERT: *Theoretische und Numerische Studien zur Lösung von Randund
Anfangswertproblemen in der Theorie Poröser Medien*, Abschlußbe-
richt zum DFG-Forschungsvorhaben Eh 107/6-2, Bericht Nr. 99-II-1,
Institut für Mechanik (Bauwesen), Universität Stuttgart, 1999.

[29] W. EHLERS, J. KUBIK: On finite dynamic equations for fluid-saturated porous media, *Acta Mechanica* **105**, pp. 101–317, 1994.

[30] W. EHLERS, B. MARKERT: On the viscoelastic behaviour of fluid-saturated materials, *Granular Matter* **2**, pp. 153–161, 2000.

[31] W. EHLERS, B. MARKERT: A linear viscoelastic biphasic model for
soft tissues based on the theory of porous media, *ASME J. Biomech.
Engng.* **123**, pp. 418-424, 2001.

[32] W. EHLERS, H. MÜLLERSCHÖN: Stress-strain behaviour of cohesionless soils: experiments, theory and numerical computations, in: A. CIVIDINI (ed.): Application of numerical methods to geotechnical problems, *CISM Courses and Lectures* **397**, Springer-Verlag, Wien, 1998, pp. 675–684.

[33] W. EHLERS, W. VOLK: On shear band localization phenomena of liquid-saturated granular elasto-plastic porous solid materials accounting for fluid viscosity and micropolar solid rotations, *Mech. Coh.-Frict. Materials* **2**, pp. 301–320, 1997.

[34] W. EHLERS, W. VOLK: On theoretical and numerical methods in the theory of porous media based on polar and non-polar elasto-plastic solid materials, *Int. J. Solids Struct.* **35**, pp. 4597–4617, 1998.

[35] P. ELLSIEPEN: *Zeit- und Ortsadaptive Verfahren Angewandt auf Mehrphasenprobleme Poröser Medien*, Dissertation, Bericht Nr. II-3, Institut für Mechanik (Bauwesen), Universität Stuttgart, 1999.

[36] A.C. ERINGEN: On non-local plasticity, *Int. J. Engng. Sci.* **19**, pp. 1461–1474, 1981.

[37] L. GALLIMARD, P. LADEVÈZE, J. P. PELLE: Error estimation and adaptivity in elastoplasticity, *Int. J. Numer. Methods Engng.* **39**, pp. 129–217, 1996.

[38] E. HAIRER, C. LUBICH, M. ROCHE: *The Numerical Solution of Differential-Algebraic Equations by Runge-Kutta Methods*, Springer-Verlag, Berlin, 1989.

[39] E. HAIRER, G. WANNER: *Solving Ordinary Differential Equations, Vol. 2: Stiff and Differential-Algebraic Problems*, Springer-Verlag, Berlin, 1991.

[40] S.M. HASSANIZADEH, W.G. GRAY: General conservation equations for multi-phase systems: 1. Averaging procedure, *Adv. Water Resour.* **2**, pp. 131–144, 1979.

[41] S.M. HASSANIZADEH, W.G. GRAY: General conservation equations for multi-phase systems: 2. Mass, momentum, energy and entropy equations, *Adv. Water Resour.* **2**, pp. 191–203, 1979.

[42] I. KOSSACZKÝ: A recursive approach to local mesh refinement in two and three dimensions, *J. Comp. Appl. Math.* **55**, pp. 275–288, 1994.

[43] R. KRAUSE, E. RANK: A fast algorithm for point-location in a finite element mesh, *Computing* **57**, pp. 849–862, 1996.

[44] P. LADEVÈSE, J. P. PELLE, P. ROUGEOT: Error estimation and mesh optimization for classic finite elements, *Eng. Comp.* **8**, pp. 69–80, 1991.

[45] R.W. LEWIS, B.A. SCHREFLER: *The Finite Element Method in the Static and Dynamic Deformation and Consolidation of Porous Media*, Wiley, Chichester, 1st–edition, 1998.

[46] W.F. MITCHELL: Adaptive refinement for arbitrary finite-element spaces with hierarchical bases, *J. Comp. Appl. Math.* **36**, pp. 65–78, 1991.

[47] V.C. MOW, M.C. GIBBS, W.M. LAI, W.B. ZHU, K.A. ATHANASIOU: Biphasic indentation of articular cartilage - II, *J. Biomechanics* **22**, pp. 853–861, 1989.

[48] W. NOWACKI: *Theory of Asymmetric Elasticity*, Pergamon Press, Oxford, 1986.

[49] P. PERZYNA: Fundamental problems in viscoplasticity *Adv. Appl. Mech. Eng.* **9**, pp. 243–377, 1966.

[50] B.A. SCHREFLER, C.E. MAJORNA, L. SANAVIA: Shear band localization in saturated porous media, *Arch. Mech.* **47**, pp. 577–599, 1995.

[51] J.R. SHEWCHUK: *Triangle: A Two-Dimensional Quality Mesh Generator and Delaunay Triangulator*, School of Computer Science, Carnegie Mellon University, Pittsburgh, Pennsylvania, 1996; http://www.cs.cmu.edu/~quake/triangel.html.

[52] P. STEINMANN: A micropolar theory of finite deformation and finite rotation multiplicative elasto-plasticity, *Int. J. Solids Struct.* **31**, pp. 1063–1084, 1994.

[53] K. TERZAGHI: *Erdbaumechanik auf Bodenphysikalischer Grundlage*, Franz Deuticke, Leipzig, 1925.

[54] K. TERZAGHI, R. JELINEK: *Theoretische Bodenmechanik*, Springer-Verlag, Berlin, 1954.

[55] C. TRUESDELL, R.A. TOUPIN: The classical field theories, in: S. FLÜGGE (ed.): *Handbuch der Physik*, Springer-Verlag, Berlin **III/1**, pp. 226–902, 1960.

[56] M.T. VAN GENUCHTEN: A closed-form equation for predicting the hydraulic conductivity of unsaturated soils, *Soil Sci. Soc. America J.* **44**, pp. 892–898, 1980.

[57] W. VOLK: *Untersuchung des Lokalisierungsverhaltens Mikropolarer Poröser Medien mit Hilfe der Cosserat-Theorie*, Dissertation, Bericht Nr. II-2, Institut für Mechanik (Bauwesen), Universität Stuttgart, 1999.

[58] O.C. ZIENKIEWICZ, J.Z. ZHU: A simple error estimator and adaptive procedure for practical engineering analysis, *Int. J. Numer. Meth. Engng.* **24**, pp. 337–357, 1987.

WOLFGANG EHLERS
Institute of Applied Mechanics (CE)
University of Stuttgart
Pfaffenwaldring, 7 D-70569 Stuttgart, GERMANY
E-mail: Ehlers@mechbau.uni-stuttgart.de
http://www.mechbau.uni-stuttgart.de/ls2

Chapter 10

A Mathematical and Numerical Model for Finite Elastoplastic Deformations in Fluid Saturated Porous Media

Lorenzo Sanavia, Bernhard A. Schrefler and Paul Steinmann

ABSTRACT Finite elastic or elastoplastic strains in fluid saturated soils are studied. Isothermal and quasi-static loading conditions are considered. The governing equations at the macroscopic level are derived in a spatial and a material setting. The constituents are assumed to be materially incompressible at the microscopic level. The elasto-plastic behaviour of the solid skeleton is described by the multiplicative decomposition of the deformation gradient into an elastic and a plastic part; the elasto-plastic evolution laws are developed in the spatial setting. The Kirchhoff effective stress tensor and logarithmic principal strains are used in conjunction with an hyperelastic free energy function. The effective stress state is limited by the von Mises or the Drucker–Prager yield surface with isotropic hardening. Algorithmically, a particular "apex formulation" is advocated for the latter case. The fluid is assumed to obey Darcy's law. The consistent linearisation of the fully non-linear coupled system of equations is derived. A spatial finite element formulation is presented. Numerical examples highlight the developments.

10.1   Introduction

Fluid saturated porous materials like soils, biological tissues, polymer foams or metal foams often exhibit large elastic or elastoplastic strains, both in field situations and in laboratory tests. For soils, such a behavior typically results when an ultimate or serviceability limit state is reached, as for example during slope instability or during the consolidation process in compressible clays. In laboratory, this can be the case of drained or undrained biaxial tests of sands, where axial logarithmic strains of the order of 0.12–0.15 are reached [22, 31], or the case of triaxial tests of peats, where axial strains of the order of 0.15 are measured. Among others, the phenomenon of strain localisation (*i.e.*, strain accumulation in well-defined narrow zones, also called shear bands) is most typical whereby large inelastic strains develop, at least inside the bands [22, 31, 24]. Moreover, it reveals also the strong coupling which occurs between the solid skeleton and the fluids filling the voids of the porous material.

The central aim of this work is the development of a mathematical and a finite element model for fully coupled fluid saturated media undergoing large elastic or elastoplastic strains. In particular the approach proposed in the present contribution belongs to the field of hyperelastoplastic models. The inelastic behaviour of the solid skeleton is hence based on the multiplicative decomposition of the deformation gradient into an elastic and a plastic part [17], using an hyperelastic free energy function.

From a micro-mechanical point of view this multiplicative decomposition describes the plastic slip in crystals; for cohesive-frictional soils its validity has been suggested by Nemat–Nasser [23], where the plastic part of the deformation gradient is viewed as an internal variable related to the amount of slipping, crushing, yielding and, for plate-like particles, plastic bending of the granules comprising the soil.

Mechanics of porous materials has a wide spectrum of engineering applications and hence, in recent years, several porous media models and their numerical solutions have appeared in the literature. Most of these models are restricted to fluid saturated materials and have been developed using small strain assumptions. More recently, fluid saturated models have been extended to large strains, first in the framework of hypoelasticity [1] and hypoplasticity [8] or hypoelastoplasticity [20], with the use of an updated Lagrangian approach, Eulerian strain rate tensor and Jaumann stress rate. Models based on an hyperelastic free energy function have been developed for saturated porous media in [9] and [11] in the dynamic and static case, respectively. After the work of J.C. Simo for single-phase materials [28, 27], multiplicative elastoplasticity was developed in [4, 5] in conjunction with the Cam–Clay model to study consolidation problems and in [2] using an associated flow rule for the Drucker–Prager constitutive model.

The approach proposed in the present contribution for binary porous materials is the first step in the development of a partially saturated model. It differs from that proposed in [2] because we have not chosen to divide the fluid content into elastic and plastic parts and because the coupled system of governing equations are here solved with a direct approach (monolithic solution) and not with a staggered procedure (undrained-drained split), allowing for a better description of the strong coupling between the solid skeleton and the fluid, which is of importance, *e.g.*, in case of strain localisation. Moreover, a peculiar aspect is related to the mathematical model, which is derived in the framework of hybrid mixture theory [18] and not the extended Biot theory as in [4, 5, 2] or the theory of porous media [9, 11].

This chapter presents in Section 2 the general model of thermo-hydro-mechanical transient behaviour of geomaterials, from which the balance equations for isothermal water saturated media with incompressible constituents and quasi-static loading conditions are derived and presented in Section 3. The governing equations of Section 3 are written in terms of both spatial and material quantities. Fluid and solid constitutive equations are presented in Subsections 2.6 and 3.3, respectively. The weak form of the governing equations, the temporal integration of the mixture mass balance equation and the linearisation are described in Sections 4, 5 and 6. The return mapping algorithm and the algorithmic tangent moduli for the Drucker–Prager yield function with isotropic linear hardening and non-associated flow rule is presented in Subsection 6.1, where, because of the presence of the apex, an algorithmic solution is adopted using the concept of multisurface plasticity. Finite element discretisation in space and some numerical examples highlight the developments in Sections 7 and 8, respectively. In particular, the classical consolidation of a soil column and an example of strain localisation of contractant/dilantant and isochoric material close the present chapter. For the aspects of the regularization properties of the multiphase model at localisation, the interested reader can see [26] and [33].

As far as notation and symbols are concerned, bold-face letters denote tensors; capital or lower case letters are used for tensors in the reference or in actual configuration. The symbol '·' denotes the scalar product between two vectors (*e.g.* $\boldsymbol{a} \cdot \boldsymbol{b} = a_i b_i$), while the symbol ':' denotes a double contraction of (adjacent) indices of two tensors of rank two or/and higher (*e.g.* $\boldsymbol{c} : \boldsymbol{d} = c_{ij} d_{ij}$, $\boldsymbol{e} : \boldsymbol{f} = e_{ijkl} f_{kl}$). Cartesian coordinates are used throughout.

10.2 General mathematical model of thermo-hydro-mechanical transient behaviour of geomaterials

The full mathematical model necessary to simulate thermo-hydro-mechanical transient behaviour of fully and partially saturated porous media is developed in [18] using averaging theories following Hassanizadeh and Gray [14, 15]. The underlying physical model, thermodynamic relations and constitutive equations for the constituents, as well as governing equations are briefly summarised for sake of completeness in the present section. The governing equations of the binary model used in the finite element discretisation are described in Section 3.

The partially saturated porous medium is treated as a multiphase system (composed of $\pi = 1, \ldots, k$ constituents) with the voids of the solid skeleton (s) filled with water (w) and gas (g). The latter is assumed to behave as an ideal mixture of two species: dry air (non-condensable gas, ga) and water vapour (condensable one, gw). Using spatial averaging operators defined over a representative elementary volume REV (of volume $dv(\boldsymbol{x}, t)$ in the deformed configuration, B_t, see Figure 10.1, where $\boldsymbol{x}$ is the vector of the spatial coordinates and t is the current time), the microscopic equations are integrated over the REV giving the macroscopic balance equations. At the macroscopic level the porous media material is hence modeled by a substitute continuum of volume B_t with boundary ∂B_t that fills the entire domain simultaneously, instead of the real fluids and the solid which fill only a part of it. In this substitute continuum each constituent π has a reduced density which is obtained through the volume fraction $\eta^\pi(\boldsymbol{x},t) = \frac{dv^\pi(\boldsymbol{x},t)}{dv(\boldsymbol{x},t)}$ with the constraint

$$\sum_{\pi=1}^{k} \eta^\pi = 1 \,, \tag{10.1}$$

where $dv^\pi(\boldsymbol{x},t)$ is the π-phase volume inside the REV in the actual placement $\boldsymbol{x}$.

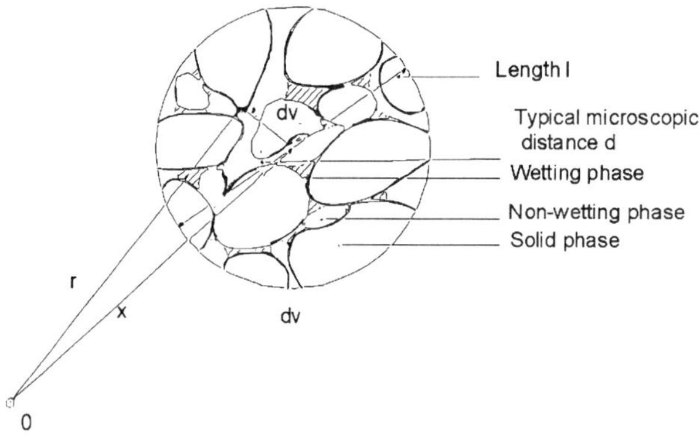

FIGURE 10.1. Typical averaging volume dv of a porous medium consisting of three constituents [18].

In this formulation heat conduction, vapour diffusion, heat convection, water flow due to pressure gradients or capillary effects and latent heat transfer due to water phase change (evaporation and condensation) inside the pores are taken into account. The solid is deformable and non-polar, and the fluid, the solid and the thermal fields are coupled. All fluids are in contact with the solid phase. The constituents are assumed to be isotropic, homogeneous, immiscible, except for dry air and vapour, and chemically non-reacting. Local thermal equilibrium between solid matrix, gas and liquid phases is assumed, so that the temperature is the same for all the constituents. The state of the medium is described by water pressure p^w, gas pressure p^g, temperature θ and displacement vector of the solid matrix $\boldsymbol{u}$.

Before summarizing the macroscopic balance equations, we specify the kinematics introducing the notion of initial and current configuration (Figure 10.2). In the following, the stress is defined as tension positive for the solid phase, while pore pressure is defined as compressive positive for fluids.

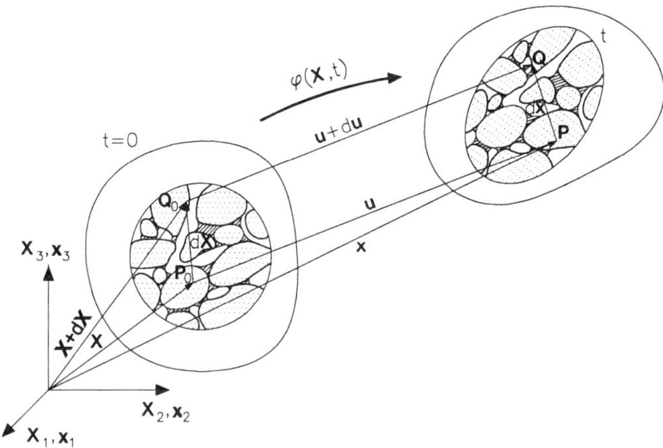

FIGURE 10.2. Initial and current configuration of a multiphase medium.

10.2.1 *Kinematic equations*

At the macroscopic level the multiphase medium is described as the superposition of all π-phases, whose material points X^π with coordinates $\boldsymbol{X}^\pi$ in the reference configuration B_0^π at time $t = t_0$ can occupy simultaneously each spatial point $\boldsymbol{x}$ in the deformed configuration B_t at time t. In the Lagrangian description of the motion in terms of material coordinates the position of each material point in the actual configuration $\boldsymbol{x}$ is a function of its placement $\boldsymbol{X}^\pi$ in a chosen reference configuration B_0^π and of the current time t,

$$\boldsymbol{x} = \boldsymbol{\chi}^\pi(\boldsymbol{X}^\pi, t), \quad \text{with} \quad \boldsymbol{x} = \boldsymbol{x}^\pi, \tag{10.2}$$

or it is given by the sum of the reference position $\boldsymbol{X}^\pi$ and the displacement $\boldsymbol{u}^\pi = (\boldsymbol{X}^\pi, t)$ at time t,

$$\boldsymbol{x} = \boldsymbol{X}^\pi + \boldsymbol{u}^\pi(\boldsymbol{X}^\pi, t). \tag{10.3}$$

In equation (10.2), $\boldsymbol{\chi}^\pi(\boldsymbol{X}^\pi, t)$ is a continuous and bijective motion function (deformation map) of each phase because the jacobian J^π of each motion function

$$J^\pi = \det \frac{\partial \boldsymbol{\chi}^\pi}{\partial \boldsymbol{X}^\pi} > 0 \tag{10.4}$$

is restricted to be a positive value. The deformation gradient F^π is defined as

$$F^\pi(X^\pi, t) = \text{Grad}^\pi \chi^\pi(X^\pi, t), \tag{10.5}$$

where the differential operator Grad^π denotes partial differentiation with respect to the reference position X^π. Hence, from equation (10.4), $J^\pi = \det F^\pi$.

The velocity and the acceleration of each constituent are given as

$$V^\pi = \frac{\partial \chi^\pi(X^\pi, t)}{\partial t}, \qquad A^\pi = \frac{\partial^2 \chi^\pi(X^\pi, t)}{\partial t^2}. \tag{10.6}$$

Due to the non-singularity of the Lagrangian relationship (10.2), the existence of its inverse function leads to the description of the motion in terms of spatial coordinates,

$$X^\pi = [\chi^\pi]^{-1}(x, t). \tag{10.7}$$

The inverse $[F^\pi]^{-1}(x, t)$ of the deformation gradient is given by

$$[F^\pi]^{-1} = \text{grad} X^\pi(x, t), \tag{10.8}$$

where the differential operator 'grad' is now referred to spatial coordinates x. The spatial parametrization of the velocity is given by

$$v^\pi = v^\pi(x, t) = V^\pi \circ [\chi^\pi]^{-1}, \tag{10.9}$$

where 'o' denotes the composition of functions. The parametrization of the spatial acceleration is related to the spatial velocity by application of the chain rule to (10.9)

$$a^\pi = a^\pi(x, t) = \frac{\partial v^\pi}{\partial t} + \text{grad} v^\pi \cdot v^\pi = A^\pi \circ [\chi^\pi]^{-1}. \tag{10.10}$$

Since the individual constituents follow in general different motions, different material time derivatives must be formulated. For an arbitrary scalar-valued function $f^\pi(x, t)$, its material time derivative following the velocity of the constituents π is defined by [18]

$$\frac{D^\pi f^\pi}{Dt} = \frac{\partial f^\pi}{\partial t} + \text{grad} f^\pi \cdot v^\pi, \tag{10.11}$$

where f^π must be replaced by f^π in case of a vector- or tensor-valued function $f^\pi(x, t)$. Thus $a^\pi = \frac{D^\pi v^\pi}{Dt}$.

In multiphase materials theory it is common to assume the motion of the solid as a reference and to describe the fluids in terms of motion relative to the solid. This means that a fluid relative velocity and the material time derivative with respect to the solid are introduced. The solid motion can be described in terms of either material or spatial coordinates. The second approach is now presented because the most natural numerical formulation of the elasto-plastic initial boundary value problem is based on the weak form of the balance equations in a spatial setting.

The fluid relative velocity $v^{\pi s}(x,t)$ in spatial parametrization or diffusion velocity is given by

$$v^{\pi s}(x,t) = v^{\pi}(x,t) - v^{s}(x,t), \tag{10.12}$$

and the material time derivative of $f^{\pi}(x,t)$ with respect to the moving solid phase (s) is given by

$$\frac{D^{s} f^{\pi}}{Dt} = \frac{D^{\pi} f^{\pi}}{Dt} + \mathrm{grad} f^{\pi} \cdot v^{s\pi}, \quad \text{with} \quad v^{s\pi} = -v^{\pi s}. \tag{10.13}$$

For the section closure, the material and the spatial velocity gradient of the solid will be recalled. The first one is given as

$$L^{s} = \mathrm{Grad}^{s} V^{s} = \frac{\partial F^{s}}{\partial t} = D^{s} + W^{s}, \tag{10.14}$$

where D^{s} and W^{s} are the symmetric and the skew-symmetric part of L^{s}, while the spatial velocity gradient l^{s} is defined as the gradient of the velocity (10.9) respect to the spatial coordinates, $i.e.$,

$$l^{s} = \mathrm{grad} v^{s} = \frac{\partial F^{s}}{\partial t} [F^{s}]^{-1} = d^{s} + w^{s}, \tag{10.15}$$

where d^{s} and w^{s} are the symmetric and the skew-symmetric part of l^{s}, also called spatial rate of deformation tensor and spin tensor, respectively. All strain measures and strain rates for each constituent follow similarly to classical non-linear continuum mechanics.

10.2.2 Mass balance equations

The averaged macroscopic balance equation for the solid phase is

$$\frac{D^{s} \rho_{s}}{Dt} + \rho_{s} \,\mathrm{div} v^{s} = \frac{\partial \rho_{s}}{\partial t} + \mathrm{div}(\rho_{s} v^{s}) = 0, \tag{10.16}$$

where $v^{s}(x,t)$ is the mass averaged solid velocity, $\rho_{s}(x,t)$ is the averaged density of the solid related to the intrinsic averaged density $\rho^{s}(x,t)$ by the volume fraction $\eta^{s}(x,t)$. For the generic π-phase the relationship between the averaged density and the intrinsic averaged density is

$$\rho_{\pi}(x,t) = \eta^{\pi}(x,t) \rho^{\pi}(x,t), \tag{10.17}$$

where the intrinsic density $\rho^{\pi}(x,t)$ is also named real or true density in the so called theory of porous media (TPM) (see, $e.g.$, [7]).

The mass balance equation for water is

$$\frac{D^w \rho_w}{Dt} + \rho_w \operatorname{div} \boldsymbol{v}^w = \frac{\partial}{\partial t}(nS_w \rho^w) + \operatorname{div}(nS_w \rho^w \boldsymbol{v}^w) = \rho_w e^w, \quad (10.18)$$

where $\rho_w e^w(\boldsymbol{x}, t)$ is the quantity of water per unit time and volume lost through evaporation. The corresponding equations for dry air and vapour are respectively

$$\frac{D^{ga} \rho_{ga}}{Dt} + \rho_{ga} \operatorname{div} \boldsymbol{v}^{ga} = \frac{\partial}{\partial t}(nS_g \rho^{ga}) + \operatorname{div}(nS_g \rho^{ga} \boldsymbol{v}^{ga}) = 0, \quad (10.19)$$

$$\frac{D^{gw} \rho_{gw}}{Dt} + \rho_{gw} \operatorname{div} \boldsymbol{v}^{gw}$$

$$= \frac{\partial}{\partial t}(nS_g \rho^{gw}) + \operatorname{div}(nS_g \rho^{gw} \boldsymbol{v}^{gw}) = \rho_{gw} e^{gw}, \quad (10.20)$$

where $n(\boldsymbol{x}, t)$ is the porosity of the medium defined as

$$n = \frac{dv^w + dv^g}{dv} = \frac{dv^{voids}}{dv} = 1 - \eta^s \quad (10.21)$$

and S_w and S_g the water and gas degree of saturation. The relationships

$$\eta^w = nS_w, \quad \text{with } S_w = \frac{dv^w}{dv^w + dv^g},$$

$$\eta^g = nS_g, \quad \text{with } S_g = \frac{dv^g}{dv^w + dv^g}, \quad (10.22)$$

hold with the saturation constraint $S_w + S_g = 1$.

10.2.3 Linear momentum balance equations

The linear momentum balance equation for the solid and the π-fluid are

$$\operatorname{div} \boldsymbol{t}^s + \rho_s[\boldsymbol{g} - \boldsymbol{a}^s] + \rho_s \hat{\boldsymbol{t}}^s = \boldsymbol{0} \quad (10.23)$$

and

$$\operatorname{div} \boldsymbol{t}^\pi + \rho_\pi[\boldsymbol{g} - \boldsymbol{a}^\pi] + \rho_\pi[\boldsymbol{e}^\pi + \hat{\boldsymbol{t}}^\pi] = \boldsymbol{0}, \quad (10.24)$$

respectively, where $\boldsymbol{t}^\pi(\boldsymbol{x}, t)$ is the partial Cauchy stress tensor defined via the constitutive equation presented in Section 2.6. $\hat{\boldsymbol{t}}^\pi(\boldsymbol{x}, t)$ accounts for the exchange of momentum due to mechanical interaction with other phases, $\rho_\pi \boldsymbol{a}^\pi$ the volume density of the inertial force, $\rho_\pi \boldsymbol{g}$ the volume density of gravitational force and $\boldsymbol{e}^\pi(\boldsymbol{x}, t)$ takes into account the momentum exchange due to averaged mass supply or mass exchange between the fluid and the gas phases and the change of density. The linear momentum balance equation of the multiphase medium is subjected to the constraint [18]

$$\sum_{\pi=1}^{k} \rho_\pi[\boldsymbol{e}^\pi + \hat{\boldsymbol{t}}^\pi] = \boldsymbol{0}. \quad (10.25)$$

10.2.4   Angular momentum balance equation

All the phases are considered microscopically non-polar and hence at macroscopic level the angular momentum balance equation states that the partial stress tensor is symmetric [18]

$$t^\pi = [t^\pi]^T. \tag{10.26}$$

10.2.5   Energy balance equation and entropy inequality

These two relationships are simply quoted from [18]; the second one is useful for development of the constitutive equations. The energy balance equation for the π-phase may be written as

$$\rho_\pi \frac{D^\pi E^\pi}{Dt} = t^\pi : d^\pi + \rho_\pi h^\pi - \mathrm{div} q^\pi + \rho_\pi R^\pi \,, \tag{10.27}$$

where $\rho_\pi R^\pi$ represents the exchange of energy between the π-phase and other phases of the medium due to phase change and mechanical interaction, q^π is the internal heat flux, h^π results from the heat sources and d^π is the spatial rate of the deformation tensor. E^π accounts for the specific internal energy of the volume element.

The entropy inequality for the mixture is

$$\sum_\pi \left[\rho_\pi \frac{D^\pi \lambda^\pi}{Dt} + \rho_\pi e^\pi \lambda^\pi + \mathrm{div} \frac{q^\pi}{\theta^\pi} - \frac{\rho_\pi h^\pi}{\theta^\pi} \right] \geq 0 \,, \tag{10.28}$$

where θ^π is the absolute temperature, λ^π is the specific entropy of the constituent π and $e^\pi \lambda^\pi$ the entropy supply due to mass exchange.

10.2.6   Constitutive equations

The momentum exchange term $\rho_\pi \hat{t}^\pi$ of the linear momentum balance equation of the fluid can be expressed as [18]

$$\rho_\pi \hat{t}^\pi = -\mu^\pi \eta^{\pi^2} k^{-1} v^{\pi s} + p^\pi \,\mathrm{grad}\, \eta^\pi \,, \qquad \text{with } \pi = g, w \,. \tag{10.29}$$

Here $k = k^{r\pi} k^\pi$ and $k^\pi(x, t) = k^\pi(\rho^\pi, \eta^\pi, T)$ is the intrinsic permeability tensor of dimension $[L^2]$ depending in the isotropic case on the porosity of the medium, $k^{r\pi}(S_\pi)$ is the relative permeability parameter function of the π-phase degree of saturation S_π determined in laboratory tests [18] and μ^π is the dynamic viscosity.

The partial stress tensor in the fluid phase of linear momentum balance equation (10.24) is related to the macroscopic pressure $p^\pi(x, t)$ of the π phase

$$t^\pi = -\eta^\pi p^\pi \mathbf{1} \,, \tag{10.30}$$

where $\mathbf{1}$ is the second-order identity tensor.

From the entropy inequality it can be also shown that the spatial solid stress tensor $t^s(\mathbf{x}, t)$ of the linear momentum balance equation (10.23) is decomposed as

$$t^s = \eta^s[t_e^s - p^s \mathbf{1}] \tag{10.31}$$

and that the effective Cauchy stress tensor $\sigma'(\mathbf{x}, t)$, which is responsible for all major deformation in the solid skeleton, is

$$\sigma' = \eta^s t_e^s . \tag{10.32}$$

In equation (10.31), t_e^s is the dissipative part [13] or effective stress tensor of the solid phase, while p^s is the equilibrium part, also called solid pressure, with $p^s = S_w p^w + S_g p^g$.

From the previous equations, it follows that the total Cauchy stress tensor $\sigma = t^s + t^w + t^g$ can be written in the usual form used in soil mechanics

$$\sigma = \sigma' - [S_w p^w + S_g p^g]\mathbf{1} . \tag{10.33}$$

The constitutive law for the solid skeleton used in this chapter will be discussed in Section 3.

The pressure p^g is given in the sequel. For a gaseous mixture of dry air and water vapour the ideal gas law is introduced because the moist air is assumed to be a perfect mixture of two ideal gases. The equation of state of perfect gas (the Clapeyron equation) and Dalton's law applied to dry air (ga), water vapour (gw) and moist air (g), yield

$$p^{ga} = \rho^{ga} R\theta/M_a , \qquad p^{gw} = \rho^{gw} R\theta/M_w , \tag{10.34}$$
$$p^g = p^{ga} + p^{gw} , \qquad \rho^g = \rho^{ga} + \rho^{gw} . \tag{10.35}$$

In the partially saturated zones, water is separated from its vapour by a concave meniscus (capillary water). Due to the curvature of this meniscus the sorption equilibrium equation gives the relationship between the capillary pressure p^c and the gas p^g and water pressure p^w [13],

$$p^c = p^g - p^w . \tag{10.36}$$

The equilibrium water vapour pressure p^{gw} can be obtained from the Kelvin–Laplace equation

$$p^{gw} = p^{gws}(\theta) \exp\left(\frac{p^c M_w}{\rho^w R\theta}\right) , \tag{10.37}$$

where the water vapour saturation pressure p^{gws}, depending only upon temperature θ, can be calculated from the Clausius–Clapeyron equation or from empirical correlation.

The saturation S_π is an experimentally determined function of capillary pressure p^c and temperature θ:

$$S_\pi = S_\pi(p^c, \theta). \tag{10.38}$$

For the binary gas mixture of the dry air and water vapour, Fick's law gives the relative (average) velocities v_g^π of the diffusing species as

$$v_g^{ga} = -\frac{M_a M_w}{M_g^2} D_g \operatorname{grad}\left(\frac{p^{ga}}{p^g}\right) = -v_g^{gw}, \tag{10.39}$$

where D_g is the effective diffusivity tensor and M_g is the molar mass of the gas mixture

$$\frac{1}{M_g} = \frac{\rho^{gw}}{\rho^g}\frac{1}{M_w} + \frac{\rho^{ga}}{\rho^g}\frac{1}{M_a}. \tag{10.40}$$

10.2.7 Initial and boundary conditions

For model closure it is necessary to define the initial and boundary conditions. The initial conditions specify the full fields of gas pressure, water pressure, temperature, displacements and velocity:

$$p^g = p_0^g, \quad p^w = p_0^w, \quad \theta = \theta_0,$$
$$\tag{10.41}$$
$$u = u_0, \quad \dot{u} = \dot{u}_0, \quad \text{at} \quad t = t_0.$$

The boundary conditions can be imposed values on ∂B_π or fluxes on ∂B_π^q, where the boundary is $\partial B = \partial B_\pi \cup \partial B_\pi^q$. The imposed values on the boundary for gas pressure, water pressure, temperature and displacements are

$$p^g = \hat{p}^g \quad \text{on} \quad \partial B_g, \quad p^w = \hat{p}^w \quad \text{on} \quad \partial B_w,$$
$$\tag{10.42}$$
$$\theta = \hat{\theta} \quad \text{on} \quad \partial B_\theta, \quad u = \hat{u} \quad \text{on} \quad \partial B_u \quad \text{for} \quad t \geq t_0.$$

The volume average flux boundary conditions for water species and dry air conservation equations and the energy equation to be imposed at the interface between the porous media and the surrounding fluid (the natural boundary conditions) are the following:

$$[\rho^{ga}v^g - \rho^g v_g^{gw}] \cdot n = q^{ga} \quad \text{on} \quad \partial B_g^q, \tag{10.43}$$

$$[\rho^{gw}v^g + \rho^w v^w + \rho^g v_g^{gw}] \cdot n = \beta_c[\rho^{gw} - \rho_\infty^{gw}] + q^{gw} + q^w \quad \text{on} \quad \partial B_c^q,$$

$$-[\rho^w v^w \Delta h_{vap} - \lambda_{eff}\nabla\theta] \cdot n = \alpha_c[\theta - \theta_\infty] + q^\theta \quad \text{on} \quad \partial B_\theta^q,$$

for $t \geq t_0$, where $n(x,t)$ is the vector perpendicular to the surface of the porous medium, pointing toward the surrounding gas, ρ_∞^{gw} and θ_∞

are, respectively, the mass concentration of water vapour and temperature in the undisturbed gas phase distant from the interface, α_c and β_c are convective heat and mass transfer coefficients, while q^{ga}, q^{gw}, q^w and q^θ are the imposed dry air flux, imposed vapour flux, imposed liquid flux and imposed heat flux, respectively.

The traction boundary conditions for the displacement field related to the total Cauchy stress tensor $\boldsymbol{\sigma}$ are

$$\boldsymbol{\sigma} \cdot \boldsymbol{n} = \bar{\boldsymbol{t}} \quad \text{on} \quad \partial B_u^q \,, \tag{10.44}$$

where $\bar{t}(\boldsymbol{x}, t)$ is the imposed Cauchy traction vector.

10.3   Macroscopic balance equations for an isothermal saturated medium, with incompressible constituents

In this section the macroscopic balance equations for mass and linear momentum of a simplified model that we shall use in the sequel are obtained. The constitutive equations for finite elastoplasticity as well as their algorithmic counterpart will close the present section.

The following assumptions are now introduced in the general model previously presented:

- All the processes are isothermal. This means that the energy balance equation, equation (10.27) is no longer necessary.

- The porous medium is assumed to be constituted of incompressible solid and water constituents at the micro level (binary porous media). The averaged intrinsic density $\rho^\pi(\boldsymbol{x}, t)$ is hence constant, while the density of the mixture $\rho(\boldsymbol{x}, t)$, equation (10.57), and the averaged density $\rho_\pi(\boldsymbol{x}, t)$ can vary due to the volume fraction $\eta^\pi(\boldsymbol{x}, t)$. Consequently, also the porosity $n(\boldsymbol{x}, t)$ can change during the deformation of the porous medium.

  Within this formulation, a second fluid phase (vapour phase) can occur in case of a dilatant behaviour of the solid skeleton due to only mechanical effects; the vapour pressure p^{gw} is then considered constant and it is neglected because of its small value (monospecies approach, presented in [26]). This means that the mass balance equations for dry air (10.19) and vapour (10.20) are neglected.

  Further, equation (10.36) is simplified as

$$p^c \cong -p^w \,, \tag{10.45}$$

  which states that capillary pressures can be approximated as pore-water tractions. Hence the water pressure can change in sign, which means that a partially saturated zone is developing and requires a proper modeling (*e.g.*, [26, 12, 34] in case of strain localisation analysis). In the present model, the effect of such a pressure on the soil stiffness is not taken into account. A partially saturated formulation is currently under investigation.

- The process is considered as quasi-static, so the solid and fluid accelerations are neglected.

As a consequence of the above assumptions, the independent fields of the model are the solid displacements u and the water pressure p^w.

The formulation in terms of spatial coordinates is now presented.

10.3.1 Mass balance equation

Taking into account the incompressibility constraint of the constituents in (10.16) and (10.18), the mass balance equations for the solid and water phase become

$$\frac{\partial}{\partial t}[1 - n] + \operatorname{div}([1 - n]v^s) = 0, \tag{10.46}$$

$$\frac{\partial n}{\partial t} + \operatorname{div}(nv^w) = 0, \tag{10.47}$$

where the RHS term of the last equation is zero because the water phase change has been neglected, and the definition of the phase averaged density (10.17) has been introduced, thus eliminating the intrinsic (constant) averaged density $\rho^\pi(x, t)$. Using the concept of material time derivative (10.11), equation (10.46) is rewritten as

$$\frac{D^s}{Dt}[1 - n] + [1 - n]\operatorname{div} v^s = 0, \tag{10.48}$$

where the classical relationship

$$\operatorname{div} v^s = \frac{D^s J^s}{Dt}[J^s]^{-1} \tag{10.49}$$

can be introduced [19] for the solid deformation. The time integration of (10.48) gives the evolution law for the porosity $n(x, t)$ related to the determinant $J^s(X^s, t)$ of the deformation gradient $F^s(X^s, t)$,

$$n = 1 - [1 - n_0][J^s]^{-1}, \tag{10.50}$$

where $n_0(X^s)$ is the porosity in the reference configuration at $t = t_0$ (or initial porosity). Because of the relation $\eta^s(x, t) = 1 - n(x, t)$, equation (10.50) can be rewritten as

$$\eta^s = \eta_0^s[J^s]^{-1}, \tag{10.51}$$

where $\eta_0^s(\boldsymbol{X}^s)$ is the solid volume fraction in the reference configuration at $t = t_0$.

The sum of the mass balance equation of the two constituents (10.46) and (10.47) produces the following mass balance equation for the binary mixture under consideration, in which the water velocity relative to the solid, $\boldsymbol{v}^{ws} = \boldsymbol{v}^w - \boldsymbol{v}^s$, has been introduced:

$$\mathrm{div}(\boldsymbol{v}^s + n\boldsymbol{v}^{ws}) = 0\,. \tag{10.52}$$

The term $n\boldsymbol{v}^{ws}(\boldsymbol{x}, t)$ represents the filtration water velocity. It is related to the water pressure by the linear momentum balance equation for water phase after introduction of the constitutive law (10.29), as will be demonstrated in the sequel.

10.3.2 Linear momentum balance equation

Neglecting the inertial term in (10.23) and (10.24) and the mass exchange term $\boldsymbol{e}^w$, the linear momentum balance equation for the solid and water phase are, respectively,

$$\mathrm{div}\boldsymbol{t}^s + [1-n]\rho^s\boldsymbol{g} + [1-n]\rho^s\hat{\boldsymbol{t}}^s = \boldsymbol{0}\,, \tag{10.53}$$
$$\mathrm{div}\boldsymbol{t}^w + n\rho^w\boldsymbol{g} + n\rho^w\hat{\boldsymbol{t}}^w = \boldsymbol{0}\,. \tag{10.54}$$

The linear momentum balance equation for the mixture

$$\mathrm{div}(\boldsymbol{t}^s + \boldsymbol{t}^w) + \rho\boldsymbol{g} = \boldsymbol{0} \tag{10.55}$$

is obtained by summation of (10.53) and (10.54) because of the constraint

$$\sum_\pi \rho_\pi\hat{\boldsymbol{t}}^\pi = \boldsymbol{0}. \tag{10.56}$$

In equation (10.55) $\rho(\boldsymbol{x}, t)$ is the density of the binary mixture

$$\rho = [1-n]\rho^s + n\rho^w \tag{10.57}$$

and $\boldsymbol{t}^s + \boldsymbol{t}^w = \boldsymbol{\sigma}$ is the total Cauchy stress. The principle of effective stress (10.33) assumes the form

$$\boldsymbol{\sigma} = \boldsymbol{\sigma}' - p^w\mathbf{1}\,. \tag{10.58}$$

Using the constitutive equation (10.29) for $\rho_\pi\hat{\boldsymbol{t}}^\pi$ and the definition (10.30) of $\boldsymbol{t}^w$, the linear momentum balance equation for water (10.54) gives Darcy's law

$$n\boldsymbol{v}^{ws} = \frac{\boldsymbol{k}}{\mu^w}[-\mathrm{grad}p^w + \rho^w\boldsymbol{g}]\,. \tag{10.59}$$

This law is valid for the transport of water in slow phenomena when the thermal effects are negligible.

The Lagrangian counterpart of the mixture balance equations (10.52) and (10.55) in terms of material coordinates is now presented. The linear momentum balance equation is obtained using the properties that the total first Piola–Kirchhoff stress tensor $\boldsymbol{P}(\boldsymbol{X}^s, t)$ can be viewed as the Piola Transform of the second leg of the total Cauchy stress tensor $\boldsymbol{\sigma}(\boldsymbol{x}, t)$ [19],

$$P^{aB} = J^s \left[F^{s-1} \right]_b^B \sigma^{ab} . \tag{10.60}$$

Multiplying (10.55) by the jacobian J^s, and using the Piola identity $\mathrm{Div}^s \boldsymbol{P} = J^s \mathrm{div}\boldsymbol{\sigma}$ and the relation $J^s \rho = \rho_0$, the linear momentum balance equation of the mixture in material setting is

$$\mathrm{Div}^s \boldsymbol{P} + \rho_0 \boldsymbol{g} = \boldsymbol{0} , \tag{10.61}$$

where 'Div^s' is the divergence operator with respect to material coordinates of the solid, and

$$\rho_0(\boldsymbol{X}^s, t) = \rho_0^i(\boldsymbol{X}^s, t_0) + \rho^w(\boldsymbol{X}^s, t_0)[J^s(\boldsymbol{X}^s, t) - 1] \tag{10.62}$$

is the pull-back of the mass density of the mixture $\rho(\boldsymbol{x}, t)$ (see equation (A.2) in the Appendix).

The total first Piola–Kirchhoff stress tensor $\boldsymbol{P}(\boldsymbol{X}^s, t)$ results in an additive decomposition into effective $\boldsymbol{P}'(\boldsymbol{X}^s, t)$ and water pressure parts using Terzaghi's principle in the form

$$\boldsymbol{P} = \boldsymbol{P}' - J^s p^w [\boldsymbol{F}^s]^{-T} , \qquad \text{with} \quad \boldsymbol{P}' = J^s \boldsymbol{\sigma}' [\boldsymbol{F}^s]^{-T} . \tag{10.63}$$

The mass balance equation of the mixture in material coordinates is obtained in a similar way, multiplying the spatial equation (10.52) by the jacobian J^s and making use of the Piola identity applied to the velocities $\boldsymbol{v}^s$ and $n\boldsymbol{v}^{ws}$:

$$\mathrm{Div}^s \bar{\boldsymbol{V}}^s = J^s \mathrm{div}\boldsymbol{v}^s , \qquad \mathrm{Div}^s (N\bar{\boldsymbol{V}}^{ws}) = J^s \mathrm{div}(n\boldsymbol{v}^{ws}) , \tag{10.64}$$

where $\bar{\boldsymbol{V}}^{ws}$ is the Piola transform of $\boldsymbol{v}^{ws}$ and $N(\boldsymbol{X}^s, t)$ is the pull-back of $n(\boldsymbol{x}, t)$, $n(\boldsymbol{x}, t) = n(\boldsymbol{x}(\boldsymbol{X}^s), t) = N(\boldsymbol{X}^s, t)$. The equation (10.52) becomes hence

$$\mathrm{Div}^s (\bar{\boldsymbol{V}}^s + N\bar{\boldsymbol{V}}^{ws}) = 0. \tag{10.65}$$

10.3.3 *Constitutive equation for the solid skeleton*

The elasto-plastic behaviour of the solid skeleton at finite strain is based on the multiplicative decomposition of the deformation gradient $\boldsymbol{F}^s(\boldsymbol{X}^s, t)$ into an elastic and plastic part, originally proposed by Lee [17] for crystals

$$\boldsymbol{F} = \boldsymbol{F}^e \boldsymbol{F}^p. \tag{10.66}$$

This decomposition states the existence of an intermediate stress-free configuration (Figure 10.3) and its validity has been suggested for cohesive-frictional soils by Nemat-Nasser [23].

In this section the subscript s will be neglected and the symbol ' $\cdot$ ' will be used for the material time derivative with respect to the solid skeleton instead of $\frac{D^s}{Dt}$ (as well as in the remaining part of the chapter).

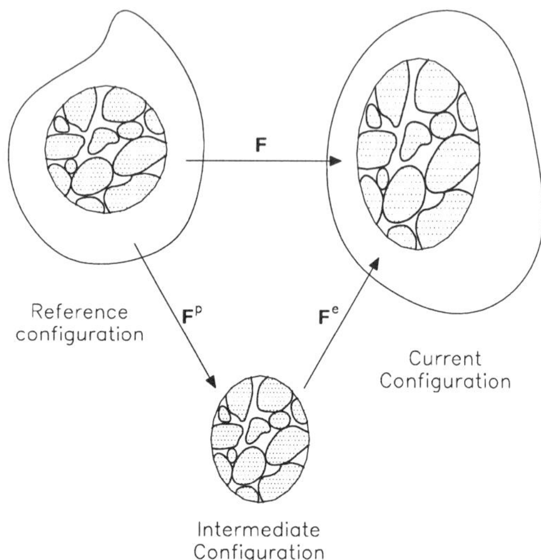

FIGURE 10.3. Illustration of the local multiplicative decomposition of the solid deformation gradient F.

The spatial formulation is used in this section. The treatment of isotropic elasto-plastic behaviour for the solid skeleton, based on the product formula algorithm proposed by Simo [27] for single phase material, is now briefly summarised.

The effective Kirchhoff stress tensor $\tau'(x, t) = J\sigma'(x, t)$ and the logarithmic principal values of the elastic left Cauchy–Green strain tensor ϵ_A are used. In the present subsection also the prime ' $'$ ' for the effective stress tensor will be neglected. The yield function restricting the stress state is developed in the form of von Mises and Drucker–Prager [10], to take into account the behaviour of clays under undrained conditions and the dilatant/contractant behaviour of dense or loose sands, respectively. The return mapping and the consistent tangent operator is developed, taking into account the singular behaviour of the Drucker–Prager yield surface in the zone of the apex, using the concept of multisurface plasticity.

The elastic behaviour of the solid skeleton is assumed to be governed by an hyperelastic free energy function ψ in the form

$$\psi = \psi(b^e, \xi) \tag{10.67}$$

dependent on the elastic left Cauchy–Green strain tensor, $b^e(x,t) = F^e[F^e]^T$, and the internal strain-like variable, $\xi(x,t)$, an equivalent plastic strain. The second law of thermodynamics yields, under the restriction of isotropy, the constitutive relations

$$\tau = 2\frac{\partial\psi}{\partial b^e}b^e\,, \qquad q = -\frac{\partial\psi}{\partial\xi} \tag{10.68}$$

and the remaining dissipation inequality

$$-\frac{1}{2}\tau : \left[[L_v b^e][b^e]^{-1}\right] + q\dot{\xi} \geq 0\,, \tag{10.69}$$

where $L_v b^e = \dot{b}^e - lb^e - b^e l^T$ is the Lie derivative of the elastic left Cauchy–Green strain tensor b^e.

The evolution equations for the rate terms of the dissipation inequality (10.69) can be derived from the postulate of maximum plastic dissipation in the case of associative flow rules [27]

$$-\frac{1}{2}L_v b^e = \dot{\gamma}\frac{\partial F}{\partial\tau}b^e\,, \tag{10.70}$$

$$\dot{\xi} = \dot{\gamma}\frac{\partial F}{\partial q}\,, \tag{10.71}$$

subjected to the classical loading-unloading conditions in Kuhn–Tucker form

$$\dot{\gamma} \geq 0\,, \quad F = F(\tau,q) \leq 0\,, \quad \dot{\gamma}F = 0\,, \tag{10.72}$$

where $\dot{\gamma}$ is the plastic multiplier and $F = F(\tau,q)$ the isotropic yield function.

For the computation, classical elastoplastic models have been selected. In particular, the Drucker–Prager and the von Mises yield functions with linear isotropic hardening have been used in the form, respectively,

$$F(p,s,\xi) = 3\alpha_F p + \|s\| - \beta_F\sqrt{\frac{2}{3}}[c_0 + h\xi] \tag{10.73}$$

and

$$F(s,\xi) = \|s\| - \sqrt{\frac{2}{3}}[\sigma_0 + h\xi] \tag{10.74}$$

in which $p = \frac{1}{3}[\tau : 1]$ is the mean effective Kirchhoff pressure, $\|s\|$ is the L_2 norm of the deviator effective Kirchhoff stress tensor τ, c_0 is the initial apparent cohesion of the Drucker–Prager model, α and β are two parameters related to the friction angle ϕ of the soil

$$\alpha_F = 2\frac{\sqrt{\frac{2}{3}}\sin\phi}{3 - \sin\phi}\,, \qquad \beta_F = \frac{6\cos\phi}{3 - \sin\phi}\,, \tag{10.75}$$

h the hardening/softening modulus and σ_0 is the yield stress in the von Mises law.

10.3.4 Algorithmic formulation for elastoplasticity

The problem of the calculation of b^e, ξ and τ is solved by an operator split into an elastic predictor and plastic corrector [28]. The calculation of the trial elastic state $(\bullet)^{tr}$ is based on the freezing of the plastic flow at time t_{n+1}. The $[b^e_{n+1}]^{tr}$ is hence the push forward of b^e_n by means of the relative deformation gradient $f_{n+1} = \frac{\partial \chi_{n+1}}{\partial x_n} = 1 + \text{grad}\Delta u_{n+1}$, i.e.,

$$[b^e_{n+1}]^{tr} = f_{n+1} b^e_n f^T_{n+1} \tag{10.76}$$

with $\xi^{tr}_{n+1} = \xi_n$, where Δu_{n+1} is the incremental displacement in the time interval $[t_n, t_{n+1}]$.

The same value can also be obtained from the reference configuration by the push-forward of $[C^p_n]^{-1}$ by means of F_{n+1}:

$$[b^e_{n+1}]^{tr} = F_{n+1} [C^p_n]^{-1} F^T_{n+1}. \tag{10.77}$$

The corresponding trial elastic stress is obtained from the hyperelastic free energy function as

$$\tau^{tr}_{n+1} = 2 \left[\frac{\partial \psi}{\partial b^e} b^e \right]_{b^e = [b^e_{n+1}]^{tr}} = 2 \frac{\partial \psi}{\partial b^e} \bigg|_{b^e = [b^e_{n+1}]^{tr}} [b^e_{n+1}]^{tr}. \tag{10.78}$$

If this trial state is admissible, it does not violate the inequality $F^{tr}_{n+1} = F(\tau^{tr}_{n+1}, q^{tr}_{n+1}) \leq 0$ and the stress state is hence already computed.

Otherwise the return mapping or plastic corrector algorithm is applied to satisfy the condition $F_{n+1} = 0$. Since during this phase the spatial position $\chi = \chi^{tr}$ is held fixed and thus $l \equiv 0$, the evolution equation for the elastic left Cauchy–Green strain tensor becomes

$$L_v b^e = -2\dot{\gamma} \frac{\partial F}{\partial \tau} b^e \quad \text{with} \quad L_v b^e \bigg|_{\chi = \chi^{tr}} = \dot{b}^e. \tag{10.79}$$

This first-order differential equation is solved by the product formula algorithm (exponential approximation of the solution, having first-order accuracy) during the time interval $[t_n, t_{n+1}]$ [27]:

$$b^e_{n+1} \cong \exp\left(-2\Delta\gamma \frac{\partial F}{\partial \tau}\right)\bigg|_{n+1} [b^e_{n+1}]^{tr}. \tag{10.80}$$

It should now be noted that b^e_{n+1} commutes with τ_{n+1} due to the assumption of isotropy and that $[b^e_{n+1}]^{tr}$ and its principal axis are held fixed during the return mapping; the spectral decomposition of $[b^e_{n+1}]^{tr}, b^e_{n+1}$

and τ_{n+1} can hence be written with the same eigenbases,

$$[b^e]^{tr} = \sum_{A=1}^{3} [\lambda_{Ae}^{tr}]^2 n_A^{tr} \otimes n_A^{tr} , \qquad b^e = \sum_{A=1}^{3} [\lambda_{Ae}]^2 n_A^{tr} \otimes n_A^{tr} ,$$

$$\tau = \sum_{A=1}^{3} \tau_A n_A^{tr} \otimes n_A^{tr} . \tag{10.81}$$

Using (10.81) the product formula (10.80) can be written in principal values in the form

$$[\lambda_{Ae}]^2 = \exp\left(-2\Delta\gamma \frac{\partial F}{\partial \tau_A}\right)\bigg|_{n+1} [\lambda_{Ae}^{tr}]^2 . \tag{10.82}$$

Taking the logarithm of (10.82) the following important *additive* decomposition of the *log* strain measure in elastic and plastic parts is obtained [27]:

$$\varepsilon_{Ae_{n+1}}^{tr} = \varepsilon_{Ae_{n+1}} + \Delta\gamma \frac{\partial F}{\partial \tau_A}\bigg|_{n+1} , \tag{10.83}$$

in which ε_{Ae} are the principal logarithmic elastic strain $\varepsilon_{Ae} = \ln \lambda_A$. This is a very important consequence of the utilised model because it permits us to use the return mapping of the elasto-plasticity developed for the linear case [28]. From the knowledge of $\Delta\gamma$ the equivalent plastic strain is computed by the backward Euler integration of equation (10.71)

$$\xi_{n+1} \cong \xi_n + \Delta\gamma \frac{\partial F}{\partial q}\bigg|_{n+1} . \tag{10.84}$$

The principal Kirchhoff stress components are then computed by the hyperelastic constitutive law

$$\tau_A = 2\lambda_{Ae} \frac{\partial \psi}{\partial \lambda_{Ae}} = \frac{\partial \psi}{\partial \varepsilon_{Ae}} , \tag{10.85}$$

where the free energy $\psi = \hat{\psi}(\varepsilon_{Ae}, \xi)$ is now written as a function of the principal elastic logarithmic strain components and the equivalent plastic strain (for isotropic linear hardening)

$$\hat{\psi} = \frac{L}{2}[\varepsilon_{1e} + \varepsilon_{2e} + \varepsilon_{3e}]^2 + G[\varepsilon_{1e}^2 + \varepsilon_{2e}^2 + \varepsilon_{3e}^2] + \frac{1}{2}h\xi^2 , \tag{10.86}$$

where L and G are the elastic Lamé constants and h the linear hardening modulus.

The strain energy function $W = \rho_{s0}\hat{\psi}$ of the solid skeleton associated to $\hat{\psi}$ and used in this chapter is the original one proposed by Simo [27]. It is valid only for moderately large elastic strain because it does not satisfy the convexity condition for very large elastic strains [6], but it has been used here because the soils behave with moderate elastic strains. Some useful remarks related to the use of this strain energy function can be found, *e.g.*, in [25], in which it has been shown that the finite element computation loses stability at finite elastic strains.

10.4   Weak form: variational approach

The weak form of the spatial governing equations presented in the previous section is now derived obtaining the variational equations formally equivalent to the initial boundary value problem given by the governing equation and the boundary conditions. This means that the governing equations (10.52) and (10.55) are multiplied by independent weighting functions that vanish on the boundary in which Dirichlet boundary conditions are applied and are then integrated over the spatial domain, $B = B_s$, with boundary, $\partial B = \partial B_s$. The linear momentum balance equation of the binary porous media (10.55) is hence weighted on the domain by the test function δu_s corresponding to the solid displacement (or virtual displacement) in the form

$$\int_B [\operatorname{div}\boldsymbol{\sigma} + \rho\boldsymbol{g}] \cdot \delta\boldsymbol{u}_s dv = 0, \quad \forall \delta\boldsymbol{u}_s. \tag{10.87}$$

Applying partial integration and Green's theorem in the form (see, *e.g.*, [19, 6])

$$\int_B \operatorname{div}\boldsymbol{\sigma} \cdot \delta\boldsymbol{u}_s dv = -\int_B \boldsymbol{\sigma} : \operatorname{grad}\delta\boldsymbol{u}_s dv + \int_{\partial B} \bar{\boldsymbol{t}} \cdot \delta\boldsymbol{u}_s ds \tag{10.88}$$

to the divergence part of (10.87) and taking into account the boundary conditions, this equation is transformed into the weak form

$$-\int_B [\boldsymbol{\sigma}' - p^w\boldsymbol{1}] : \operatorname{grad}\delta\boldsymbol{u}_s dv$$
$$+\int_B \rho\boldsymbol{g} \cdot \delta\boldsymbol{u}_s dv + \int_{\partial B} \bar{\boldsymbol{t}} \cdot \delta\boldsymbol{u}_s ds = 0, \quad \forall \delta\boldsymbol{u}_s, \tag{10.89}$$

where the effective stress principle (10.58) has been introduced. Using the relation $\operatorname{div}\delta\boldsymbol{u}_s = \operatorname{grad}\delta\boldsymbol{u}_s : \boldsymbol{1}$, the previous weak form is transformed in

$$-\int_B \boldsymbol{\sigma}' : \operatorname{grad}\delta\boldsymbol{u}_s dv + \int_B p^w \operatorname{div}\delta\boldsymbol{u}_s dv$$
$$+\int_B \rho\boldsymbol{g} \cdot \delta\boldsymbol{u}_s dv + \int_{\partial B} \bar{\boldsymbol{t}} \cdot \delta\boldsymbol{u}_s ds = 0, \quad \forall \delta\boldsymbol{u}_s. \tag{10.90}$$

The weak form of the mixture mass balance equation (10.52) is obtained in a similar way, introducing Darcy's law (10.59) and using the test function δp^w corresponding to p^w (or virtual water pressure)

$$\int_B \operatorname{div} v^s \, \delta p^w \, dv + \int_B \operatorname{div} \left(\frac{k}{\mu^w} [-\operatorname{grad} p^w + \rho^w g] \right) \delta p^w \, dv = 0, \quad \forall \delta p^w. \quad (10.91)$$

Applying Green's theorem to the underlined term of the previous equation, the following is obtained,

$$\int_B \operatorname{div} v^s \, \delta p^w \, dv + \int_B \left[\frac{k}{\mu^w} [\operatorname{grad} p^w - \rho^w g] \right] \cdot \operatorname{grad} \delta p^w \, dv$$
$$+ \int_{\partial B} q^w \delta p^w \, ds = 0, \quad \forall \delta p^w, \quad (10.92)$$

where q^w is the water flow draining through the surface ∂B.

Remarks. It can be observed that the weak forms (10.90) and (10.92) are very similar to those of the small strain theory, *e.g.*, [18], by substituting the deformed integration domain, B, with the undeformed one, B_0. Moreover, in the small strain theory $\operatorname{div} v^s = \dot{\varepsilon} : 1$, where ε is the small strain tensor of the solid skeleton, while in finite strain $\operatorname{div} v^s = \dot{J}^s [J^s]^{-1}$. In small strain theory the additive decomposition of the strain tensor ε in elastic and plastic parts is also possible, thus rendering the computation of the constitutive tangent operator in the linearisation of the weak form particularly easy.

10.5   Time discretisation

Time integration of the weak form of the mass balance equation (10.92) over a finite time step $\Delta t = t_{n+1} - t_n$ is necessary because of the time dependent term $\operatorname{div} v^s$.

The generalised trapezoidal method is here used, as shown for instance in [18]. Because of the dependency of the integration domain on the time, we rewrite the weak forms (10.90) and (10.92) with respect to the undeformed domain as follows:

$$\int_{B_0} [\tau' - J^s p^w 1] : \operatorname{grad} \delta u_s \, dV - \int_{B_0} \rho_0 g \cdot \delta u_s \, dV$$
$$- \int_{\partial B_0} \bar{T} \cdot \delta u_s \, dA = 0, \quad \forall \delta u_s, \quad (10.93)$$

314   L. Sanavia, B.A. Schrefler and P. Steinmann

$$\int_{B_0} J^s \operatorname{div} \boldsymbol{v}^s \delta p^w dV + \int_{B_0} \left[J^s \frac{\boldsymbol{k}}{\mu^w} [\operatorname{grad} p^w - \rho^w \boldsymbol{g}] \right] \operatorname{grad} \delta p^w dV$$

$$+ \int_{\partial B_0} Q^w \delta p^w dA = 0, \quad \forall \delta p^w, \tag{10.94}$$

where τ' is the effective Kirchhoff stress tensor and $\bar{\boldsymbol{T}} = \boldsymbol{P} \cdot \boldsymbol{N}$ and $Q^w = N \bar{\boldsymbol{V}}^{ws} \cdot \boldsymbol{N}$ are, respectively, the traction vector and the water flow computed with respect to the undeformed configuration. The transport theorems for density, volume and area elements of multiphase materials used to derive equations (10.93) and (10.94) can be found in the Appendix. The form of equations (10.93) and (10.94) is also useful for the subsequent linearisation, because it will be easily performed with respect to the undeformed (fixed) domain.

Equation (10.94) is now rewritten at time t_{n+1} using the relationships

$$\dot{J}^s_{n+\beta} = \frac{J^s_{n+1} - J^s_n}{\Delta t}, \tag{10.95}$$

$$(\bullet)_{n+\beta} = [1 - \beta](\bullet)_n + \beta(\bullet)_{n+1} = (\bullet)_n + \beta[(\bullet)_{n+1} - (\bullet)_n] \tag{10.96}$$

with $\beta = (0, 1]$, obtaining

$$\int_{B_0} [J^s_{n+1} - J^s_n] \delta p^w dV - \Delta t \int_{B_0} [J^s \boldsymbol{v}^D \cdot \operatorname{grad} \delta p^w]_{n+\beta} dV$$

$$+ \Delta t \int_{\partial B_0} Q^w_{n+\beta} \delta p^w dA = 0, \quad \forall \delta p^w, \tag{10.97}$$

where $\boldsymbol{v}^D = -\frac{\boldsymbol{k}}{\mu^w}(\operatorname{grad} p^w - \rho^w \boldsymbol{g})$ is Darcy's velocity of the water.

The weak form of the linear momentum balance equation (10.93) is directly written at time t_{n+1} because it is time independent:

$$\int_{B_0} \left[[\tau' - J^s p^w \boldsymbol{1}] : \operatorname{grad} \delta \boldsymbol{u}_s \right]_{n+1} dV - \int_{B_0} \rho_{0_{n+1}} \boldsymbol{g} \cdot \delta \boldsymbol{u}_s dV$$

$$- \int_{\partial B_0} \bar{\boldsymbol{T}}_{n+1} \cdot \delta \boldsymbol{u}_s dA = 0, \quad \forall \delta \boldsymbol{u}_s. \tag{10.98}$$

Linearised analysis of accuracy and stability suggest the use of $\beta \geq \frac{1}{2}$. In the examples section, implicit one-step time integration has been performed ($\beta = 1$).

The weak forms (10.97) and (10.98) represent a non-linear coupled equations system where the non-linearities are introduced by the finite kinematics and the constitutive laws.

10.6   Consistent linearisation

The non-linear equation system (10.97, 10.98) can be written in the compact form

$$G(\chi, \eta) = 0, \quad \text{where} \quad \chi = [\chi^s, p^w]^T \quad \text{and} \quad \eta = [\delta u_s, \delta p^w]^T. \quad (10.99)$$

For its numerical solution, iterative methods have to be employed and the linearisation at $\bar{\chi}$ is hence necessary:

$$G(\bar{\chi}, \eta, \Delta u) \cong G(\bar{\chi}, \eta) + DG(\bar{\chi}, \eta) \cdot \Delta u \cong 0, \qquad (10.100)$$

where $\Delta u = [\Delta u_s, \Delta p^w]^T$ and $DG \cdot \Delta u = \frac{d}{d\alpha}G(\bar{\chi} + \alpha\Delta u)|_{\alpha=0}$ is the directional derivative or Gateaux derivative of G at $\bar{\chi}$ in the direction of Δu (*e.g.*, [19], [32] for single phase material). Since the equation system G is composed of the weak form of the linear momentum balance equation (G_{LBE}) and of mass balance equation (G_{MBE}), then

$$DG \cdot \Delta u = \left[\begin{array}{c} DG_{LBE} \cdot \Delta u_s + DG_{LBE} \cdot \Delta p^w \\[2mm] DG_{MBE} \cdot \Delta u_s + DG_{MBE} \cdot \Delta p^w \end{array} \right]. \qquad (10.101)$$

Using the symbol $(\bullet)_{n+1}^{k+1}$ to indicate the current iteration in the current time step, the linearisation on the configuration $(\bullet)_{n+1}^{k}$ is written as

$$DG_{n+1}^k \cdot \Delta u_{n+1}^{k+1} = -G_{n+1}^k \qquad (10.102)$$

and the solution vector $u = [u_s, p^w]^T$ is then updated by the incremental relationship

$$u_{n+1}^{k+1} = u_{n+1}^k + \Delta u_{n+1}^{k+1}. \qquad (10.103)$$

For an efficient numerical performance of the scheme (10.102) a consistent linearisation is applied [32] in which the linearisation of the integrated constitutive equation (10.81) plays a central role (this concept was first pointed out in [29] for the small strain case).

The linearisation of equations (10.97) and (10.98), performed in the undeformed configuration, B_0, and then pushed forward in the deformed configuration, B, (see the Appendix), gives the following result:

- for the linear momentum balance equation

$$\int_B [\,\mathrm{grad}\delta u_s : c^{ep} : \mathrm{sym}(\,\mathrm{grad}\Delta u_s) + \sigma' : \mathrm{grad}^T\delta u_s\,\mathrm{grad}\Delta u_s]dv$$

$$+ \int_B p^w\,\mathrm{grad}\delta u_s : [\,\mathrm{grad}^T\Delta u_s - \mathrm{div}\Delta u_s\mathbf{1}]dv \qquad (10.104)$$

$$- \int_B \rho^w\delta u_s \cdot g\,\mathrm{div}\Delta u_s dv - \int_B \mathrm{div}\delta u_s\Delta p^w dv\,;$$

316   L. Sanavia, B.A. Schrefler and P. Steinmann

- for the mass balance equation

$$
\int_B \delta p^w \, \mathrm{div}\Delta \boldsymbol{u}_s dv + \beta \Delta t \int_B \frac{k}{\mu^w} \mathrm{grad}\delta p^w \cdot \mathrm{grad}\Delta p^w dv
$$

$$
+\beta \Delta t \int_B \mathrm{grad}\delta p^w \cdot \left[[\frac{1-n}{k}\frac{\partial k}{\partial n} + 1] \frac{k}{\mu^w}[\mathrm{grad}p^w - \rho^w \boldsymbol{g}]\mathrm{div}\Delta \boldsymbol{u}_s \right] dv
$$

$$
\tag{10.105}
$$

$$
-\beta \Delta t \int_B \mathrm{grad}\delta p^w \cdot \left[\frac{2k}{\mu^w} \mathrm{sym}(\mathrm{grad}\Delta \boldsymbol{u}_s) \mathrm{grad}p^w \right] dv
$$

$$
+\beta \Delta t \int_B \mathrm{grad}\delta p^w \cdot \left[\frac{k}{\mu^w}\rho^w \mathrm{grad}\Delta \boldsymbol{u}_s \boldsymbol{g} \right] dv .
$$

It can be observed that the term $-\int_B p^w \, \mathrm{grad}\delta \boldsymbol{u} : \mathrm{div}\Delta \boldsymbol{u}_s \mathbf{1} dv$ of the linearised LBE is not present in [4].

In the directional derivative $DG_{LBE} \cdot \Delta \boldsymbol{u}_s$ the term

$$
\int_B \mathrm{grad}\delta \boldsymbol{u}_s : \boldsymbol{c}^{ep} : \mathrm{sym}(\mathrm{grad}\Delta \boldsymbol{u}_s) + \boldsymbol{\sigma}' : \mathrm{grad}^T \delta \boldsymbol{u}_s \, \mathrm{grad}\Delta \boldsymbol{u}_s dv \tag{10.106}
$$

contains $\boldsymbol{c}^{ep}$, the spatial constitutive operator following the linearisation of (10.81)

$$
\boldsymbol{c}^{ep}_{n+1} = \sum_{A=1}^3 \sum_{B=1}^3 a^{ep}_{AB_{n+1}}[\boldsymbol{n}^{tr}_A \otimes \boldsymbol{n}^{tr}_A] \otimes [\boldsymbol{n}^{tr}_B \otimes \boldsymbol{n}^{tr}_B]
$$

$$
\tag{10.107}
$$

$$
+2 \sum_{A=1}^3 \tau_{A_{n+1}} \boldsymbol{c}^{tr(A)}_{n+1} .
$$

It is useful to remark that in (10.107) only the second-order tensor $\boldsymbol{a}^{ep} = \frac{\partial \tau_A}{\partial \varepsilon^{tr}_B}$ depends on the specific model of plasticity and the structure of the return mapping algorithm in principal stretches, while the tensors $\boldsymbol{c}^{tr(A)}_{n+1}$ and $\boldsymbol{n}^{tr}_A \otimes \boldsymbol{n}^{tr}_A$ are independent of the specific plastic model used. Moreover it is easy to proof that the moduli $\boldsymbol{a}^{ep}$ have a form identical to the algorithmic elastoplastic tangent moduli of the infinitesimal theory [27]. The expression for $\boldsymbol{c}^{tr(A)}_{n+1}$ can obtained by linearisation of the eigenbases dyadic $\boldsymbol{n}^{tr}_A \otimes \boldsymbol{n}^{tr}_A$ in the spatial setting

$$
\boldsymbol{c}^{tr(A)}_{n+1} = \frac{\partial [\boldsymbol{n}^{tr}_A \otimes \boldsymbol{n}^{tr}_A]}{\partial \boldsymbol{g}} , \tag{10.108}
$$

where $\boldsymbol{g}$ is the spatial metric, or by pull-back [19] of $\boldsymbol{n}^{tr}_A \otimes \boldsymbol{n}^{tr}_A$, subsequent to linearisation in the material setting using the relations, $e.g.$, of [21] and then by push-forward of the linearisation in spatial setting.

10.6.1 Drucker–Prager model with linear isotropic hardening: return mapping and algorithmic tangent moduli with apex solution

Originally the return mapping algorithm was developed for J_2-plasticity. Extension of this method to the Drucker–Prager model can be made taking into account a special treatment of the corner region using the concept of multi-surface plasticity, as developed in [16] in case of perfect plasticity and deviatoric non-associative plasticity. In this chapter the return mapping and the algorithmic tangent moduli will be obtained for isotropic linear hardening/softening and volumetric-deviatoric non-associative plasticity. To this end, a plastic potential function $Q(p, s, \xi)$ similar to (10.73) is defined, where the dilatancy angle φ is introduced.

The key idea is based on the fact that the return mapping algorithm developed without any special treatment of the apex region leads to physically meaningless results (*i.e.*, $\|s_{n+1}\| < 0$) for a certain range of trial elastic stress. Once the plastic consistency parameter $\Delta\gamma_{n+1}$ is computed by the return mapping, this happens when the following relationship obtained from the updated deviatoric components of the stress tensor

$$\|s_{n+1}\| = \|s_{n+1}^{tr}\| - 2G\Delta\gamma_{n+1} \geq 0 \qquad (10.109)$$

is violated. Without going into details, violation of inequality (10.109) and the consistency condition $F_{n+1} = F(\tau_{n+1}, \xi_{n+1}) = 0$ yields the inequality for which the return mapping needs to be modified, *i.e.*,

$$p_{n+1}^{tr} > \frac{3\alpha_Q K}{2G}\|s_{n+1}^{tr}\| + \frac{\beta_F\sqrt{\frac{2}{3}}}{\alpha_F}\left[\frac{\|s_{n+1}^{tr}\|}{2G}h\sqrt{1 + 3\alpha_Q^2} + c_n\right], \qquad (10.110)$$

where the indexes F and Q of α and β are referred to the yield and the plastic potential surface, respectively.

In this case, the stress region characterised by (10.110) may be treated like a corner region in non-smooth multi-surface plasticity. To this end a second yield condition F_2 is introduced in addition to (10.73) as

$$F_2(p, \xi) = 3\alpha_F p - \beta_F\sqrt{\frac{2}{3}}[c_0 + h\xi], \qquad (10.111)$$

which is derived from (10.73) with the condition $\|s\| = 0$ and the plastic evolution equations need to be modified following Koiter's generalisation introducing a second plastic consistency parameter $\dot{\gamma}_2$ related to F_2. Hence the evolution equations (10.70) will be replaced by the generalised plastic evolution laws

$$L_v b^e = -2\sum_i \dot{\gamma}_i \frac{\partial F_i}{\partial \tau} b^e, \qquad i = 1, 2, \qquad (10.112)$$

$$\dot{\xi} = \sum_i \dot{\gamma}_i \frac{\partial F_i}{\partial q}, \qquad i = 1, 2. \tag{10.113}$$

The algorithmic tangent moduli are computed by linearisation of the computed Kirchhoff stress tensor. Two tangent moduli are obtained, the first one valid for the stress state where the Drucker–Prager model is satisfied, *i.e.*, for the stress for which equation (10.110) is violated, the second one for the stress state which belongs to the corner region. The computed moduli for the two cases are respectively:

- for the non-corner zone:

$$
\begin{aligned}
\boldsymbol{a}^{ep}_{n+1} = {} & c_1 K \boldsymbol{1} \otimes \boldsymbol{1} + 2G \left[\boldsymbol{I} - \tfrac{1}{3} \boldsymbol{1} \otimes \boldsymbol{1} \right] \left[1 - \tfrac{2G\Delta\gamma_{n+1}}{\|\boldsymbol{s}^{tr}_{n+1}\|} \right] \\
& - \frac{6\alpha_Q KG}{c_2} \boldsymbol{1} \otimes \boldsymbol{n}^{tr}_{n+1} - \frac{6\alpha_F KG}{c_2} \boldsymbol{n}^{tr}_{n+1} \otimes \boldsymbol{1} \\
& - 4G^2 \left[\frac{1}{c_2} - \frac{\Delta\gamma_{n+1}}{\|\boldsymbol{s}^{tr}_{n+1}\|} \right] \boldsymbol{n}^{tr}_{n+1} \otimes \boldsymbol{n}^{tr}_{n+1},
\end{aligned} \tag{10.114}
$$

where the coefficients c_1 and c_2 are

$$c_1 = \left[1 - \frac{9\alpha_F \alpha_Q K}{c_2} \right], \quad c_2 = 9\alpha_F \alpha_Q K + 2G + \beta_F h \sqrt{\frac{2}{3}[1 + 3\alpha_Q^2]};$$

- for the corner zone:

$$\boldsymbol{a}^{ep}_{n+1} = \left[K\boldsymbol{1} \otimes \boldsymbol{1} + \frac{K}{2\alpha_Q G(\Delta\gamma_1 + \Delta\gamma_2)_{n+1}} \boldsymbol{1} \otimes \boldsymbol{s}^{tr}_{n+1} \right] c_3, \tag{10.115}$$

where the coefficient c_3 is

$$c_3 = \frac{\alpha_Q \beta_F \sqrt{\frac{2}{3}} h [\Delta\gamma_1 + \Delta\gamma_2]_{n+1}}{3\alpha_F K \sqrt{\Delta\gamma_{1\,n+1}^2 + 3\alpha_Q^2 [\Delta\gamma_1 + \Delta\gamma_2]_{n+1}^2} + \alpha_Q \beta_F \sqrt{\frac{2}{3}} h [\Delta\gamma_1 + \Delta\gamma_2]_{n+1}}.$$

It can be observed that the moduli (10.115) are non-symmetric even for associated plasticity, while (10.114) are non-symmetric only for non-associated plasticity. In case of perfect plasticity ($h = 0$) the coefficient c_3 and hence the moduli of equation (10.115) vanish. In case of geometrically non-linear analysis, this implies that only the geometrical part of the stiffness matrix is activated. Moreover, the moduli (10.115) are reduced to those of the von Mises model by selecting $\alpha = 0$ and $\beta = 1$.

10.7   Finite element discretisation in space

The suitable spatial finite element formulation is derived by applying the well known Galerkin procedure, in which the weighting functions are approximated by the same shape functions used to approximate the driving variables (isoparametric finite elements). This means that the geometry, $\boldsymbol{X}^s$, the current configuration, $\boldsymbol{x}$, the displacement field, $\boldsymbol{u}_s$, the water pressure, p^w, the incremental generalised displacement, $\Delta\boldsymbol{u} = [\Delta\boldsymbol{u}_s, \Delta p^w]^T$, and the variations, $\boldsymbol{\eta} = [\delta\boldsymbol{u}_s, \delta p^w]^T$, are interpolated within a finite element by the same type of functions. In the present setting different shape functions are chosen for quantities associated respectively to the solid and the fluid, thus satisfying the LBB condition for the locally undrained case. Standard procedures have been applied, following any textbook on FEM. With respect to the small strain case, the discretisation of the spatial form of the linearised system of equations is made taking into account that each quantity is referred to the spatial coordinates, $\boldsymbol{x}$, instead of the coordinates of the undeformed configuration, $\boldsymbol{X}^s$. (In the present formulation quadrilateral $Q_2 Q_1$ or $S_2 Q_1$ elements have been used for consolidation of the soil column and the localisation example, respectively). The solid displacement and the water pressures are hence expressed in the whole domain by global shape function matrices $\boldsymbol{N}_u(\boldsymbol{x})$ and $\boldsymbol{N}_w(\boldsymbol{x})$ and the nodal value vectors $\bar{u}(t)$ and $\bar{p}(t)$,

$$\boldsymbol{u} = \boldsymbol{N}_u\bar{u}, \qquad p^w = \boldsymbol{N}_w\bar{p}. \tag{10.116}$$

The linearised system of equations (10.102) in matrix form can be expressed as

$$\begin{bmatrix} \boldsymbol{K}_T & -\boldsymbol{Q}_{sw} \\ \boldsymbol{Q}_{ws} & \beta\Delta t\boldsymbol{H} \end{bmatrix} \begin{bmatrix} \Delta\bar{u} \\ \Delta\bar{p} \end{bmatrix} = - \begin{bmatrix} \boldsymbol{G}^u \\ \boldsymbol{G}^p \end{bmatrix}. \tag{10.117}$$

Owing to the strong coupling between the mechanical and the pore-fluid problem a monolithic solution of (10.117) is preferred using a Newton–Raphson scheme. As far as the numerical performance of the proposed implementation is concerned, it can be demonstrated that quadratic rate of convergence for the global Newton–Raphson iteration in each step has been obtained in the computation. A typical rate is reported in Table 10.1.

10.8   Numerical examples

The numerical model obtained in the previous section is now used to study a classical consolidation test and a case of soil failure where localisation takes place. The first example is the analysis of a water saturated one-dimensional soil column of $10m$ depth and $2m$ width, discretised with $Q_2 Q_1$ finite elements for the solid displacements and the water pressure, respectively. The upper boundary is drained (water pressure is hence assumed to be at atmospheric value) and loaded by a step load of $10kN$. The other boundaries are impervious. Horizontal displacements are constrained on the lateral surfaces while the bottom surface is also vertically constrained. Zero

INCREMENT NO.: 50		
Iteration n.: 0	Residuum Norm:	4.58716E-02
Iteration n.: 1	Residuum Norm:	1.91969E+02
Iteration n.: 2	Residuum Norm:	6.47109E+00
Iteration n.: 3	Residuum Norm:	3.23143E-03
Iteration n.: 4	Residuum Norm:	3.37295E-09
INCREMENT NO.: 75		
Iteration n.: 0	Residuum Norm:	4.58869E-02
Iteration n.: 1	Residuum Norm:	4.26410E+02
Iteration n.: 2	Residuum Norm:	1.51897E+01
Iteration n.: 3	Residuum Norm:	3.00819E-02
Iteration n.: 4	Residuum Norm:	1.11320E-07
Iteration n.: 5	Residuum Norm:	4.41052E-09

TABLE 10.1. Typical rate of convergence for the implemented Newton–Raphson scheme.

initial conditions are assumed and gravity acceleration is neglected during the computation. The hyperelastic constitutive law of equation (10.86) is utilized for the solid skeleton. The material parameters used in the computation are listed in Table 10.2.

Solid bulk modulus	$K = 330 kN/m^2$
Solid shear modulus	$G = 500 kN/m^2$
Porosity	$n = 0.30$
Isotropic permeability	$k = 0.1 m/s$
Water unit weight	$\gamma_w = 10 kN/m^3$
Gravity acceleration	$g = 0 m/s^2$

TABLE 10.2. Material parameters used in the computation of the consolidation test.

A consolidation process due to water drainage through the upper surface develops and a typical time history for settlement of the top surface of the soil column and dissipation of the excess of water pressure is shown in the following. First a numerical comparison of the results of the present model with the ones obtained using the linearised theory is illustrated in Figures 10.4 and 10.5 (only the first 4s are depicted). In particular, Figure 10.4 shows the time history of the vertical displacement of the top surface, while Figure 10.5 shows the excess of water pressure versus time at a node of the bottom surface. Due to the fact that the applied load leads only to small strains (axial nominal strain $\epsilon_y = 0.005$), there is no visible difference between the linear and non-linear computations (a similar example has been solved in [9]).

If the load is increased, the difference between the linear and the ge-

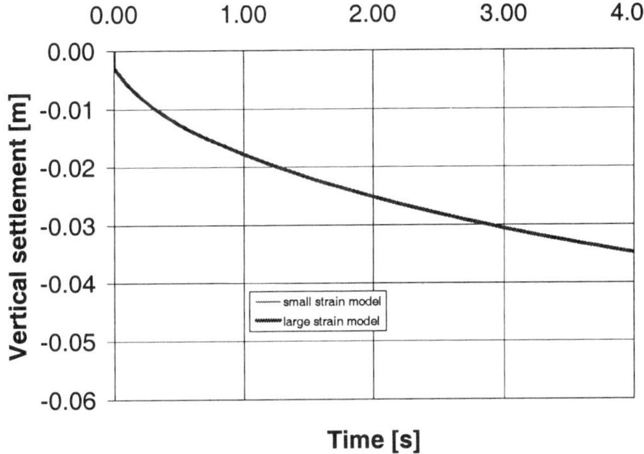

FIGURE 10.4. Vertical settlement of the top surface of the soil column versus time.

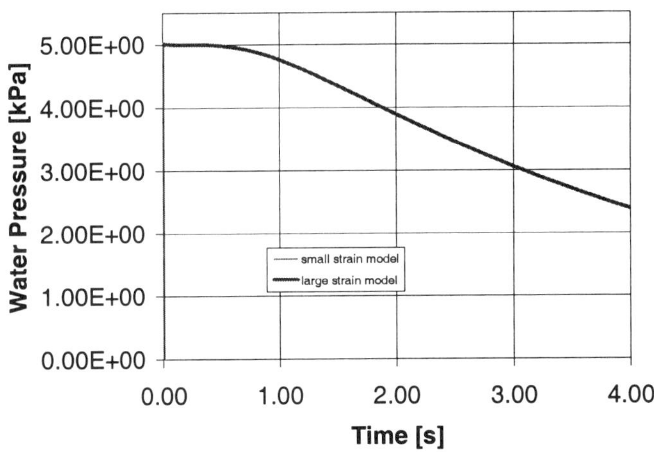

FIGURE 10.5. Excess of water pressure at a node of the bottom surface of the soil column versus time.

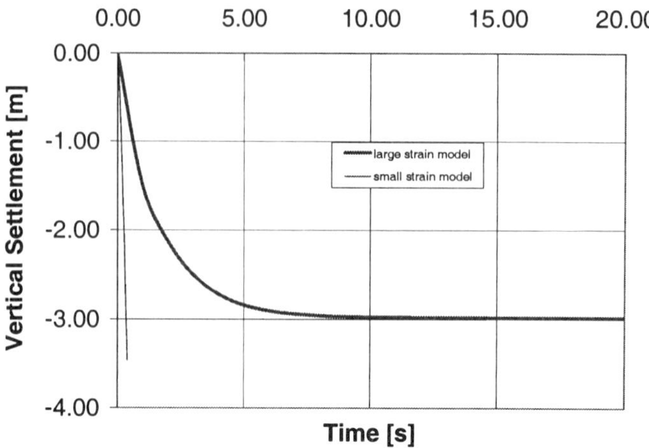

FIGURE 10.6. Vertical settlement of the top surface of the soil column versus time.

ometrically non-linear model becomes visible (*e.g.*, [20, 9]). To point out the effect of the non-linear terms, the first example is now solved applying a step load of $1000kN$ (for axial nominal strain $\epsilon_y = 0.5$). The classical consolidation behaviour in the plateau of the vertical settlement (Figure 10.6) due to dissipation of the excess of water pressure (Figure 10.7) can be noted. Moreover, the small strain model used for comparison is not suitable to describe the phenomenon from a quantitative point of view, because the maximum settlement is limited by the void closure, which is not captured with the small strain model.

The second example deals with the analysis of a representative square domain of water saturated porous material loaded by a rigid footing (Figure 10.8).

This example was solved in [30] using the linear theory. The BVP is now solved numerically using the formulation developed in the previous sections. Plane-strain conditions are assumed. The homogeneous soil domain has side length of $l = 10m$, with the rigid footing spanning over $5m$ at the right part of the top surface. On the boundary, the horizontal and vertical displacements are constrained respectively on the left and bottom surface. Drainage of the water is allowed only through the unloaded part of the top surface of the domain. The solid skeleton is assumed to obey the elastoplastic von Mises or the Drucker–Prager constitutive model, both with isotropic linear softening behaviour as a phenomenological description of damage effects. The material parameters used in the computation are listed in Table 10.3.

The von Mises law has been selected as a reference material law used to test the implementation and to describe qualitatively the mechanical behaviour of clays under undrained conditions, while the Drucker–Pra-

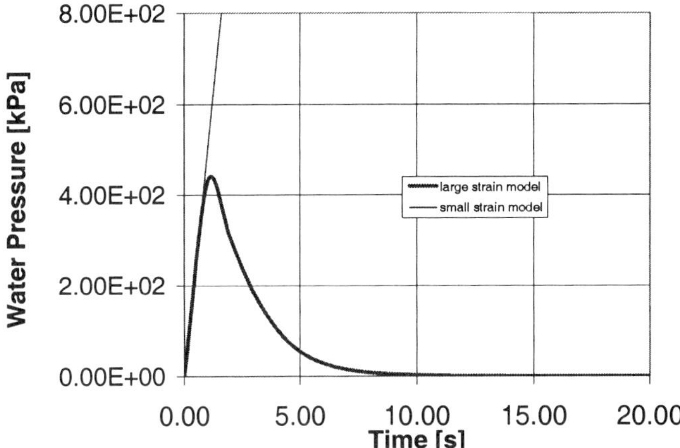

FIGURE 10.7. Excess of water pressure at a node of the bottom surface of the soil column versus time.

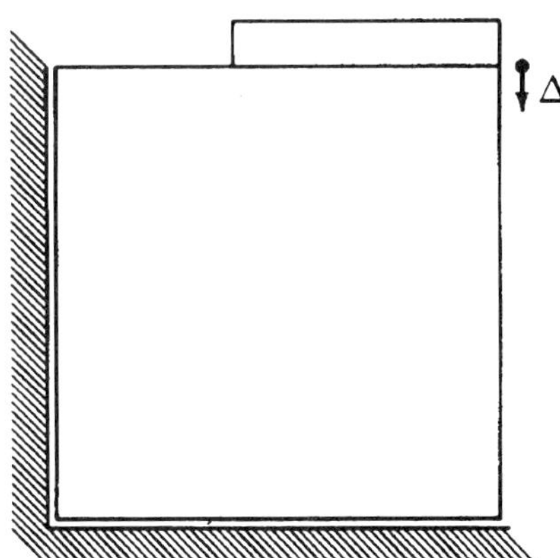

FIGURE 10.8. Rigid footing on a square domain of porous elastoplastic material.

324   L. Sanavia, B.A. Schrefler and P. Steinmann

Solid bulk modulus	$K = 8333kN/m^2$
Solid shear modulus	$G = 3486kN/m^2$
initial yield stress	$y_0 = 100kN/m^2$
linear softening modulus	$h = -10kN/m^2$
initial apparent cohesion	$c_0 = 100kN/m^2$
angle of internal friction	$\phi = 20^0$
angle of dilatancy	$\varphi = -10^0, 0^0, +5^0, +10^0, +20^0$
isotropic permeability	$k = 0.0001m/s$
water unit weight	$\gamma_w = 10kN/m^3$
gravity acceleration	$g = 0m/s^2$

TABLE 10.3. Material parameters used in the computation of the localisation example.

ger law has been chosen to simulate the behaviour of dilatant/contractant geomaterials such as dense and loose sands, respectively. The loading is applied quasi-statically to the rigid footing by displacement control with a constant vertical velocity of $5e - 3m/s$ until the maximum displacement is obtained. The domain has been discretised using 20 x 20 S_2Q_1 elements for the solid and the fluid mesh (*i.e.*, 8 nodes for the solid, 4 nodes for the fluid).

Figure 10.9 shows the distribution of the equivalent plastic strain at the end of the load history ($0.5m$ in this case) on the deformed configuration using the von Mises material model. No magnification of the displacements has been used in this or any of the following figures. The plastic zone indicates the pronounced accumulation of inelastic strains in a narrow band, while the deformed configuration outlines the classical slip of a part of the domain on the other. Figure 10.10 shows the excess water pressures at the end of the load history (values expressed in kPa), where only positive pressure values can be observed due to the compressive load in a solid skeleton with isochoric plastic flow.

The effect of the plastic dilatancy/contractancy is shown by analysing the square panel using the Drucker–Prager material law. The sample has been solved using different values of the angle of dilatancy: $0^0, +5^0, +10^0, +20^0$ and -10^0. In particular, Figures 10.11 and 10.12 show the equivalent plastic strain and the excess water pressure distribution at the end of the load history ($1m$) in case of zero dilatancy (which means isochoric plastic flow). The resulting shear band and the deformation pattern are hence very similar to those obtained using the von Mises material law, as can be observed by comparison with Figures 10.9 and 10.10.

Increasing the value of the angle of dilatancy ($\varphi = 5^0, 10^0$ and 20^0, see Figures 10.13, 10.14 and 10.15, respectively), an increase of the slope of the shear band can be observed. Moreover, the same pictures reveal the increasing value of the horizontal displacement of the right side of

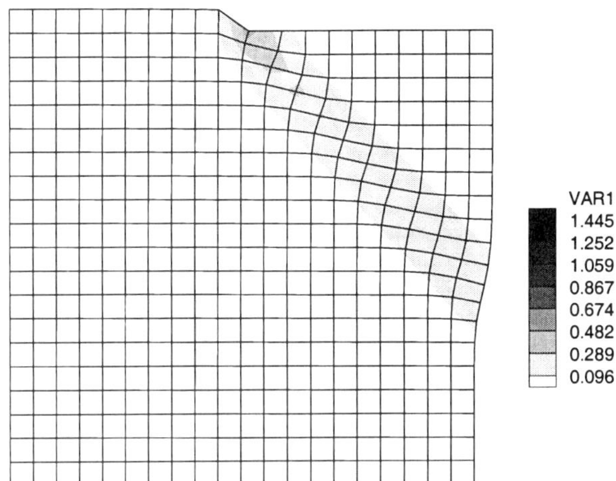

FIGURE 10.9. Equivalent plastic strain contour using von Mises law.

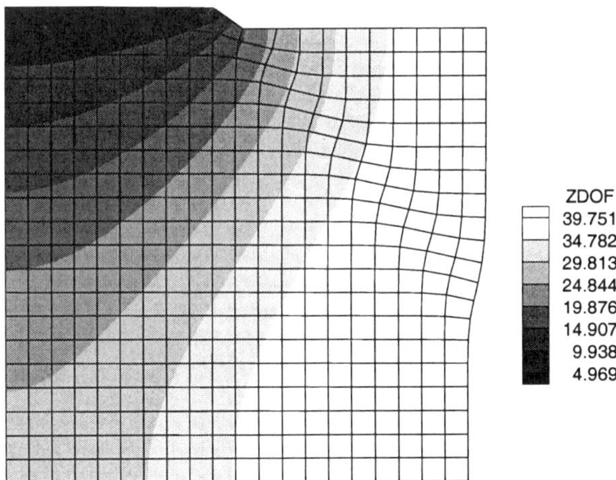

FIGURE 10.10. Excess water pressure contour using von Mises law (kPa).

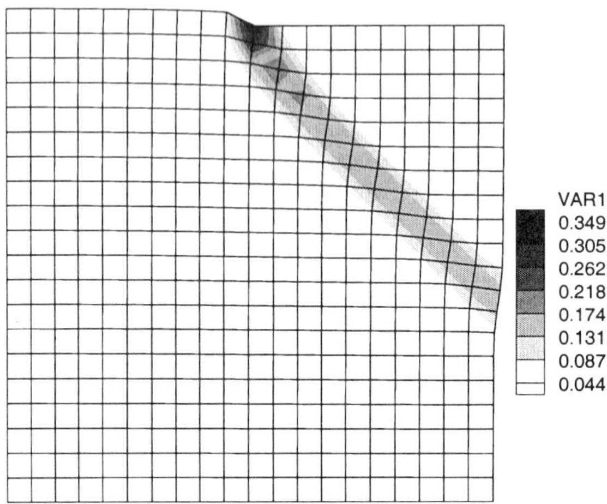

FIGURE 10.11. Equivalent plastic strain contour using Drucker–Prager law with $\varphi = 0^0$.

the panel, due to the increase of volumetric plastic strain with dilatancy. The opposite behaviour appears in case of negative value of the dilatancy angle, see Figure 10.16, where the equivalent plastic strain contour in case of dilatancy angle of -10^0 is depicted.

The effect of the plastic dilatancy/contractancy is evidenced also in the contour of the excess water pressures. In fact the variation of the porosity with the deformation of the medium, see equation (10.50) and the localisation of the plastic deformation imply the presence of negative water pressure, with the lowest values inside the plastic zones in case of dilatant plastic strains (Figures 10.17, 10.18 and 10.19), as opposed to the case of contractant plastic flow (Figure 10.20). The presence of negative pressures is not surprising. In fact, it was experimentally observed at localisation by [22] and [31] during biaxial tests of globally undrained dense sands under imposed displacements. In particular, the values of -80 and $-91 kPa$ were measured by the two authors, respectively. At those pressures, cavitation of the pore fluid was observed, which means the presence of the vapour phase separated from the liquid phase by a meniscus. The low values of negative excess of water pressures computed in the numerical examples of this chapter (e.g., $-280 kPa$ in Figure 10.19) suggest the presence of cavitation phenomenon, which should be modeled by introducing the effect of the partial saturation (see [26] and [12] in case of small strains). These improvements will be further pursued.

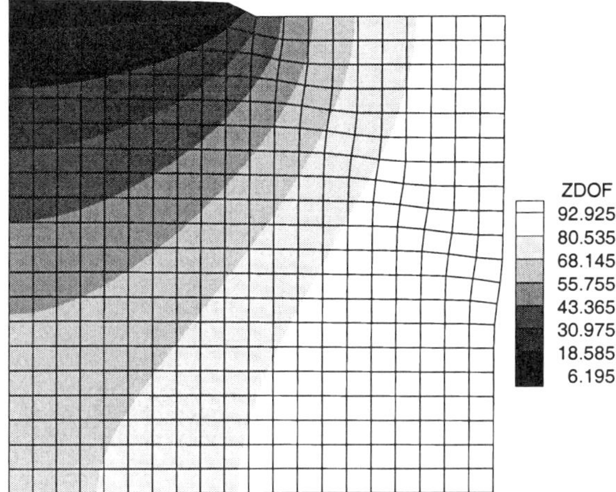

FIGURE 10.12. Excess water pressure contour using Drucker–Prager law with $\varphi = 0^0$ (kPa).

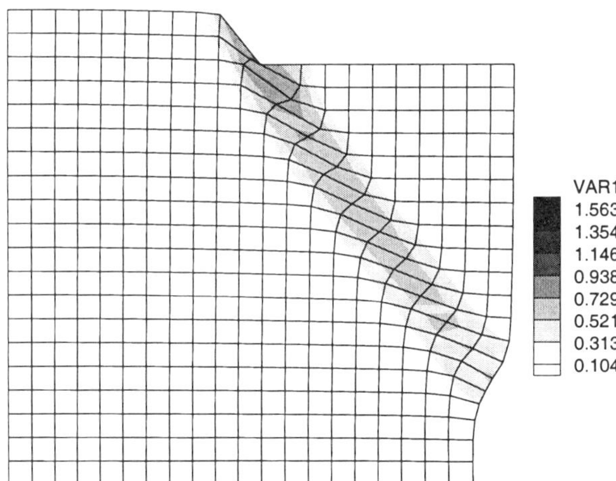

FIGURE 10.13. Equivalent plastic strain contour using Drucker–Prager law with $\varphi = 5^0$.

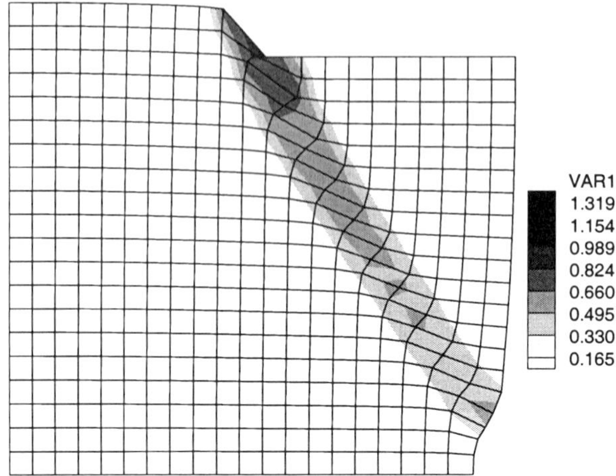

FIGURE 10.14. Equivalent plastic strain contour using Drucker–Prager law with $\varphi = 10^0$.

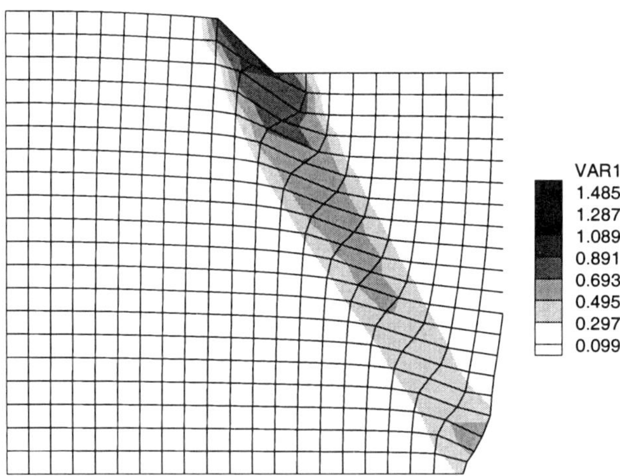

FIGURE 10.15. Equivalent plastic strain contour using Drucker–Prager law with $\varphi = 20^0$.

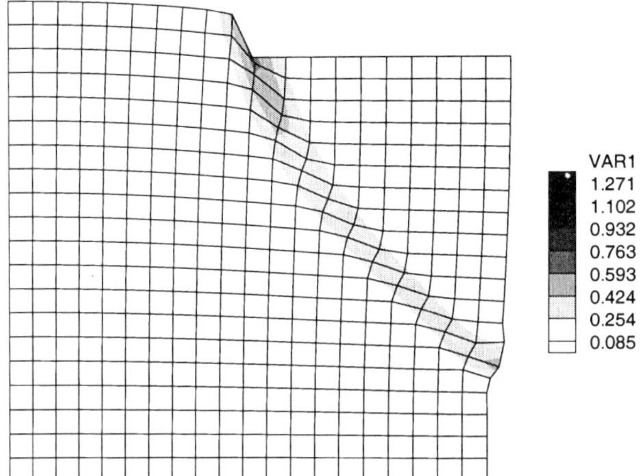

FIGURE 10.16. Equivalent plastic strain contour using Drucker–Prager law with $\varphi = -10^0$.

Water velocity in case of dilatant and contractant material is shown in Figures 10.21 and 10.22, respectively, where the different directions of the fluid can be noted, flowing into the band in case of dilatant material, out of it in the other case.

Remarks. Negative water pressures start from the top surface, close to the left corner of the foundation, where the dilatant shear band first appears, and propagate inside the plastic zone following the evolution of the shear band. Then, they propagate also outside the band due to the consolidation generated by the high pressure gradient in conjunction with the quasi-static process, until they occupy all the domain, as can be observed in Figure 10.19. This phenomenon is dependent on the value of the time step: decreasing its value, a reduced zone with negative water pressure has been observed, but it is always present. The localisation of the negative pressures could be probably easily captured in the dynamic case, as shown in [26, 12, 34] using a small strain model.

10.9   Conclusions

This chapter shows a mathematical model and the related finite element discretisation of quasi-static and isothermal inelastic two phase geomateri-

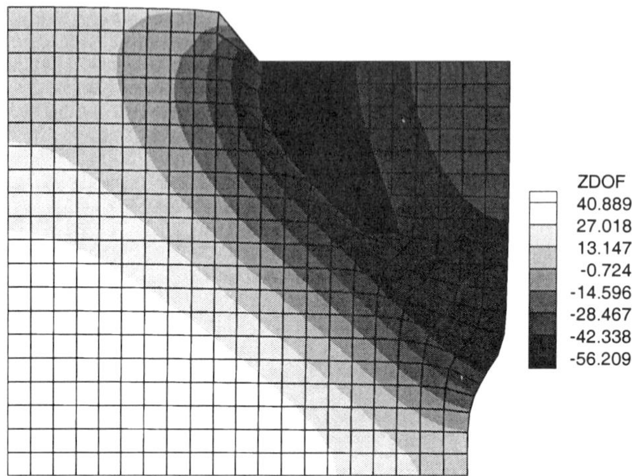

FIGURE 10.17. Excess water pressure contour using Drucker–Prager law with $\varphi = 5^0$ (kPa).

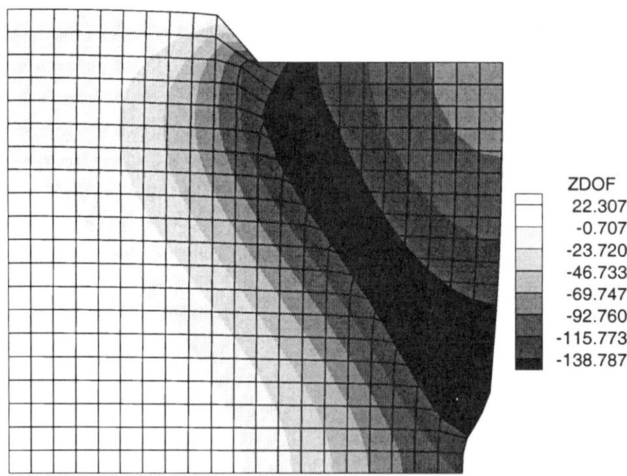

FIGURE 10.18. Excess water pressure contour using Drucker–Prager law with $\varphi = 10^0$ (kPa).

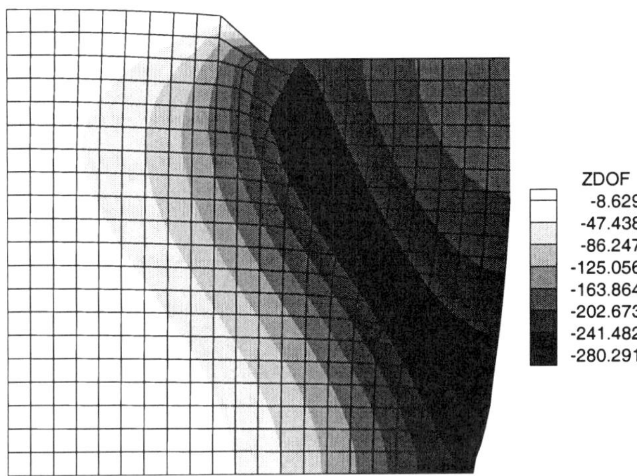

FIGURE 10.19. Excess water pressure contour using Drucker–Prager law with $\varphi = 20^0$ (kPa).

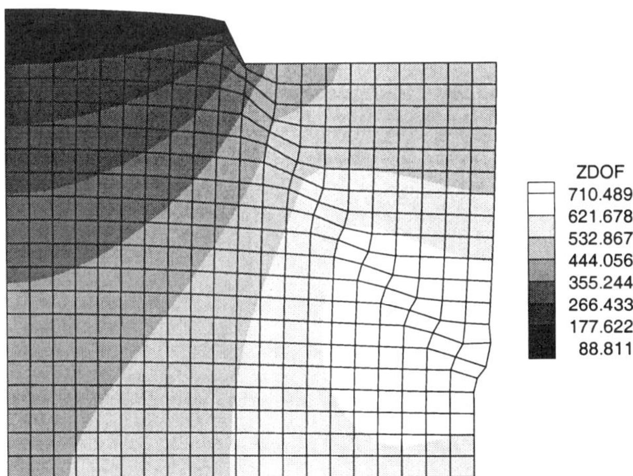

FIGURE 10.20. Excess water pressure contour using Drucker–Prager law with $\varphi = -10^0$ (kPa) .

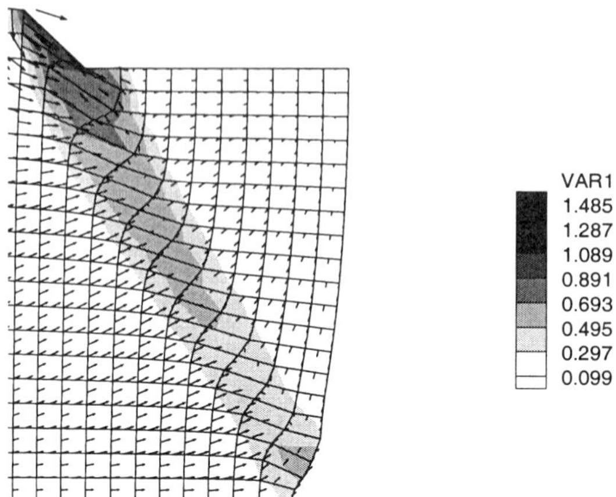

FIGURE 10.21. Water velocity close to the shear band using Drucker–Prager law with $\varphi = 20^0$.

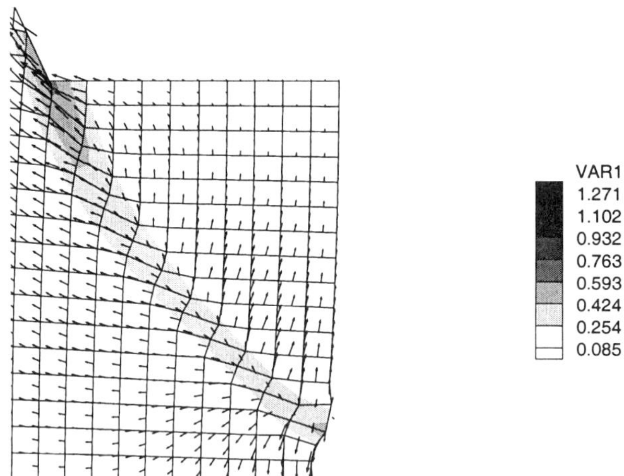

FIGURE 10.22. Water velocity close to the shear band using Drucker–Prager law with $\varphi = -10^0$.

als, assuming incompressible constituents at microscopic level. The governing equations are derived in material and spatial formulation. The elastoplastic behaviour of the solid skeleton is based on the multiplicative decomposition of the deformation gradient in an elastic and plastic part and is developed in spatial setting. The solid effective stress is hyperelastic or limited by the von Mises or the Drucker–Prager yield criterion with isotropic linear softening. A particular apex formulation is developed for the later case. The water behaves following Darcy's law. Consistent linearisation of the non-linear equation system and finite element formulation are derived. Numerical results of this research in progress on large elastic or inelastic strains are shown.

10.10   Appendix: Transport theorems and linearisation

Transport theorems for multiphase materials and some useful details related with the linearisation (10.101) are reported in this appendix.

10.10.1   Transport theorems

1. For mixture density (binary porous media with incompressible constituents).

   The pull-back of the spatial density of mixture $\rho(\boldsymbol{x}, t)$ in the material setting reveals the intrinsic nature of the porous media. In fact the material density of the mixture is not constant in time because the change of the volume of the voids and the consequent variation of the fluid content. The classical transport relationship for single phase material

   $$J(\boldsymbol{X}, t)\rho(\boldsymbol{x}, t) = \rho_0(\boldsymbol{X}, t) = \rho_0(\boldsymbol{X}, t_0) \tag{A.1}$$

   is hence not valid and must be replaced by

   $$\begin{aligned} J^s(\boldsymbol{X}, t)\rho(\boldsymbol{x}, t) &= \rho_0(\boldsymbol{X}^s, t)\,, \\ \rho_0(\boldsymbol{X}^s, t) &= \rho_0^i(\boldsymbol{X}^s, t_0) + [J^s(\boldsymbol{X}^s, t) - 1]\rho^w(\boldsymbol{X}^s, t_0)\,, \end{aligned} \tag{A.2}$$

   where $\rho_0^i(\boldsymbol{X}^s, t_0)$ is the mixture density at $t = t_0$.

   The proof of (A.2) follows easily using the definition of averaged density (10.17). In fact the spatial density of the mixture is

   $$\rho = \rho_s + \rho_w = \eta^s \rho^s + \eta^w \rho^w = \eta^s \rho^s + [1 - \eta^s]\rho^w. \tag{A.3}$$

   By introducing equation (10.51), valid for incompressible constituents at microscopic level, summing and subtracting the term $[1 - \eta_0^s]\rho^w$, in which $\eta_0^s(\boldsymbol{X}^s, t_0)$ is the solid volume fraction in the reference configuration, it can be obtained that

$$
\begin{aligned}
J^s \rho &= \eta_0^s \rho^s + [J^s - \eta_0^s]\rho^w + [1 - \eta_0^s]\rho^w - [1 - \eta_0^s]\rho^w \\
&= \eta_0^s \rho^s + [1 - \eta_0^s]\rho^w + [J^s - 1]\rho^w = \rho_o^i + [J^s - 1]\rho^w.
\end{aligned}
\tag{A.4}
$$

2. For area and volume elements.

From the definition of deformation gradient $\boldsymbol{F}^\pi$, equation (10.5), follows that a vector $d\boldsymbol{X}^\pi$ in the reference configuration is transformed to a vector $d\boldsymbol{x}$ in the current configuration. Since we have assumed that the motion of the fluid phase is referred to the moving solid, we can write (10.5) in terms of the quantities referred to the solid skeleton

$$
d\boldsymbol{x} = \boldsymbol{F}^s d\boldsymbol{X}^s.
\tag{A.5}
$$

The transport theorems for area and volume elements follow computing the change of area and volume as in classical non-linear continuum mechanics (see, *e.g.*, [19, 6]).
Let $d\boldsymbol{X}^s, d\boldsymbol{Y}^s$ and $d\boldsymbol{Z}^s$ be three non-coplanar vectors in the reference configuration and $d\boldsymbol{x} = \boldsymbol{F}^s d\boldsymbol{X}^s, d\boldsymbol{y} = \boldsymbol{F}^s d\boldsymbol{Y}^s$ and $d\boldsymbol{z} = \boldsymbol{F}^s d\boldsymbol{Z}^s$ their spatial counterpart. Hence the area and the volume element can be computed as

$$
\begin{aligned}
d\boldsymbol{a} = \boldsymbol{n}da = d\boldsymbol{x} \times d\boldsymbol{y} &= \boldsymbol{F}^s d\boldsymbol{X}^s \times \boldsymbol{F}^s d\boldsymbol{Y}^s \\
&= J^s \boldsymbol{F}^{s^{-T}} \boldsymbol{N} dA = J^s [\boldsymbol{F}^s]^{-T} d\boldsymbol{A},
\end{aligned}
\tag{A.6}
$$

$$
\begin{aligned}
dv = d\boldsymbol{x} \times d\boldsymbol{y} \times d\boldsymbol{z} &= \boldsymbol{F}^s d\boldsymbol{X}^s \times \boldsymbol{F}^s d\boldsymbol{Y}^s \times \boldsymbol{F}^s d\boldsymbol{Z}^s \\
&= \boldsymbol{F}^s d\boldsymbol{X}^s \cdot [\boldsymbol{F}^s d\boldsymbol{Y}^s \times \boldsymbol{F}^s d\boldsymbol{Z}^s] = J^s dV.
\end{aligned}
\tag{A.7}
$$

It can be observed that $J^s = \det \boldsymbol{F}^s$ is related to the transformation law for volume element of the mixture and not for the volume element occupied by the solid phase, $dV_s = \eta_0^s dV^s$, at microscopic level.

The interested reader can find in [3] some additional remarks on the transport theorems and the volume fraction concept.

10.10.2 *Linearisation with respect to the undeformed domain*

Linearised equations (10.104) and (10.105) are referred to the deformed domain, B. They have been obtained from the linearisation of the weak form of the LBE and of the MBE, equations (10.97) and (10.98), computed with respect to the undeformed domain B_0, because it is time independent. These governing equations are now rewritten for sake of completeness:

$$\int_{B_0} \left[[\boldsymbol{\tau}' - J^s p^w \mathbf{1}] : \mathrm{grad}\delta\boldsymbol{u}_s \right]_{n+1} dV$$
$$- \int_{B_0} \rho_{0_{n+1}} \boldsymbol{g} \cdot \delta\boldsymbol{u}_s dV - \int_{\partial B_0} \bar{\boldsymbol{T}}_{n+1} \cdot \delta\boldsymbol{u}_s dA = 0\,, \quad \forall \delta\boldsymbol{u}_s\,, \tag{A.8}$$

$$\int_{B_0} \Delta J^s_{n+1} \delta p^w dV + \Delta t \int_{B_0} \left[J^s \frac{k}{\mu^w} [\mathrm{grad}p^w - \rho^w \boldsymbol{g}] \cdot \mathrm{grad}\delta p^w \right]_{n+\beta} dV$$
$$+ \Delta t \int_{\partial B_0} Q^w_{n+\beta} \delta p^w dA = 0\,, \qquad \forall \delta p^w\,, \tag{A.9}$$

where $\boldsymbol{\tau}'$ is the effective Kirchhoff stress tensor and $\bar{\boldsymbol{T}} = \boldsymbol{P} \cdot \boldsymbol{N}$ and $Q^w = N\bar{\boldsymbol{V}}^{ws} \cdot \boldsymbol{N}$ are, respectively, the traction vector and the water flow computed with respect to the undeformed configuration. $\Delta J^s_{n+1} = J^s_{n+1} - J^s_n$.

In the following, a (consistent) linearisation of the previous equations in the reference configuration is performed. Applying the definition of directional derivative, the linearised system of equations with respect to the volume element dV is:

- for the linear momentum balance equation:

$$\int_{B_0} [\,\mathrm{grad}\delta\boldsymbol{u}_s : \boldsymbol{c}^{ep} : \mathrm{sym}(\mathrm{grad}\Delta\boldsymbol{u}_s) + \boldsymbol{\sigma}' : \mathrm{grad}^T\Delta\boldsymbol{u}_s\, \mathrm{grad}\delta\boldsymbol{u}_s] J^s dV$$
$$+ \int_{B_0} p^w \,\mathrm{grad}\delta\boldsymbol{u}_s : [\,\mathrm{grad}^T\Delta\boldsymbol{u}_s - \mathrm{div}\Delta\boldsymbol{u}_s \mathbf{1}] J^s dV \tag{A.10}$$
$$- \int_{B_0} \rho^w \delta\boldsymbol{u}_s \cdot \boldsymbol{g} \,\mathrm{div}\Delta\boldsymbol{u}_s J^s dV - \int_{B_0} \mathrm{div}\delta\boldsymbol{u}_s \Delta p^w J^s dV\,;$$

- for the mass balance equation (A.9):

$$\int_{B_0} \delta p^w \,\mathrm{div}\Delta\boldsymbol{u}_s J^s dV$$
$$+ \beta\Delta t \int_{B_0} \mathrm{grad}\delta p^w \cdot \left[dkdn \frac{k}{\mu^w} [\,\mathrm{grad}p^w - \rho^w \boldsymbol{g}]\,\mathrm{div}\Delta\boldsymbol{u}_s \right] J^s dV$$
$$- \beta\Delta t \int_{B_0} \mathrm{grad}\delta p^w \cdot \left[\frac{2k}{\mu^w} \mathrm{sym}(\mathrm{grad}\Delta\boldsymbol{u}_s)\,\mathrm{grad}p^w \right] J^s dV \tag{A.11}$$
$$+ \beta\Delta t \int_{B_0} \frac{k}{\mu^w} \mathrm{grad}\delta p^w \cdot [\,\mathrm{grad}\Delta p^w + \rho^w \,\mathrm{grad}\Delta\boldsymbol{u}_s \boldsymbol{g}] J^s dV,$$

336   L. Sanavia, B.A. Schrefler and P. Steinmann

with $dkdn = \frac{1-n}{k} \frac{\partial k}{\partial n} + 1$.

In (A.11) an isotropic permeability variable with the void ratio, *i.e.*, $\boldsymbol{k} = k(n)\boldsymbol{1}$, has been assumed. The case of constant permeability can be obtained by setting $dkdn = 1$.

In the following, the linearisation of each term of equations (A.10) and (A.11) is developed. The symbols D_u and D_p for the directional derivatives $D(\ldots) \cdot \Delta\boldsymbol{u}_s$ and $D(\ldots) \cdot \Delta p^w$ with respect to the material coordinates are used.

- For the linear momentum balance equation:

$$DG_{LBE} \cdot \Delta\boldsymbol{u}_s;$$

$$D(\boldsymbol{\tau}' : \mathrm{grad}\delta\boldsymbol{u}_s) \cdot \Delta\boldsymbol{u}_s = D_u(\boldsymbol{\tau}' : \mathrm{Grad}^s\delta\boldsymbol{u}_s[\boldsymbol{F}^s]^{-1})$$

$$= D_u(\boldsymbol{\tau}'[\boldsymbol{F}^s]^{-T} : \mathrm{Grad}^s\delta\boldsymbol{u}_s) = D_u(\boldsymbol{S}' : [\boldsymbol{F}^s]^T \mathrm{Grad}^s\delta\boldsymbol{u}_s)$$

$$= (D_u\boldsymbol{S}') : [\boldsymbol{F}^s]^T \mathrm{Grad}^s\delta\boldsymbol{u}_s + \boldsymbol{S}' : (D_u[\boldsymbol{F}^s]^T) \mathrm{grad}_s\delta\boldsymbol{u}_s$$

$$= [\boldsymbol{F}^s]^T \mathrm{Grad}^s\delta\boldsymbol{u}_s : \boldsymbol{C}^{ep} : \mathrm{sym}([\boldsymbol{F}^s]^T \mathrm{Grad}^s\Delta\boldsymbol{u}_s) \qquad (A.12)$$

$$+\boldsymbol{S}' : [\,\mathrm{Grad}^s]^T\Delta\boldsymbol{u}_s \mathrm{Grad}^s\delta\boldsymbol{u}_s$$

$$= J^s \mathrm{grad}\delta\boldsymbol{u}_s : \boldsymbol{c}^{ep} : [\,\mathrm{sym}\,(\,\mathrm{grad}\Delta\boldsymbol{u}_s)] + \boldsymbol{\tau}' : \mathrm{grad}^T\delta\boldsymbol{u}_s \mathrm{grad}\Delta\boldsymbol{u}_s\,,$$

where $\boldsymbol{S}'$ is the effective second Piola–Kirchhoff stress tensor and $\boldsymbol{C}^{ep}$ is the material elasto-plastic constitutive tangent operator.

$$D(-J^s p^w \boldsymbol{1} : \mathrm{grad}\delta\boldsymbol{u}_s) \cdot \Delta\boldsymbol{u}_s$$

$$-[(D_u J^s)\,p^w\boldsymbol{1} : \mathrm{grad}\delta\boldsymbol{u}_s + J^s p^w\boldsymbol{1} : D_u \mathrm{grad}\delta\boldsymbol{u}_s]$$

$$-\left[J^s \mathrm{div}\Delta\boldsymbol{u}_s p^w\boldsymbol{1} : \mathrm{grad}\delta\boldsymbol{u}_s + J^s p^w\boldsymbol{1} : D_u\left(\mathrm{Grad}^s\delta\boldsymbol{u}_s[\boldsymbol{F}^s]^{-1}\right)\right]$$

$$-[J^s \mathrm{div}\Delta\boldsymbol{u}_s p^w\boldsymbol{1} : \mathrm{grad}\delta\boldsymbol{u}_s - J^s p^w\boldsymbol{1} : \mathrm{grad}\delta\boldsymbol{u}_s \mathrm{grad}\Delta\boldsymbol{u}_s]$$

$$= J^s p^w \mathrm{grad}\delta\boldsymbol{u} : \left[\mathrm{grad}^T\Delta\boldsymbol{u}_s - \mathrm{div}\Delta\boldsymbol{u}_s\boldsymbol{1}\right]. \qquad (A.13)$$

$$D(\rho_0\boldsymbol{g} \cdot \delta\boldsymbol{u}_s) \cdot \Delta\boldsymbol{u}_s$$

$$= D_u\left\{\left[\rho_0^i + \rho^w\left(J^s - 1\right)\right]\boldsymbol{g} \cdot \delta\boldsymbol{u}_s\right\} = J^s \rho^w \mathrm{div}\Delta\boldsymbol{u}_s\boldsymbol{g} \cdot \delta\boldsymbol{u}_s. \qquad (A.14)$$

$$DG_{LBE} \cdot \Delta p^w;$$

$$D\left(J^s p^w\boldsymbol{1} : \mathrm{grad}\delta\boldsymbol{u}_s\right) \cdot \Delta p^w = J^s \Delta p^w\boldsymbol{1} : \mathrm{grad}\delta\boldsymbol{u}_s. \qquad (A.15)$$

- For the mass balance equation:

$DG_{MBE} \cdot \Delta u_s;$

$$D\left(J^s \delta p^w\right) \cdot \Delta u_s = J^s \, \mathrm{div}\Delta u_s \delta p^w, \tag{A.16}$$

$$D\left(J^s \frac{k}{\mu^w}\left[\mathrm{grad}p^w - \rho^w g\right] \cdot \mathrm{grad}\delta p^w\right) \cdot \Delta u_s$$
$$= (D_u k)\,\frac{J^s}{\mu^w}\left[\mathrm{grad}p^w - \rho^w g\right] \cdot \mathrm{grad}\delta p^w$$
$$+ (D_u J^s)\,\frac{k}{\mu^w}\left[\mathrm{grad}p^w - \rho^w g\right] \cdot \mathrm{grad}\delta p^w$$
$$+ J^s \frac{k}{\mu^w}\left(D_u \,\mathrm{grad}p^w\right) \cdot \mathrm{grad}\delta p^w \tag{A.17}$$
$$+ J^s \frac{k}{\mu^w}\left[\mathrm{grad}p^w - \rho^w g\right] \cdot D_u \,\mathrm{grad}\delta p^w$$
$$= \left[\left[\frac{1-n}{k}\frac{\partial k}{\partial n} + 1\right]\frac{k}{\mu^w}\left[\mathrm{grad}p^w - \rho^w g\right] \mathrm{div}\Delta u_s\right] \cdot J^s \,\mathrm{grad}\delta p^w$$
$$- \left[\frac{2k}{\mu^w}\,\mathrm{sym}\left(\mathrm{grad}\Delta u_s\right)\mathrm{grad}p^w - \frac{k}{\mu^w}\rho^w \,\mathrm{grad}\Delta u_s g\right] \cdot J^s \,\mathrm{grad}\delta p^w,$$

where:

$$D_u k\left(n\right) = \frac{\partial k}{\partial n}\frac{1-n_0}{J^s}\,\mathrm{div}\Delta u_s = \frac{1-n}{k}\frac{\partial k}{\partial n}\,\mathrm{div}\Delta u_s$$

in which $n = n(J^s)$ and equation (10.50) has been introduced. The term $D_u \,\mathrm{grad}p^w$ of (A.17) is computed as follows:

$$D_u \,\mathrm{grad}p^w = D_u\left([F^s]^{-T}\,\mathrm{Grad}^s p^w\right) = \left(D_u [F^s]^{-T}\right)\mathrm{Grad}^s p^w$$
$$-[F^s]^{-T}\,\mathrm{Grad}^s \Delta u_s [F^s]^{-T}\,\mathrm{Grad}^s p^w = -\,\mathrm{grad}^T \Delta u_s \,\mathrm{grad}p^w; \tag{A.18}$$

$D_u \,\mathrm{grad}\delta p^w$ is easily obtained by the substitution of δp^w for p^w in (A.18).

$DG_{MBE} \cdot \Delta p^w;$

$$D\left\{J^s \frac{k}{\mu^w}\left[\mathrm{grad}p^w - \rho^w g\right] \cdot \mathrm{grad}\delta p^w\right\} \cdot \Delta p^w$$
$$= \frac{k}{\mu^w}\,\mathrm{grad}\Delta p^w \cdot \mathrm{grad}\delta p^w J^s. \tag{A.19}$$

Acknowledgments: The authors would like to thank *Programma Vigoni 2000* from CRUI, Italy, for the financial support of the stay of the first author at the University of Kaiserslautern (D).

References

[1] D.H.ADVANI, T.S. LEE, J.K. LEE, C.S. KIM: Hygrothermomechanical evaluation of porous media under finite deformation. Part. I - Finite element formulations, *Int. J. Num. Meth. Eng.* **36**, pp. 147–160, 1993.

[2] F. ARMERO: Formulation and finite element implementation of a multiplicative model of coupled poro-plasticity at finite strains under fully saturated conditions, *Comput. Methods Appl. Mech. Eng.* **171**, pp. 205–241, 1999.

[3] J. BLUHM, R. DE BOER: The volume fraction concept in the porous media theory, *ZAMM Z. Angew. Math. Mech.* **77**, pp. 563–577, 1997.

[4] R.I. BORJA, E. ALARCON: A mathematical framework for finite strain elastoplastic consolidation. Part 1: balance laws, variational formulation and linearization, *Computer Meth. Appl. Mech. Eng.* **122**, pp. 145–171, 1995.

[5] R.I. BORJA, C. TAMAGNINI: Numerical implementation of a mathematical model for finite strain elastoplastic consolidation, in: D.R.J. OWEN, E. ONATE, E. HINTON (eds.): *Computational Plasticity - Fundamentals and Applications*, CIMNE, Barcelona, 1997, pp. 1631–1640.

[6] P.G. CIARLET: *Mathematical Elasticity. Volume I: Three-Dimensional Elasticity*, Elsevier Science Publishers, Amsterdam, 1988.

[7] R. DE BOER: *Theory of Porous Media: Highlights in Historical Development and Current State*, Springer, Berlin, 2000.

[8] J. DESRUES, R. CHAMBON, W. HAMMAD, R. CHARLIER: Soil modelling with regard to consistency: Cloe, a new rate type constitutive model, in: C.S. DESAI, S. KREMPL (eds.): 3rd *Int. Conf. Constitutive Laws for Eng. Materials, Tucson*, ASME press, 1991, pp. 399–402.

[9] S. DIEBELS, W. EHLERS: Dynamic analysis of a fully saturated porous medium accounting for geometrical and material non-linearities, *Int. J. Numer. Meth. Eng.* **39**, pp. 81–97, 1996.

[10] D.C. DRUCKER, W. PRAGER: Soil mechanics and plastic analysis or limit design, *Quart. Appl. Math.* **10**, pp. 157–165, 1952.

[11] W. EHLERS, G. EIPPER: Finite elastic deformation in liquid-saturated and empty porous solids, *Trans. Porous Media* **34**, pp. 179–191, 1999.

[12] D. GAWIN, L. SANAVIA, B.A. SCHERFLER: Cavitation modelling in saturated geomaterials with application to dynamic strain localisation, *Int. J. Num. Meth. Fluids* **27**, pp. 109–125, 1998.

[13] W.G. GRAY, M. HASSANIZADEH: Unsaturated flow theory including interfacial phenomena, *Water Resour. Res.* **27**, pp. 1855–1863, 1991.

[14] M. HASSANIZADEH, W.G. GRAY: General conservation equations for multi-phase system: 1. Averaging technique; 2. Mass, Momenta, Energy and Entropy Equations, *Adv. Water Res.* **2**, pp. 191–201/131–144, 1979.

[15] M. HASSANIZADEH, W.G. GRAY: General conservation equations for multi-phase system: 3. Constitutive theory for porous media flow, *Adv. Water Res.* **3**, pp. 25–40, 1980.

[16] G. HOFSTETTER, R.L. TAYLOR: Non-associative Drucker–Prager plasticity at finite strains, *Comm. Appl. Num. Meth. Engrg.* **6**, pp. 583–589, 1990.

[17] E.H. LEE: Elastic-plastic deformation at finite strains, *J. Appl. Mech.* **36**, pp. 1–6, 1969.

[18] R.W. LEWIS, B.A. SCHREFLER: *The Finite Element Method in the Static and Dynamic Deformation and Consolidation of Porous Media*, John Wiley and Sons, New York, 1998.

[19] J.E. MARSDEN, T.J.R. HUGHES: *Mathematical Foundations of Elasticity*, Prentice Hall Inc., Upper Saddle River, New Jersey, 1983.

[20] E. MEROI, B.A. SCHREFLER, O.C. ZIENKIEWICZ: Large strain static and dynamic semi-saturated soil behaviour, *Int. J. Num. Anal. Meth. Geomech.* **19**, pp. 81–106, 1995.

[21] C. MIEHE: Computation of isotropic tensor functions, *Comm. Num. Meth. Eng.* **9**, pp. 889–896, 1993.

[22] M. MOKNI, J. DESRUES: Strain localisation measurements in undrained plane-strain biaxial tests on Hostun RF sand, *Mech. Coh.-Frict. Mater.* **4**, pp. 419–441, 1998.

[23] S. NEMAT-NASSER: On finite plastic flow of crystalline solids and geomaterials, *Trans. ASME* **50**, pp. 1114–1126, 1983.

[24] J.F. PETERS, P.V. LADE, A. BRO: Shear band formation in triaxial and plane strain tests, in: R.T. DONAGHE, R.C. CHANEY, M.L. SILVER (eds.): *Advanced Triaxial Testing of Soil and Rock*, Am. Soc. Testing and Materials **977**, Philadelphia, 1988, pp. 604–615.

[25] S. REESE: Elastopastic material behaviour with large elastic and large plastic deformation, *ZAMM Z. Angew. Math. Mech.* **77**, pp. S277–S278, 1997.

[26] B.A. SCHREFLER, L. SANAVIA, C.E. MAJORANA: A multiphase medium model for localisation and postlocalisation simulation in geomaterials, *Mech. Coh.-Frict. Mater.* **1**, pp. 95–114, 1996.

[27] J.C. SIMO: Numerical analysis and simulation of plasticity, in: P.G. CIARLET, J.L. LIONS (eds.): Numerical methods for solids (Part 3), *Handbook of Numerical Analysis* **6**, North-Holland, Amsterdam, 1998.

[28] J.C. SIMO, T.J.R. HUGHES: *Computational Inelasticity*, Springer, Berlin, 1998.

[29] J.C. SIMO, R. TAYLOR: Consistent tangent operators for rate-independent elastoplasticity, *Comp. Meth. Applied Mech. Eng.* **48**, pp. 101–118, 1985.

[30] P. STEINMANN: A finite element formulation for strong discontinuities in fluid-saturated porous media, *Mech. Coh.-Frict. Mater.* **4**, pp. 133–152, 1999.

[31] I. VARDOULAKIS, J. SULEM: *Bifurcation Analysis in Geomechanics*, Blakie Academic and Professional, London, 1995.

[32] P. WRIGGERS: Continuum mechanics, non-linear finite element techniques and computational stability, in: E. STEIN (ed.): *Progress in Computational Analysis of Inelastic Structures, CISM* **321**, Springer-Verlag, Wien-New York, 1993.

[33] H.W. ZHANG, L. SANAVIA, B.A. SCHREFLER: An internal length scale in strain localisation of multiphase porous media, *Mech. Coh.-Frict. Mater.* **4**, pp. 433–460, 1999.

[34] H.W. ZHANG, L. SANAVIA, B.A. SCHREFLER: Numerical analysis of dynamic strain localisation in initially water saturated dense sand with a modified generalised plasticity model, *Comp. and Struct.* **79**, pp. 441–459, 2001.

LORENZO SANAVIA and
BERNHARD A. SCHREFLER
Dipartimento di Ingegneria
Strutturale e dei Trasporti
Università di Padova
Via F. Marzolo, 9
I-35131 Padova, ITALY

E-mail:
Sanavia@caronte.dic.unipd.it
Bas@caronte.dic.unipd.it

PAUL STEINMANN
Chair of Applied Mechanics
University of Kaiserslautern
PO-Box 3049
D-67653 Kaiserlautern
GERMANY

E-mail: Ps@rhrk.uni-kl.de

Chapter 11

Numerical Modeling of Initiation and Propagation Phases of Landslides

Manuel Pastor, Manuel Quecedo, Pablo Mira, José A. Fernández-Merodo, Li Tongchun and Liu Xiaoqing

ABSTRACT This paper deals with initiation and propagation of landslides, for which suitable numerical models are presented. Concerning the initiation phase, we present a coupled displacement-pore pressure formulation (u-pw) proposed by Zienkiewicz and co-workers. Particular attention is paid to capture of the failure surface where strain localizes. An example is presented where the triggering mechanism is the pore pressure changes induced by rainfall. Once failure has been triggered, propagation is analyzed using an Eulerian formulation of the balance of mass and momentum equations. Two simplified, one-phase models are presented for the two extreme cases of dry granular flows and mudflows. The first model uses a level set algorithm to track the free surface, and is suitable for length scales of 100 m. For longer distances of propagation, we propose a depth integrated model which is discretized using a Taylor–Galerkin technique.

11.1   Introduction

Landslides are one of natural catastrophes causing important losses of human lives and damage to property.

There is a wide variety of types of landslides, depending on the materials involved and triggering mechanism. The time scale involved can range from minutes to years. For a classification, it is worth consulting the reference [9] where a detailed description is provided.

It could be said that a slope fails either because the material strength decreases (degradation, chemical attack, erosion) or because the effective stresses change. In the latter case, the variation can be induced by a change in total stresses (for instance, load is added at the top of the slope), or by a change in the pore pressures. Here we could mention the rise of the water table induced by reservoir filling or the change caused by heavy rain. Of course, there could also exist a combination of causes such as occurs during earthquakes, where both the total stresses and the pore pressures change.

The study of landslides and their consequences has become a multidisciplinary subject, where geographical, pedagogical and urban planning aspects are important. Here we will deal only with engineering aspects and will focus on the *prediction* of the initiation and propagation of landslides. The first aspect is important not only to know what has really happened but also to avoid possible landslides. Prediction tools can be applied, for instance, to tailing dams, to propose the rate at which they can be built, their slopes and their maximum allowable height.

Concerning the propagation phase, once the landslide has taken place, it is important to know the velocity of the flow, how far it will reach and what will be the path followed. In this way, it is possible to propose strategies based on channeling and protection structures. Another interesting example is that of landslide affecting a reservoir. Here, we will need to know both the speed and the mass of soil involved in the problem in order to feed the hydrodynamic model with proper data.

The prediction tools which will be presented in this work are based on mathematical and constitutive models for which there are extremely few analytical solutions. Therefore, numerical models such as the finite element model are required to produce suitable numerical models.

It is possible to describe the whole process (initiation and propagation) using a single mathematical model. However, there exists a difficulty in the cases where the landslide evolves to flow type phenomena, where the problem of changing from solid-like to fluid-like type of behaviour presents important difficulties. This is why we will deal separately with both phases, using different approaches for each of them.

The purpose of the paper is twofold: (i) we will present the models which have been developed by the authors, describing the basic aspects of the models, and (ii) we will concentrate on some important numerical difficulties such as the influence of element type on the computations.

The paper is structured as follows:

- The first part deals with the initiation mechanism, where we will consider both diffuse and localized mechanisms. After describing the model for the coupling of pore fluid and the solid structure, we will propose a discretization technique based on the finite element method. Then we will concentrate on the numerical difficulties in obtaining the failure conditions and mechanisms, providing some solutions. Finally, we will present an application case where a slope fails due to heavy rain.

- Once the landslide has been triggered, we will study the propagation of flowslides and debris flows. Depending on the scale at which the problem needs to be solved, we will consider two cases. The first consists of problems for which we require a detailed structure of the flow, as it happens when designing a dam or a protection structure.

It is necessary to describe how the contact between the fluid and the air evolves, and we will use a technique recently developed called the level set. The main difficulty of this approach is the computer effort required, which makes it uneconomical for propagation over long distances. In this case, it is convenient to use equations integrated in depth. Again, we find the problem of a moving boundary, but it can be solved without having to use special and expensive techniques.

11.2   Mathematical model for the initiation phase

Geomaterials (soil, rocks, concrete and other similar materials) are porous materials with voids which can be filled with water, air and other fluids. They are, therefore, multiphase materials, exhibiting a mechanical behaviour governed by the coupling between all the phases. Pore pressure plays a paramount role in the behaviour of a soil structure and, indeed, its variations can induce failure.

The first mathematical model describing this coupling was proposed by Biot [2, 3] for linear elastic materials. This work was followed by further development at Swansea University, where Zienkiewicz and coworkers [31, 36, 32, 37, 33] extended the theory to non-linear materials and large deformation problems. The mathematical model of solid skeleton and pore fluid interaction which will be recalled here is that proposed by Zienkiewicz *et al.* and described in [33].

In what follows we will assume that the voids are filled with air and water, and introduce S_w and S_a, the degrees of saturation of water and air, defined as the volume percents of each phase. It is possible to define an intersticial pressure $\bar{p}$ as $\bar{p} = S_w\, p_w + S_a\, p_a$, where p_w and p_a are the pressures of the water and air respectively.

11.2.1   The effective stress tensor

The effective stress σ' is defined from

$$d\sigma' = d\sigma + \mathbf{I}\, d\bar{p}\,,$$

where $\mathbf{I}$ is the second order identity tensor δ_{ij}. We have used positive values of the stress for tension and negative for extension.

We will assume also that the air is at atmospheric pressure, *i.e.*, $p_a = 0$ and, therefore, $\bar{p} = S_w\, p_w$. This approach is valid for many engineering cases of interest. A more complete description can be found in [15]

The increments of stresses which we will consider are of Jaumann–Zaremba type and they are related to the increments of Cauchy stress tensor by

$$d\hat{\sigma}' = d\sigma' - d\sigma'^{R}\,,$$

where the co-rotational increment $d\sigma'^R$ is given by

$$d\sigma_{ij}'^R = \sigma_{ik}d\Omega_{kj} + \sigma_{jk}d\Omega_{ki} \, .$$

The term $d\Omega_{ij}$ is the antisymmetric gradient of the displacement of the solid skeleton u:

$$d\Omega_{ij} = \frac{1}{2}\left(du_{i,j} - du_{j,i}\right) \, .$$

Increments of strain are then related to $d\hat{\sigma}'$ by the constitutive equation

$$d\hat{\sigma}' = D' : d\varepsilon' \, .$$

Another point worth mentioning is that the effective stress above introduced has to be modified to account for compressibility of soil grains relative to that of solid skeleton. The total increment of strain $d\varepsilon$ can be decomposed into three parts: $d\varepsilon^g$ which accounts for the deformation of the solid grains, $d\varepsilon^0$ which describes changes of strain caused by other mechanisms than mechanical (thermal, for instance) and $d\varepsilon'$ which corresponds to the soil skeleton. It is possible then to write the constitutive equation as

$$d\hat{\sigma}' = D' : \left(d\varepsilon - d\varepsilon^0 - d\varepsilon^g\right) \, .$$

The strain induced in the grains by a change in the pore pressure $\bar{p}$ is

$$d\varepsilon^g = -\frac{1}{3K_g}\mathbf{I}\,d\bar{p} \, ,$$

where K_g is the volumetric stiffness of the soil grains. From here, we obtain

$$d\sigma' = D' : \left(d\varepsilon - d\varepsilon^0\right) + (D' : \mathbf{I})\frac{d\bar{p}}{3K_g} + d\sigma^R \, .$$

The increment of total stress (Cauchy) will be given by

$$d\sigma = D' : \left(d\varepsilon - d\varepsilon^0\right) - \left[\mathbf{I} - \frac{D' : \mathbf{I}}{3K_g}\right]d\bar{p} + d\sigma^R \, . \tag{11.1}$$

As far as the term $D' : \mathbf{I}$ is concerned, if we assume elastic behaviour of the soil skeleton, we will have

$$\begin{aligned}
D'_{ijkl}\delta_{kl} &= \left(\lambda\delta_{ij}\delta_{kl} + \mu\delta_{ik}\delta_{jl} + \mu\delta_{il}\delta_{jk}\right)\delta_{kl} \\
&= \left(3\lambda' + 2\mu'\right)\delta_{ij} \\
&= 3K_s\delta_{ij} \, ,
\end{aligned}$$

where K_s is the volumetric stiffness of the skeleton. Therefore,

$$\left[\mathbf{I} - \frac{D' : \mathbf{I}}{3K_g}\right]d\bar{p} = \left(1 - \frac{K_s}{K_g}\right)d\bar{p} = (1 - \alpha)\,d\bar{p} \, ,$$

where we have introduced the stiffness ratio $\alpha = \frac{K_s}{K_g}$. Using this result in the expression (11.1), we obtain

$$\begin{aligned} d\sigma &= D' : \left(d\varepsilon - d\varepsilon^0\right) - \alpha \mathbf{I}\, d\bar{p} + d\sigma^R \,, \\ d\sigma' &= D' : \left(d\varepsilon - d\varepsilon^0\right) + \left(1 - \alpha\right)\mathbf{I}\, d\bar{p} + d\sigma^R \,. \end{aligned} \qquad (11.2)$$

The value of α ranges from very small values in the case of soils, $\alpha \sim 0$, to values close to 0.7 in the case of rocks. Therefore, the definition of effective stress has to be modified in the case of soils having a stiff skeleton introducing the parameter α.

11.2.2 Kinematics

We will describe the displacement of soil skeleton by the vector field $u(r, t)$ and the displacement of the intersticial water by $U(r,t)$, which will be decomposed into two parts, corresponding to the displacements of the soil skeleton and the relative movement with respect to it as

$$\dot{U} = \dot{u} + \frac{\dot{w}}{n} \,,$$

where n is the material porosity. Above, $\dot{w}$ is a fictitious velocity which provides the same debit when we consider flow over the total section (and not only the pores).

11.2.3 Balance of mass

The purpose of this section is to relate the change of volume of an infinitesimal element of the mixture to the changes of volume of the constituents and the net balance of fluid flowing through it.

We will characterize the rates of volume change caused by (i) deformation of soil skeleton, $\dot{\theta}_1$, (ii) volumetric deformation of solid particles caused by changes in pore pressure, $\dot{\theta}_2$, (iii) volumetric deformation of pore fluid caused by changes in pore pressure, $\dot{\theta}_3$, (iv) deformation of solid particles caused by changes in effective stresses, $\dot{\theta}_4$, and (v) water storage induced by change of degree of saturation, $\dot{\theta}_5$.

The first term is given by

$$\dot{\theta}_1 = S_r \mathrm{tr}\left(\dot{\varepsilon}\right) \,.$$

The second can be written as

$$\dot{\theta}_2 = \frac{1-n}{K_s}\dot{\bar{p}} = \frac{1-n}{K_s}\left\{S_w + p_w \frac{C_s}{n}\right\}\dot{\bar{p}}_w \,,$$

where we have introduced C_s, the specific storage coefficient, as

$$C_s = n\frac{\partial S_w}{\partial t} \,.$$

To derive the above equation, we have used the relationships

$$
\begin{aligned}
\frac{\partial \overline{p}}{\partial t} &= \frac{\partial}{\partial t}\left(S_w p_w\right) = S_w \frac{\partial p_w}{\partial t} + p_w \frac{\partial S_w}{\partial t}, \\
\frac{\partial S_w}{\partial t} &= \frac{\partial S_w}{\partial p_w}\frac{\partial p_w}{\partial t} = \frac{C_s}{n}\frac{\partial p_w}{\partial t}.
\end{aligned}
\tag{11.3}
$$

The volume change due to fluid compressibility is given by

$$
\dot{\theta}_3 = n\frac{S_w}{K_w}\frac{\partial p_w}{\partial t},
$$

where K_w is the volumetric stiffness of the fluid.

The fourth term, $\dot{\theta}_4$, accounts for the effect of the effective stress on the solid particles:

$$
\dot{\theta}_4 = -\frac{S_w}{3K_g}\operatorname{tr}\left(\dot{\sigma}\right).
$$

If we substitute in the above the value of $\dot{\sigma}$ given by (11.2), we obtain

$$
\dot{\theta}_4 = -\frac{S_w}{3K_g}\operatorname{tr}\left\{D' : \dot{\varepsilon} - D' : \dot{\varepsilon}_0 + (1-\alpha)\,\mathbf{I}\,\dot{p} + \dot{\sigma}^R\right\}.
$$

The first term can be simplified using the definition of α :

$$
\frac{1}{3K_s}\operatorname{tr}\left\{D' : \dot{\varepsilon}\right\} = \dot{\varepsilon} : \left\{\frac{D' : \mathbf{I}}{3K_s}\right\} = \dot{\varepsilon} : (1-\alpha)\,\mathbf{I} = (1-\alpha)\operatorname{tr}\dot{\varepsilon}.
$$

Making use of (11.3) we finally arrive at

$$
\begin{aligned}
\dot{\theta}_4 = &-S_w\left(1-\alpha\right)\operatorname{tr}\left(\dot{\varepsilon}\right) \\
&-\frac{S_w}{K_s}\left(1-\alpha\right)\left\{S_w + p_w\frac{C_s}{n}\right\}\frac{\partial p_w}{\partial t} + \frac{S_w}{3K_s}\operatorname{tr}\left(D' : \dot{\varepsilon}_0\right).
\end{aligned}
$$

Finally, the term $\dot{\theta}_5$ is

$$
\dot{\theta}_5 = n\frac{\partial S_w}{\partial t} = C_s\frac{\partial p_w}{\partial t}.
$$

The sum of all five contributions should be equal to the rate of fluid entering (or exiting) the domain:

$$
-\nabla^T \dot{w} = \dot{\theta}_1 + \dot{\theta}_2 + \dot{\theta}_3 + \dot{\theta}_4 + \dot{\theta}_5
$$

from which

$$
\nabla^T \dot{w} + S_w \alpha \operatorname{tr}\left(\dot{\varepsilon}\right) + C_s \dot{p}_w + \frac{1}{Q}\dot{p}_w + \frac{S_w}{3K_s}\operatorname{tr}\left(D' : \dot{\varepsilon}_0\right) = 0,
\tag{11.4}
$$

where we have introduced the mixed volumetric stiffness Q as

$$
\frac{1}{Q} = \left\{n\frac{S_w}{K_w} + \frac{\alpha - n}{K_s}\left(S_w + p_w\frac{C_s}{n}\right)\right\}.
$$

11.2.4 Balance of linear momentum

The balance of linear momentum for the mixture can be written as

$$S^T \sigma + \rho b - \rho \ddot{u} - n\rho_w S_w \frac{D}{Dt}\left(\frac{\dot{w}}{nS_w}\right) = 0, \tag{11.5}$$

where the first three terms account for the joint movement of the solid and fluid phases and the last for the relative displacement of the fluid with respect to the solid. The transpose of the strain operator S coincides with the vector representation of the divergence operator

$$S = \begin{bmatrix} \dfrac{\partial}{\partial x} & 0 & 0 \\[6pt] 0 & \dfrac{\partial}{\partial y} & 0 \\[6pt] 0 & 0 & \dfrac{\partial}{\partial z} \\[6pt] \dfrac{\partial}{\partial y} & \dfrac{\partial}{\partial x} & 0 \\[6pt] 0 & \dfrac{\partial}{\partial z} & \dfrac{\partial}{\partial y} \\[6pt] \dfrac{\partial}{\partial z} & 0 & \dfrac{\partial}{\partial x} \end{bmatrix}, \tag{11.6}$$

where we have represented the stress tensor in vector form:

$$\sigma = (\sigma_{xx}, \sigma_{yy}, \sigma_{zz}, \sigma_{xy}, \sigma_{yz}, \sigma_{zx})^T. \tag{11.7}$$

It is also useful to introduce m, the vector representation of the identity tensor δ_{ij},

$$m^T = (1, 1, 1, 0, 0, 0).$$

In above, b are the body forces, ρ_m the density of the mixture and $\ddot{u}$ the acceleration of the solid skeleton.

The balance of momentum of the fluid phase is given by the expression

$$-\nabla p_w + \rho_w b - k_w^{-1}\dot{w} - \rho_w\left[\ddot{u} + \frac{D}{Dt}\left(\frac{\dot{w}}{nS_w}\right)\right] = 0. \tag{11.8}$$

11.2.5 A note on permeability and degree of saturation

In the proposed approach we have assumed that the air is always at atmospheric pressure. Concerning the pore pressure and the degree of saturation, it can be assumed that they are related by a given empirical law. Further simplification can be made by assuming that the evolution of both variables does not present hysteresis effects. Still, it is important to recall that degree of saturation will depend on pore pressure.

Concerning the permeability tensor, it will also depend on the degree of saturation and will change also with the volumetric strains induced in the soil.

11.2.6 The u-p_w formulation

Under certain conditions [31], we can neglect the relative accelerations of the fluid phase and eliminate the relative displacements of the fluid phase w. This can be done by substituting the value of $\nabla^T \dot{w}$ obtained from (11.8)

$$\nabla^T \dot{w} = \nabla^T \left[k_w \left(-\nabla p_w + \rho_w b - \rho_w \ddot{u} \right) \right]$$

into the balance of mass of the fluid phase (11.4)

$$\nabla^T \dot{w} + S_w \alpha \operatorname{tr}(\dot{\varepsilon}) + C_s \dot{p}_w + \frac{1}{Q} \dot{p}_w + \frac{S_w}{3K_s} \operatorname{tr}(D' : \dot{\varepsilon}_0) = 0$$

from which we obtain

$$\nabla^T \left\{ k_w \left(-\nabla p_w + \rho_w b - \rho_w \ddot{u} \right) \right\} + \left(C_s + \frac{1}{Q} \right) \dot{p}_w$$

$$+ S_w \alpha \operatorname{tr}(\dot{\varepsilon}) + \frac{S_w}{3K_s} \operatorname{tr}(D' : \dot{\varepsilon}_0) = 0 . \tag{11.9}$$

The main advantage of this formulation is that the field variables reduce to two, displacements u and pressures p_w.

Therefore, the basic equations of this $u - p_w$ formulation are (11.5) and (11.9), together with the constitutive equation (11.2) and the kinematic relations between strain and displacement.

11.2.7 Boundary and initial conditions

The equations presented so far have to be complemented by suitable boundary and initial conditions for the problem variables. In the $u - p_w$ the conditions are the following:

(i) u prescribed on Γ_u;

(ii) tractions prescribed on Γ_t, $\sigma.n = \bar{t}$, $\Gamma_u \cup \Gamma_t = \Gamma$, $\Gamma_u \cap \Gamma_t = \{\emptyset\}$;

(iii) p_w prescribed on Γ_{pw};

(iv) flux $-k_w \nabla p_w n$ prescribed along Γ_q, $\Gamma_{pw} \cup \Gamma_q = \Gamma$, $\Gamma_{pw} \cap \Gamma_q = \{\emptyset\}$.

In the case of a non-saturated slope under rain, the boundary conditions to be applied can be simplified to a prescribed pressure (atmospheric) on the surface, but care should be taken as the inflow cannot be larger than the amount of percolating water. In this way, the saturation will increase within the material and the effective stresses will decrease.

The initial conditions will be:

(i) solid displacements and velocities at $t = 0$,

(ii) pore pressures at $t = 0$.

11.2.8 A note on the incompressible undrained limit

The $u - p_w$ version of Biot equations can be further simplified according to whether accelerations are small, leading to what is known as "consolidation", or "slow consolidation phenomena", which in the case of saturated materials are

$$S^T \sigma + \rho b = 0$$

and

$$\nabla^T \{k_w (-\nabla p_w + \rho_w b)\} + \frac{1}{Q} \dot{p}_w + \operatorname{tr}(\dot{\varepsilon}) = 0 . \tag{11.10}$$

If we now consider a material of very small permeability and very large volumetric stiffness, the above equations can be written as:

$$\begin{aligned} S^T (\sigma' + m\, p_w) + \rho b &= 0, \\ \nabla^T \dot{u} &= 0, \end{aligned}$$

which are similar to the equations found in solid mechanics for incompressible materials. Here, important difficulties regarding both the interpolation spaces which can be used for displacements and pressures and simulation of failure processes are found, and will be discussed later.

11.3   Numerical model

11.3.1   Discretization of the u-p_m model

The mathematical model consists of the following system of PDE's:

$$S^T \sigma + \rho b - \rho \ddot{u} - n \rho_w S_w \frac{D}{Dt}\left(\frac{\dot{w}}{n S_w}\right) = 0 , \tag{11.11}$$

$$\nabla^T \{k_w (-\nabla p_w + \rho_w b - \rho_w \ddot{u})\} + \left(C_s + \frac{1}{Q}\right) \dot{p}_w$$

$$+ S_w \alpha \operatorname{tr}(\dot{\varepsilon}) + \frac{S_w}{3K_s} \operatorname{tr}(D' : \dot{\varepsilon}_0) = 0,$$

or

$$S_w \alpha \operatorname{tr}(\dot{\varepsilon}) - \nabla^T (k_w \nabla p_w) + \left(\tfrac{1}{Q^*}\right) \dot{p}_w$$

$$+ \nabla^T (k_w \rho_w b) - \nabla^T (k_w \rho_w \ddot{u}) + s_0 = 0, \tag{11.12}$$

where

$$\begin{aligned} \frac{1}{Q^*} &= C_s + \frac{1}{Q}, \\ s_0 &= \frac{S_w}{3K_s} \operatorname{tr}(D' : \dot{\varepsilon}_0) . \end{aligned}$$

The system of partial differential equations can be discretized using standard Galerkin techniques. After approximating the fields u and p_w as $u = \mathbf{N}_u \bar{\mathbf{u}}$, $p_w = \mathbf{N}_p \bar{\mathbf{p}}_w$, it results in two ordinary differential equations:

$$\mathbf{M}\ddot{\mathbf{u}} + \int_\Omega \mathbf{B}^T . \sigma' d\Omega - \mathbf{Q}\bar{\mathbf{p}}_w - \mathbf{f}_u = \mathbf{0} \,, \tag{11.13}$$

where $\mathbf{B} = \mathbf{S}.\mathbf{N}_u$ and

$$\mathbf{Q}^T . \dot{\mathbf{u}} + \mathbf{H}.\bar{\mathbf{p}}_w + \mathbf{C}.\dot{\bar{\mathbf{p}}}_w - \mathbf{f}_p = \mathbf{0} \,. \tag{11.14}$$

The matrices given above are defined as

$$\begin{aligned}
\mathbf{M} &= \int_\Omega \rho \mathbf{N}_u^T \mathbf{N}_u d\Omega \,, \\[4pt]
\mathbf{C} &= \int_\Omega \tfrac{1}{Q^*} \mathbf{N}_p^T \mathbf{N}_p d\Omega \,, \\[4pt]
\mathbf{Q} &= \int_\Omega S_w \alpha \mathbf{B}^T m \mathbf{N}_p d\Omega \,, \\[4pt]
\mathbf{H} &= \int \nabla \mathbf{N}_p^T k_w \nabla \mathbf{N}_p d\Omega
\end{aligned} \tag{11.15}$$

and

$$\begin{aligned}
\mathbf{f}_u &= \int_\Omega \mathbf{N}_u^T b\, d\Omega + \int_{\Gamma_t} \mathbf{N}_u^T \bar{t}\, d\Gamma \,, \\[4pt]
\mathbf{f}_p &= \int_{\Gamma_q} \mathbf{N}_p^T k_w \frac{\partial p}{\partial n} d\Gamma + \int_\Omega \nabla \mathbf{N}_p^T k_w \rho_w\, b\, d\Omega \\
&\quad - \left(\int_\Omega \nabla \mathbf{N}_p k_w \mathbf{N}_u d\Omega \right) \ddot{\mathbf{u}} - \int_\Omega \mathbf{N}_p^T s_0 d\Omega \,.
\end{aligned} \tag{11.16}$$

The term including accelerations in $\mathbf{f}_p$ is usually disregarded.

The time derivatives of u and p_w are approximated in a typical step of computation using the Generalized Newmark GN22 scheme for displacements and a GN11 for the water pressure [13, 38].

If we introduce the notation [13]

$$\begin{aligned}
\Delta \ddot{u}^n &= \ddot{u}^{n+1} - \ddot{u}^n \,, \\[4pt]
\Delta \dot{p}_w^n &= \dot{p}_w^{n+1} - \dot{p}_w^n \,,
\end{aligned}$$

$$\begin{aligned}
\dot{u}^{n+1} &= \dot{u}^{p,n+1} + \beta_1 \Delta t \Delta \ddot{u}^n \,, \\[4pt]
u^{n+1} &= u^{p,n} + \tfrac{1}{2}\beta_2 \Delta t^2 \Delta \ddot{u}^n \,, \\[4pt]
p_w^{n+1} &= p_w^{p,n+1} + \theta \Delta t \Delta \dot{p}_w^n \,,
\end{aligned} \tag{11.17}$$

where

$$\begin{aligned}
\dot{u}^{p,n+1} &= \dot{u}^n + \Delta t\, \ddot{u}^n \,, \\[4pt]
u^{p,n+1} &= u^n + \Delta t\, \dot{u}^n + \tfrac{1}{2}\Delta t^2\, \ddot{u}^n \,, \\[4pt]
p_w^{p,n+1} &= p_w^n + \Delta t\, \dot{p}_w^n \,,
\end{aligned} \tag{11.18}$$

we obtain the discretized system of equations valid in each time step:

$$\mathbf{M}\Delta\ddot{\mathbf{u}}^n + \int \mathbf{B}^T.\sigma'^{n+1} - \theta\Delta t\mathbf{Q}\Delta\dot{\mathbf{p}}_w^n - \mathbf{F}_u^{n+1} = \mathbf{\Phi}_u = \mathbf{0}\,, \qquad (11.19)$$

$$\beta_1\Delta t\mathbf{Q}^T.\Delta\ddot{\mathbf{u}}^n + (\theta\Delta t\,\mathbf{H} + \mathbf{C}).\Delta\dot{\mathbf{p}}_w^n - \mathbf{F}_p^{n+1} = \mathbf{\Phi}_p = \mathbf{0}\,, \qquad (11.20)$$

where the unknown values are $\Delta\ddot{\mathbf{u}}^n$ and $\Delta\dot{\mathbf{p}}_w^n$.

If this system is non-linear, it can be solved by using a Newton–Raphson method with a suitable jacobian matrix

$$\begin{bmatrix} \dfrac{\partial\mathbf{\Phi}_u}{\partial\Delta\ddot{\mathbf{u}}} & \dfrac{\partial\mathbf{\Phi}_u}{\partial\Delta\dot{\mathbf{p}}} \\[2ex] \dfrac{\partial\mathbf{\Phi}_p}{\partial\Delta\ddot{\mathbf{u}}} & \dfrac{\partial\mathbf{\Phi}_p}{\partial\Delta\dot{\mathbf{p}}} \end{bmatrix}^{(i)} \begin{bmatrix} \delta(\Delta\ddot{\mathbf{u}}) \\[1ex] \delta(\Delta\dot{\mathbf{p}}) \end{bmatrix}^{(i+1)} = - \begin{bmatrix} \mathbf{\Phi}_u \\[1ex] \mathbf{\Phi}_p \end{bmatrix}^{(i)}. \qquad (11.21)$$

Using equations (11.19) and (11.20) we can write the above step as

$$\begin{bmatrix} \mathbf{M} + \tfrac{1}{2}\Delta t^2\beta_2\mathbf{K}_T & -\theta\Delta t\mathbf{Q} \\[1ex] \beta_1\Delta t\mathbf{Q}^T & \theta\Delta t\mathbf{H} + \mathbf{S} \end{bmatrix}^{(i)} \begin{bmatrix} \delta(\Delta\ddot{\mathbf{u}}) \\[1ex] \delta(\Delta\dot{\mathbf{p}}) \end{bmatrix}^{(i+1)} = - \begin{bmatrix} \mathbf{\Phi}_u \\[1ex] \mathbf{\Phi}_p \end{bmatrix}^{(i)}, \qquad (11.22)$$

where $\mathbf{K}_T$ is the tangent stiffness matrix: $\mathbf{K}_T = \int \mathbf{B}^T\mathbf{D}_{ep}\mathbf{B}d\Omega$.

11.3.2 *Restrictions on the interpolation spaces for displacements and pressures*

The system of equations given in equation (11.22) achieves in steady state in the undrained-incompressible limit, *i.e.*, with $k_w \to 0$, $Q^* \to \infty$ a system of the type frequently found in constrained problems

$$\begin{bmatrix} \mathbf{A} & -\mathbf{B} \\ -\mathbf{B}^T & \mathbf{0} \end{bmatrix} \begin{bmatrix} \Delta\xi \\ \Delta\phi \end{bmatrix} = \begin{bmatrix} \mathbf{f}_\xi \\ \mathbf{f}_\phi \end{bmatrix}. \qquad (11.23)$$

This is in detail similar to those found in mixed formulations of incompressible solid mechanics and fluid dynamics problems, and it can be demonstrated that the system will be singular and present pressure oscillations whenever the number of $\Delta\xi$ variables $n_\xi \le n_\phi$, *i.e.*, the number of $\Delta\phi$ variables. Although this condition is not sufficient for stability (and solvability) it is necessary and it generally excludes equal order of interpolation spaces for pressure and displacements.

A complete mathematical treatment of the problem can be found in [1] and [4]. However, a much simpler explanation is provided by the patch test for mixed formulations proposed by Zienkiewicz, Taylor and Chan [38]. Therefore, the interpolation spaces for displacements and pressures cannot be the same unless stabilization is provided, such as the one proposed by the authors in references [19, 21, 35, 20].

11.4   Applications

11.4.1   Failure of a vertical slope

The first problem we will consider is that of the failure of a 45° slope
under undrained loading. The geometry and boundary conditions are given
in Figure 11.1. In the quasi-static limit, this is a simple boundary value
problem for which an analytical solution of collapse load exists [5]:

$$\frac{F}{BC_u} = 3.57 \, .$$

It is important to note that this problem becomes ill-posed from a mathematical point of view which results in non-uniqueness and mesh dependence. We will not focus on these issues, but on how simple T3P3 are able to
provide an accurate description of both limit load and failure mechanism.

We will use adaptive remeshing to improve resolution of the shear band as
introduced by Peraire, Vahdati, Morgan and Zienkiewicz [24] and described
in [34].

The soil has been assumed to be an elastoplastic material with a Tresca
yield surface and no hardening.

Applied external forces F and vertical displacement δ at the point at
which load is applied are described in terms of the non-dimensional parameters $\frac{F}{BC_u}$ and $\alpha = \frac{E\delta}{BC_u}$.

The soil domain is discretized with an initial mesh of linear T3P3. The
problem is solved using adaptive remeshing and Figure 11.2 shows the first
and third meshes together with the plastic strain contours at $\alpha = 15$.
Finally, the load *vs.* displacement curves are given in Figure 11.3.

The error in the computed limit load is below 5%, with a failure surface
clearly defined.

11.4.2   Failure of a cut slope under rain action

The second example we will present concerns the progressive failure of a
non-saturated clay cut slope due to rainfall. The height has been chosen as
1 and the slope is 2:1. Initial conditions have been simulated by assuming
an initial horizontal soil layer which has been excavated to obtain the cut
slope. The finite element mesh can be seen in Figure 11.4. The interpolation
functions are quadratic for displacements and bilinear for pore pressures.
A reduced integration rule of 2x2 points has been used.

Concerning boundary conditions, we have used $\partial_n p_w = 0$ and $u.n = 0$,
where n is the normal to the artificial boundary. In the surface, we have
assumed traction free conditions and a suction of $20 \, kPa$.

The material has been assumed elasto-plastic, with an associated Druc-
ker–Prager yield criterion and a softening law characterized by $\varepsilon^p_{peak} = 0.05$,
$\phi_{peak} = 20°$, $c_{peak} = 7 \, kPa$, $\varepsilon^p_{res} = 0.20$, $\phi_{res} = 13°$ and $c_{res} = 2 \, kPa$. The

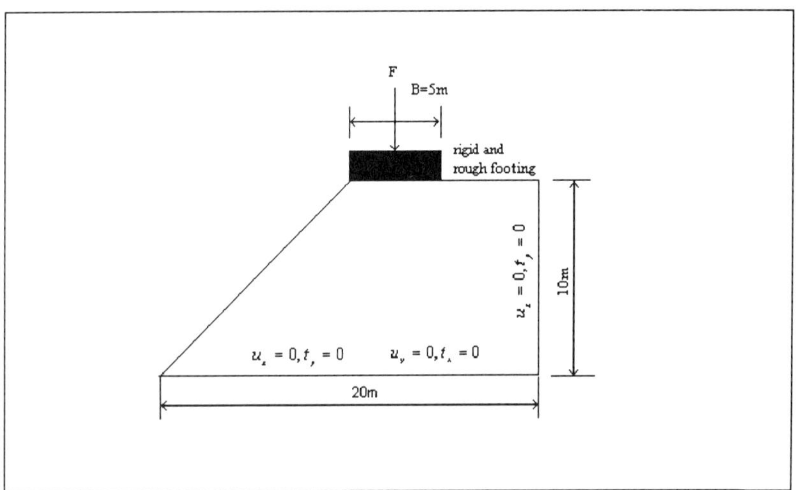

FIGURE 11.1. Slope stability problem: geometry and boundary conditions.

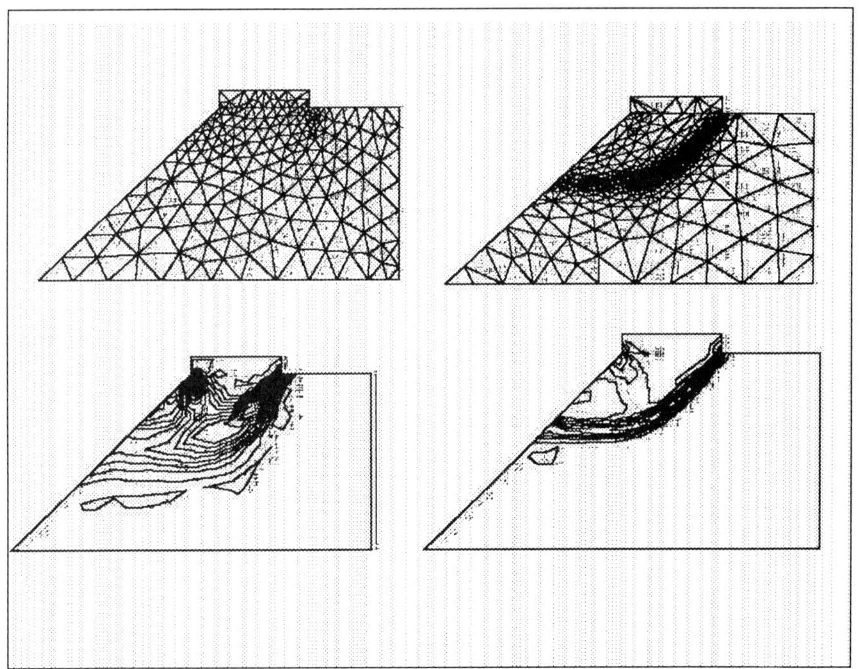

FIGURE 11.2. Adaptive meshes and plastic strain contours.

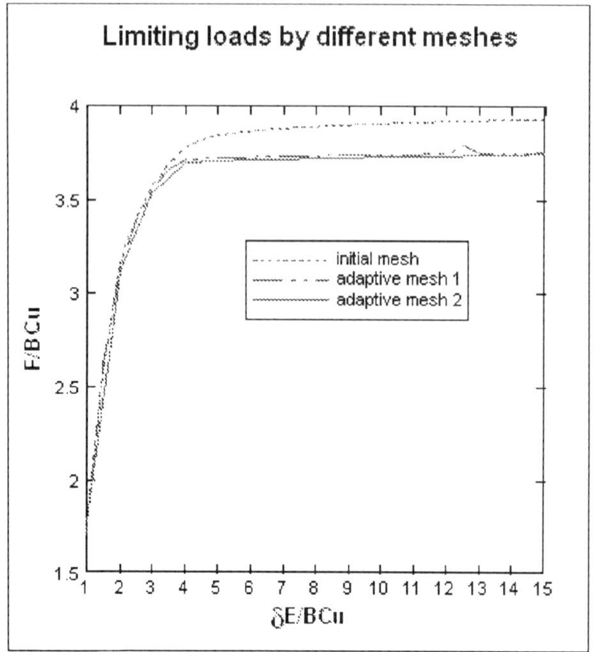

FIGURE 11.3. Slope stability problem: force vs. displacement plot.

Young modulus varies with the effective confining pressure as

$$E \;=\; \min\left(400\,kPa,\; 25\left(p' + 100\right)\right),$$
$$\nu \;=\; 0.2.$$

After checking equilibrium steady state conditions after excavation, the rain has been applied in two simplified ways: (I) Flooding, $p_w = 0$, and (II) 6 mm/day. In both cases failure is reached. Figure 11.5 shows the plastic strain contours at failure and Figure 11.6 displays the time evolution of the horizontal displacement of the midslope point.

11.5   Propagation phase

11.5.1   Introduction

Once failure has been triggered, the soil mass will move with a velocity which will depend largely on the type of problem. In some cases (flow slides, mudflows, *etc.*) the material will flow in a "fluid-like" manner. There exists a strong coupling between the solid and the fluid phases and the equations describing the movement are similar to those of the initiation phase. However, a first simplification of assuming a single phase can be made in two

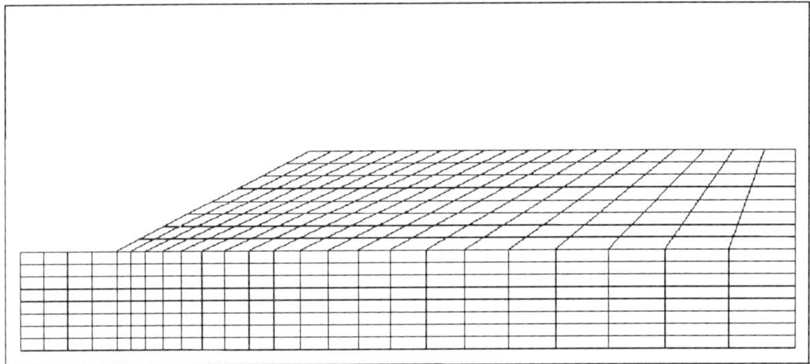

FIGURE 11.4. Finite element mesh.

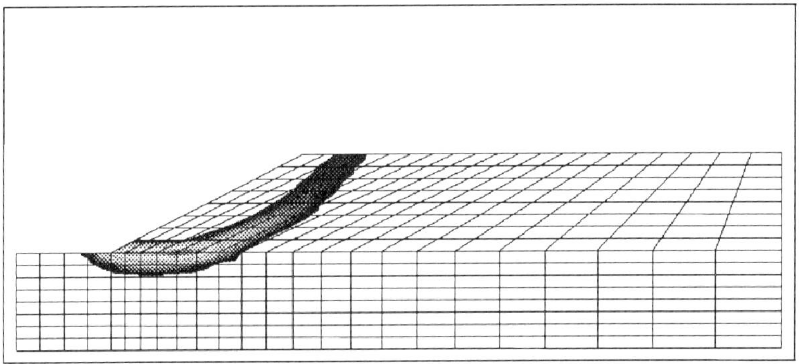

FIGURE 11.5. Isolines of equivalent deviatoric plastic strain.

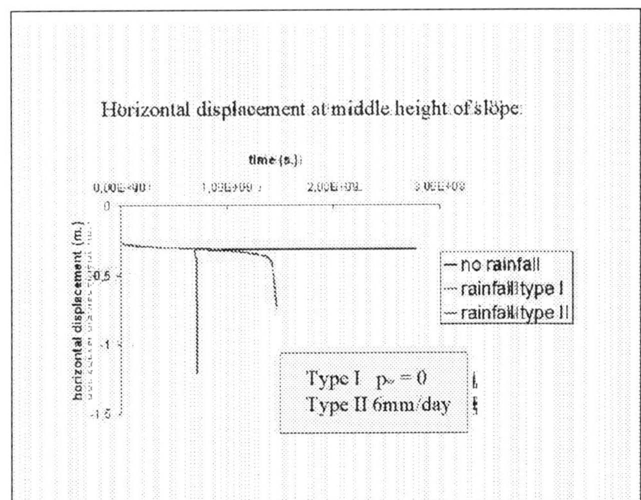

FIGURE 11.6. Horizontal displacement at middle height of slope.

limit cases (i) very permeable or dry materials, and (ii) materials for which
the time scale of consolidation is much larger than that of propagation. Examples of both situations are rock avalanches and mudflows. Care should
be taken when selecting constitutive material properties, as the apparent
friction angle can be much smaller than the effective, because of the pore
pressures. In fact, liquefied material can flow over slopes much smaller than
its effective friction angle.

11.5.2 A level-set finite element model for free surface flow of geomaterials

Description of the model

The main problem in the propagation phase is to follow the domain which is
covered by the flowing material. The approach which will be followed here
consists in considering a domain of propagation which includes two regions,
one of which is occupied by the soil mixture. Concerning the second region,
we will assume it filled by atmospheric air. Furthermore, we will assume
that both fluids are immiscible.

Both soil mixture and air will move in the domain and it would be possible to write a set of equations (balance of momentum and mass) for each
material provided we are able to determine at every point of the domain
the material which occupies it. Therefore, the key is to determine the exact
position of the interface. Among the different alternatives proposed in the
past, we can mention the so-called "front-tracking" methods, where the interface was described using segments of lines or pieces of surfaces. The main
difficulty arises when two fronts coalesce and special ad-hoc techniques were
proposed to re-structure the front in these cases [29]. The volume-tracking
algorithms use markers convected by the fluid to obtain the position of the
front. Here it is worth mentioning the MAC (marker and cell) technique
of Harlow and Welch [11], the VOF (Volume of Fluid) method of Hirt and
Nichols [12] and the SLIC (Simple Line Interface Calculation) technique
described in [17, 6].

Concerning finite elements, Dieterlen *et al.* [8] proposed a combination of
finite elements and cells, but the most effective approach was based in the
concept of the "pseudo-concentration function", introduced by Thompson
[28] and applied by Lewis [14]. This approach was applied to model debris
flows by [10].

The pseudo-concentration function is an indicator which takes a value
of unity in the domain occupied by the soil and zero in the rest. The
computation cycle consists of two steps (i) a dynamic step, where balance of
linear moment and constitutive equations of materials provide the velocity
field at t^{n+1}, and (ii) an advection step where the pseudoconcentration
function Ψ is advected with the velocity field obtained in the previous
step.

The main problem here concerns the second step, where severe oscillations can appear as we have to propagate a discontinuity. To circumvent this difficulty, Sethian introduced in the framework of finite differences the "level set" method [18], which was further refined by Sussman [27]. The method consists of transforming the discontinuous pseudo-concentration function onto a much smoother function ϕ (a distance function of unity gradient) which takes a reference value (zero, for instance) at the interface. Therefore, the interface is defined by the set of points at which ϕ is zero. It can be said that the interface is the level set of ϕ.

This method has been recently extended to finite elements by the authors [25] and will be applied here to model flow of geomaterials. The basic ingredients are:

- A suitable numerical scheme to integrate in time the equations of motion, and

- The level set algorithm to advect the pseudoconcentration function.

The balance of momentum equation will be written for both phases (air and soil mixture) as

$$\rho\frac{\partial\bar{u}}{\partial t} + \operatorname{grad}\bar{u}\,(\rho\bar{u}) = \operatorname{div}\tau - \operatorname{grad}p + \rho\bar{b}, \qquad (11.24)$$

where we have split the stress tensor into its deviatoric and hydrostatic components. This is necessary when dealing with incompressible materials for which

$$\operatorname{div}\bar{u} = 0. \qquad (11.25)$$

These equations have to be complemented by suitable constitutive or rheological equations relating stresses to rate of strains. In the case of Newtonian fluids, we have

$$\tau = 2\mu\dot{\varepsilon} = \mu\left(\operatorname{grad}\bar{u} + \operatorname{grad}^{T}\bar{u}\right).$$

Advection by the velocity field $\bar{u}$ is described by

$$\frac{\partial\phi}{\partial t} + \operatorname{grad}\phi\cdot\bar{u} = 0, \qquad (11.26)$$

where $\phi(x;t)$ is the normalized pseudoconcentration function, which fulfills $\left\|\operatorname{grad}\phi^{n+1}\right\| = 1$.

Once this function has been advected by the velocity field, it has to be transformed again into a distance function, which is achieved solving the evolution algorithm

$$\frac{\partial\phi}{\partial t^{*}} + S\left(\phi^{n}\right)|\operatorname{grad}\phi| = S\left(\phi^{n}\right) \qquad (11.27)$$

with initial conditions

$$\phi(\bar{x}, t) = \phi^{n}(\bar{x}).$$

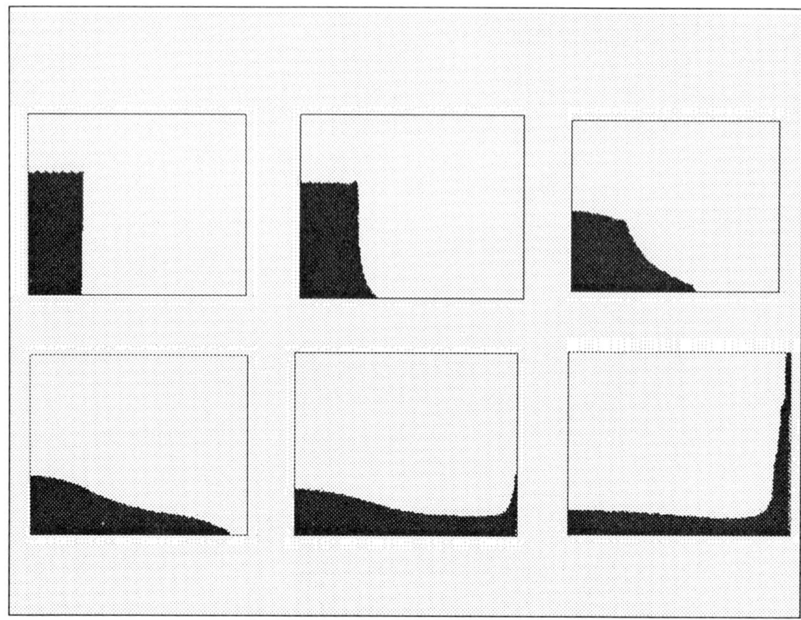

FIGURE 11.7. Dam break problem: initial conditions and propagation.

In the above, $S(\cdot)$ is the "sign" function and t^* a fictitious time variable introduced to obtain the distance function at steady state.

The equation (11.27) can be written also as

$$\frac{\partial \phi}{\partial t} + S(\phi^n) \frac{\operatorname{grad} \phi}{|\operatorname{grad} \phi|} \cdot \operatorname{grad} \phi = S(\phi^n) \,,$$

where it can be seen that the problem to be solved is the advection of ϕ by the velocity field $S(\phi_n) \frac{\operatorname{grad} \phi}{|\operatorname{grad} \phi|}$.

Applications

We will consider first the case of breaking of a dam filled with a Newtonian fluid and the effect of a downstream wall. The results obtained with the proposed model can be seen in Figure 11.7, where the initial conditions together with the propagation at different instants can be seen.

The second example concerns the propagation over obstacles such as small dams. Figure 11.8 shows the propagation of a Newtonian fluid over a small dam.

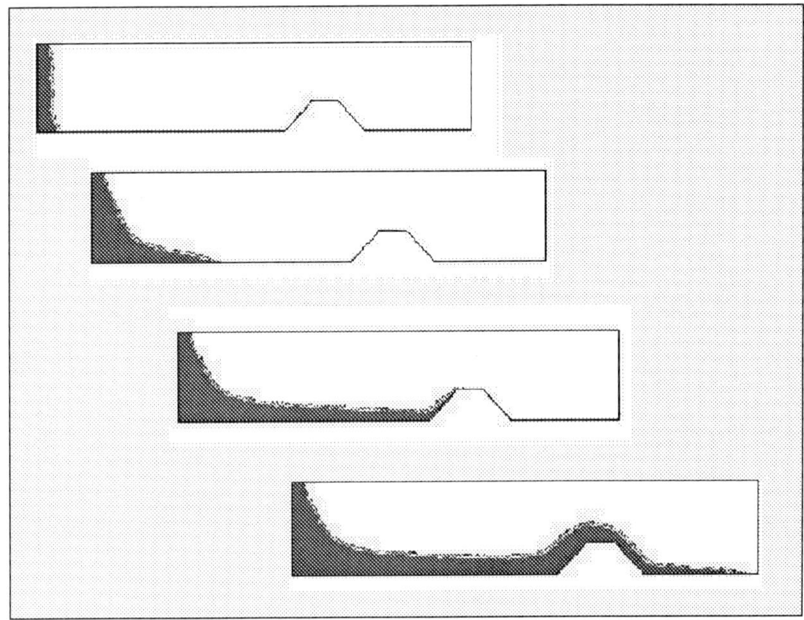

FIGURE 11.8. Propagation over a small dam.

11.5.3 Geoflow2D: a 2D depth integrated model

Description of the model

The model described in the previous section is able to provide details of the
flow such as vertical velocities and accelerations, or forces against obstacles.
However, it is costly in terms of computer time and storage. Therefore,
its optimal range of application is restricted to length scales of 100 m. To
study propagation over longer distances, it is convenient to make additional
assumptions to reduce the size of the computational model. For length
scales in the range of 1000 to 10000 m., depth integrated models can be
applied.

The equations are obtained from the balance of mass and momentum
by integration on depth, making additional assumptions such as neglecting
vertical accelerations. Following the notation introduced in Figure 11.9,
depth integrated equations can be written as

$$\frac{\partial h}{\partial t} + \frac{\partial}{\partial x}\left(uh\right) + \frac{\partial}{\partial y}\left(vh\right) \;=\; 0\,,$$

$$\frac{\partial}{\partial t}\left(uh\right) + \frac{\partial}{\partial x}\left(u^2h\right) + \frac{\partial}{\partial y}\left(uvh\right) \;=\; -gh\frac{\partial \eta}{\partial x} + \tau_{s,x} - \tau_{b,x}\,,$$

$$\frac{\partial}{\partial t}\left(vh\right) + \frac{\partial}{\partial x}\left(uvh\right) + \frac{\partial}{\partial y}\left(v^2h\right) \;=\; -gh\frac{\partial \eta}{\partial y} + \tau_{s,y} - \tau_{b,y}\,,$$

where:

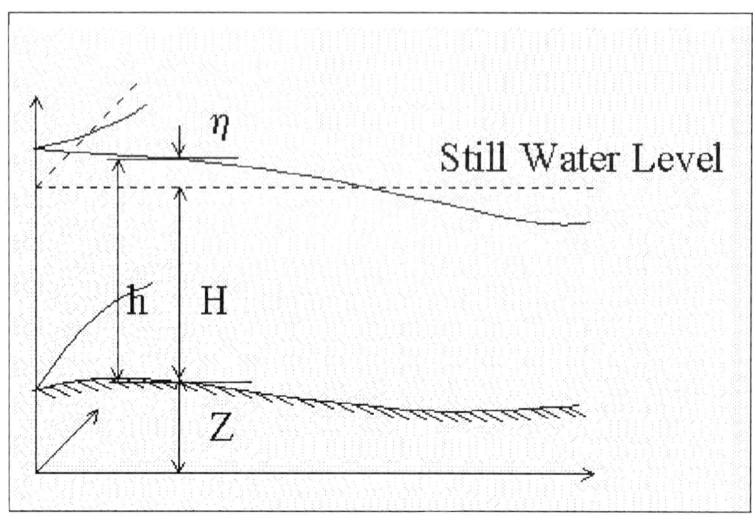

FIGURE 11.9. Depth integrated equations: notation.

- u, v are the components of the averaged velocity $\mathbf{u}$ along X and Y,

- τ_s and τ_b are shear stresses on the surface and bottom.

After introducing the change

$$\operatorname{grad} \eta = \operatorname{grad}\,(h + Z)$$

these equations can be written as

$$\frac{\partial h}{\partial t} + \operatorname{div}\,(\mathbf{U}) \;=\; 0\,, \tag{11.28}$$

$$\frac{\partial \mathbf{U}}{\partial t} + \operatorname{div}\left(\mathbf{u} \otimes \mathbf{U} + \frac{h^2}{2}\,g\,\mathbf{I}\right) \;=\; -gh\operatorname{grad} Z + \tau_b\,,$$

where $\mathbf{U} = h\mathbf{u}$. It is important to note that bottom friction depends on the rheological model used and the structure of flow. A simple case is the Chezy–Manning formula used in hydraulics:

$$\tau_b = \frac{g\,n^2\,|\mathbf{u}|\,\mathbf{u}}{h^{\frac{1}{3}}}\,. \tag{11.29}$$

We will use the alternative conservation form

$$\frac{\partial \phi}{\partial t} + \operatorname{div} \mathbf{F} = \mathbf{S}\,, \tag{11.30}$$

where we have introduced the flow vector

$$\phi = \left\{ \begin{array}{c} h \\ U_x \\ U_y \end{array} \right\}\,,$$

the fluxes along X and Y,

$$F_x = \left\{ \begin{array}{c} U_x \\ u_x U_x + g\frac{h^2}{2} \\ u_y U_x \end{array} \right\}, \qquad F_y = \left\{ \begin{array}{c} U_y \\ u_x U_y \\ u_y U_y + g\frac{h^2}{2} \end{array} \right\},$$

and the source vector

$$\mathbf{S} = \left\{ \begin{array}{c} 0 \\ g\,h\,\frac{\partial Z}{\partial x} + \tau_{b,x} \\ g\,h\,\frac{\partial Z}{\partial y} + \tau_{b,y} \end{array} \right\}.$$

These equations can be improved following the ideas presented by Savage and Hutter in [26], where both the effects of curvature and a distinction between active and passive states were included.

Discretization

The equations described in the previous section are non-linear and can exhibit shocks and discontinuities. The problem is similar to that of propagation of a flood wave caused by breaking of a dam and it has been solved using special algorithms for problems involving shocks (see for instance references [39] and [30])

However, it will be shown here that classical algorithms can provide enough accuracy with a smaller computational effort. Here we will use the two-step Taylor–Galerkin algorithm introduced by Peraire in [22] and [23] for coastal dynamics problems.

The method consists of performing a Taylor series expansion on time which allows substitution of time by space derivatives.

The first step, from t^n to $t^{n+\frac{1}{2}}$ gives

$$\phi^{n+\frac{1}{2}} = \phi^n + \frac{\Delta t}{2} \left.\frac{\partial \phi}{\partial t}\right|^n$$

from which, by using the system of PDEs results in

$$\bar{\phi}^{n+\frac{1}{2}} = \bar{\phi}^n + \frac{\Delta t}{2} \left(\bar{\mathbf{S}} - \mathrm{div}\,\bar{\mathbf{F}} \right)^n,$$

where the overbar denotes values averaged over elements

$$\bar{\phi} = \frac{1}{\Omega^e} \int_\Omega \phi\,d\Omega.$$

Once we have obtained the values of the vector of unknowns at $t^{n+\frac{1}{2}}$ we compute the fluxes and sources, which will be used in the second step:

$$\phi^{n+1} = \phi^n + \Delta t \left(\bar{\mathbf{S}} - \mathrm{div}\,\bar{\mathbf{F}} \right)^{n+\frac{1}{2}}.$$

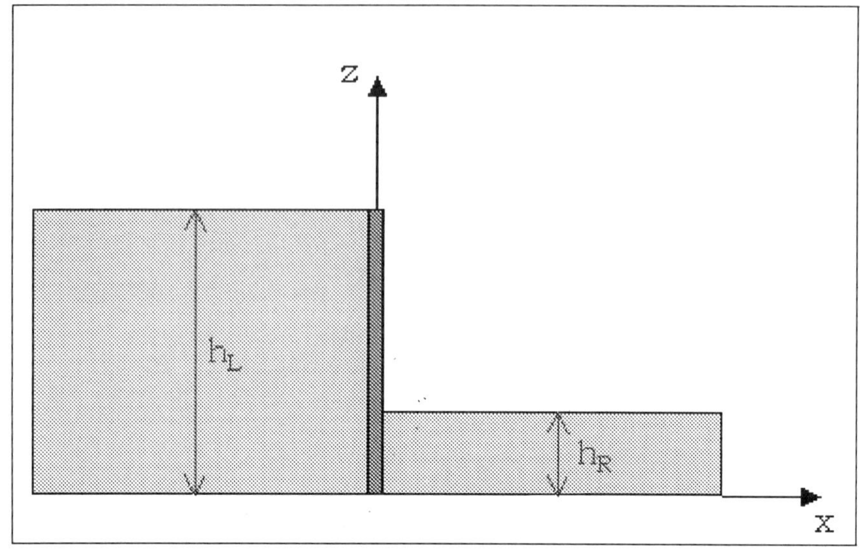

FIGURE 11.10. Breaking of a dam.

This equation is discretized using the classical Galerkin method:

$$
\bar{\mathbf{M}}\Delta\phi = \Delta t \left(\int_{\Omega} \bar{\mathbf{N}}\mathbf{S}^{n+\frac{1}{2}} d\Omega + \right.
$$
$$
\left. + \int_{\Omega} \bar{\mathbf{F}}^{n+\frac{1}{2}} \operatorname{grad}\bar{\mathbf{N}} \, d\Omega - \int_{\Gamma_N} \bar{\mathbf{N}}\bar{\mathbf{F}}^{n+\frac{1}{2}} d\Omega \right). \tag{11.31}
$$

Applications

The purpose of our first example is to show how the proposed model is able to reproduce a related case for which an analytical solution is available: the propagation of the flood wave caused by breaking of a dam over both dry and wet bottoms. The case considered here is sketched in Figure 11.10 and consists of a 10 m. high dam located at the middle of a 1000 m. long 1D channel.

The solution depends on whether propagation of the flood wave takes place over a wet or a dry bottom. In the former, there exists a shock and a rarefaction wave propagating downstream with speeds given by $\sqrt{gh_L}$ and $-2\sqrt{gh_L}$ respectively. The results are given in Figure 11.11 where it can be seen that model predictions agree reasonably well with the analytical solution.

If propagation takes place over a dry bottom, the solution consists of an expansion and a rarefaction wave, as can be seen in Figure 11.12.

The second example concerns the propagation of a landslide. We will not consider here the initiation phase, assuming for the propagation a cohesion-

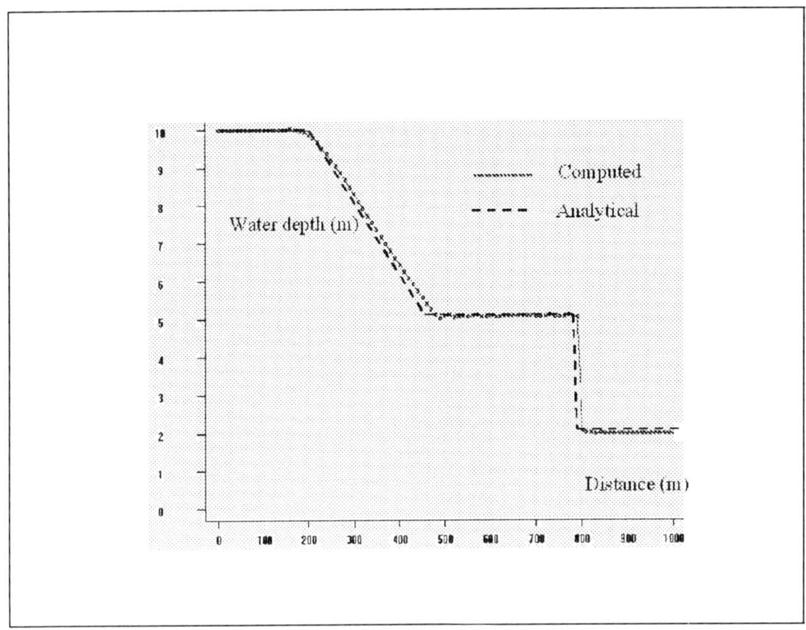

FIGURE 11.11. Propagation over wet bed.

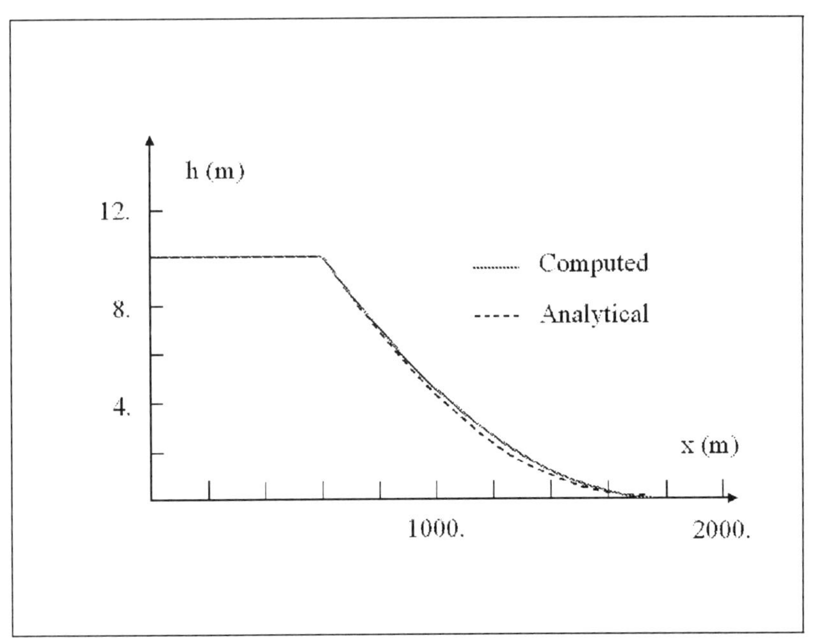

FIGURE 11.12. Propagation over dry bed.

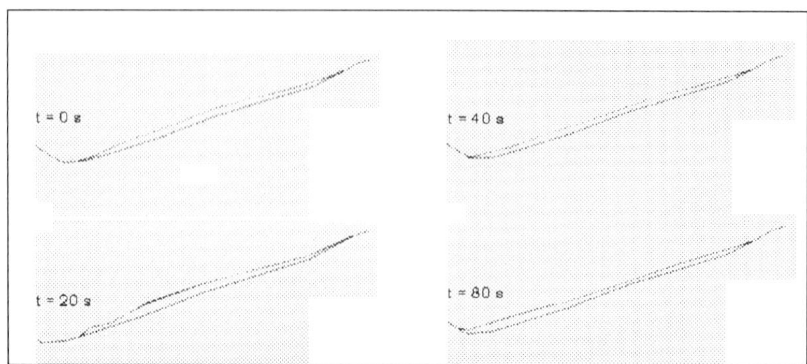

FIGURE 11.13. Propagation of a landslide.

less soil with an apparent friction angle of 20°. A series of "snapshots" of the propagation process is presented in Figure 11.13.

Finally, we will consider the failure of a tailing dam. The geometry is a 1D approximation of that of Aberfan, but the material properties have not been obtained either from back analysis nor from the analysis of the initiation phase. Figure 11.14 shows the initial situation of the sliding mass and the situation at several instants of propagation.

11.6   Conclusions

We have presented in this paper an approach to reproduce both initiation and propagation of landslides. The initiation phase can be modeled using a coupled formulation for the solid skeleton and the pore fluid, and indeed failure in some cases is due to the changes in pore water pressure induced by rain. Here, we have used a simple Drucker–Prager model incorporating softening, as it is one of the simplest able to trigger localized failure. It has to be noted that diffuse failure mechanisms such as occurs when a part of the slope liquefies require more complex constitutive models such as those described in [33].

Concerning the propagation phase, we assume a one-phase material. This approach is valid in two extreme cases: (i) when pore pressures are negligible (for instance, in dry granular flows), or (ii) when propagation time is much smaller than consolidation and pore pressures can be assumed constant. Care should be taken here when selecting parameters such as apparent friction, especially when using depth integrated models.

The performance of the proposed models is assessed using examples of failure initiation and landslide propagation.

Acknowledgments: The authors gratefully acknowledge the financial help

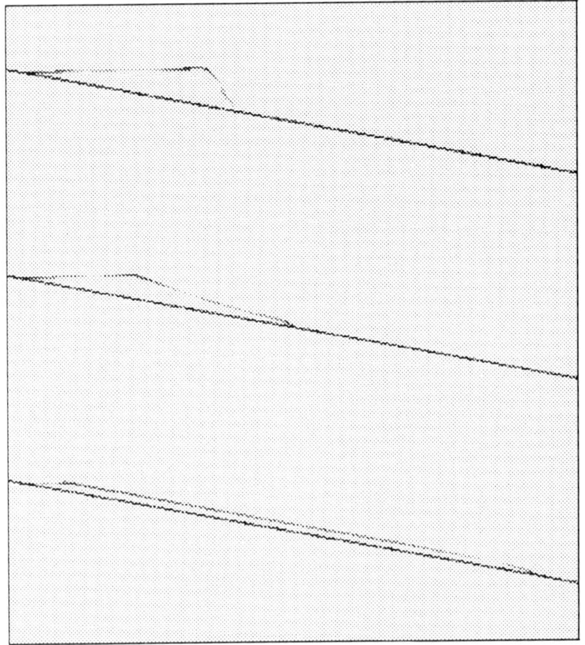

FIGURE 11.14. Failure of a tailing dam.

of both the spanish Agencia de Cooperación Internacional (AECI) and the European Union (Project Lamé, ENV4-CT97-0619).

References

[1] I. BABŬSKA: The finite element method with Lagrange multipliers, *Num. Math.* **20**, pp. 179–192, 1973.

[2] M.A. BIOT: General theory of three-dimensional consolidation, *J. Appl. Phys.* **12**, pp. 155–164, 1941.

[3] M.A. BIOT: Theory of elasticity and consolidation for a porous anisotropic solid, *J. Appl. Phys.* **26**, pp. 182–185, 1955.

[4] F. BREZZI: On the existence, uniqueness and approximation of saddle point problems arising from lagrangian multipliers, *RAIRO* **8-R2**, pp. 129–151, 1974.

[5] W.F. CHEN: *Limit Analysis and Soil Plasticity*, Elsevier Science Publishers, Amsterdam, 1975.

[6] A.J. CHORIN: Flame advection and propagation algorithms, *J. Comput. Phys.* **35**, pp. 1–11, 1980.

[7] G. DHATT, D.M. GAO, A. BEN CHEIKH: A finite element simulation of metal flow in moulds, *Int. J. Num. Meth. Eng.* **30**, pp. 821–831, 1990.

[8] R. DIETERLEN, V. MARONNIER, M. PICASSO, J. RAPPAZ: Numerical simulation of free surface flows, in: S. IDELSOHN, E. OÑATE, E. DVORKIN (eds.): *Computational Mechanics. New Trends and Application*, CIMNE, Barcelona, 1998.

[9] R. DIKAU, D. BRUNDSEN, L. SCHROTT, M.L. IBSEN: *Landslide Recognition*, John Wiley and Sons, New York, 1996.

[10] R. FRENETTE, D. EYHERAMENDI, T. IMMERMANN: Numerical modeling of dam-break type problems for Navier-Stokes and granular flows, in: C.-L. CHEN (ed.): *Debris-Flow Hazards and Mitigation: Mechanics, Prediction and Assessment*, ASCE, 1997, pp. 586–595.

[11] J.H. HARLOW, J.E. WELCH: Numerical study of large amplitude free surface motion, *Phys. Fluids* **9**, pp. 842–851, 1966.

[12] C.W. HIRT, B.D. NICHOLS: Volume of fluid (VOF) method for the dynamics of free boundaries, *J. Comput. Phys.* **39**, pp. 201–225, 1981.

[13] M.G. KATONA, O.C. ZIENKIEWICZ: A unified set of single-step algorithms. Part 3: the beta-m method, a generalisation of the Newmark scheme, *Int. J. Num. Meth. Eng.* **21**, pp. 1345–1359, 1985.

[14] R.W. LEWIS, A.S. USMANI, J.T. CROSS: Efficient mould filling simulation in castings by an explicit finite element method, *Int. J. Num. Meth. Fluids* **20**, pp. 493–506, 1995.

[15] R.L. LEWIS, B.A. SCHREFLER: *The Finite Element Method in the Static and Dynamic Deformation and Consolidation of Porous Media*, John Wiley and Sons, New York, 1998.

[16] M. MEDALE, M. JAEGER: Numerical simulation of incompressible flows with moving interfaces, *Int. J. Num. Meth. Fluids* **24**, pp. 615–638, 1997.

[17] W. NOH, P. WOODWARD: Simple line interface calculation, in: A.I. VOOREN, P.J. ZANBERGEN (eds.): *Proc. 5th Int. Conf. Num. Meth. Fluid Dynamics*, Springer-Verlag, Wien, 1976, p. 330.

[18] S. OSHER, J.A. SETHIAN: Fronts propagating with curvature–dependent speed: algorithms based on Hamilton-Jacobi formulation, *J. Comput. Phys.* **79**, pp. 12–49, 1988.

[19] M. PASTOR, T. LI, J.A. FERNÁNDEZ-MERODO: Stabilized finite elements for harmonic soil dynamics problems near the undrained-incompressible limit, *Soil Dyn. Earthquake Eng.* **16**, pp. 161–171, 1997.

[20] M. PASTOR, T. LI, X. LIU, O.C. ZIENKIEWICZ: Stabilized low order finite elements for failure and localization problems in undrained soils and foundations, *Comp. Meth. Appl. Mech. Eng.* **174**, pp. 219–234, 1999.

[21] M. PASTOR, O.C. ZIENKIEWICZ, T. LI, X. LIU, M. HUANG: Stabilized finite elements with equal order of interpolation for soil dynamics problems, *Arch. Comp. Mech.* **6**, pp. 3–33, 1999.

[22] J. PERAIRE: *A Finite Element Method for Convection Dominated Flows*, Ph.D. thesis, University of Wales, Swansea, 1986.

[23] J. PERAIRE, O.C. ZIENKIEWICZ, K. MORGAN: Shallow water problems. A general explicit formulation, *Int. J. Num. Meth. Eng.* **22**, pp. 547–574, 1986.

[24] J. PERAIRE, M. VAHDATI, K. MORGAN, O.C. ZIENKIEWICZ: Adaptive remeshing for compressible flow computations, *J. Comput. Phys.* **72**, pp. 449–466, 1987.

[25] M. QUECEDO, M. PASTOR: Application of the level set method to the finite element solution of two-phase flows, *Int. J. Num. Meth. Eng.*, 2002, in press.

[26] S.B. SAVAGE, K. HUTTER: The dynamics of avalanches of granular materials from initiation to run out. Part I: analysis, *Acta Mechanica* **86**, pp. 210–223, 1991.

[27] M. SUSSMAN, P. SMEREKA, S. OSHER: A level set approach for computing solutions to incompressible two-phase flow, *J. Comput. Phys.* **114**, pp. 146–159, 1994.

[28] E. THOMPSON: Use of pseudo-concentrations to follow creeping viscous flows during transient analysis, *Int. J. Num. Meth. Fluids* **6**, pp. 749–761, 1986.

[29] S.O. UNVERDI, G. TRYGGVASON: A front-tracking method for viscous, incompressible, multi-fluid flows, *J. Comput. Phys.* **100**, pp. 25–37, 1992.

[30] J.S. WANG, H.G. NI, Y.S. HE: Finite-difference TVD scheme for computation of dam-break problems, *J. Hyd. Eng.* **126**, pp. 253–262, 2000.

[31] O.C. ZIENKIEWICZ, C.T. CHANG, P. BETTESS: Drained, undrained, consolidating dynamic behaviour assumptions in soils, *Géotechnique* **30**, pp. 385–395, 1980.

[32] O.C. ZIENKIEWICZ, A.H.C. CHAN, M. PASTOR, D.K. PAUL, T. SHIOMI: Static and dynamic behaviour of soils: a rational approach to quantitative solutions. I. Fully saturated problems, *Proc. R. Soc. Lond.* **A 429**, pp. 285–309, 1990.

[33] O.C. ZIENKIEWICZ, A.H.C. CHAN, M. PASTOR, B. SCHREFLER, T. SHIOMI: *Computational Geomechanics*, John Wiley and Sons, New York, 2000.

[34] O.C. ZIENKIEWICZ, M. HUANG, M. PASTOR: Localization problems in plasticity using finite elements with adaptive remeshing, *Int. J. Num. Anal. Meth. Geomechs.* **19**, pp. 127–148, 1995.

[35] O.C. ZIENKIEWICZ, J. ROJEK, R.L. TAYLOR, M. PASTOR: Triangles and tetrahedra in explicit dynamic codes for solids, *Int. J. Num. Meth. Eng.* **43**, pp. 565–583, 1998.

[36] O.C. ZIENKIEWICZ, T.SHIOMI: Dynamic behaviour of saturated porous media: the generalised Biot formulation and its numerical solution. *Int. J. Num. Anal. Meth. Geomech.* **8**, pp. 71–96, 1984.

[37] O.C. ZIENKIEWICZ, Y.M. XIE, B.A. SCHREFLER, A. LEDESMA, N. BICANIC: Static and dynamic behaviour of soils: a rational approach to quantitative solutions. II. Semi-saturated problems, *Proc. R. Soc. Lond.* **A 429**, pp. 311–321, 1990.

[38] O.C. ZIENKIEWICZ, R.L. TAYLOR: *The Finite Element Method*, Vol. 2, 4th-edition, McGraw-Hill, New York, 1991.

[39] C. ZOPPOU, S. ROBERTS: Catastrophic collapse of water supply reservoirs in urban areas, *J. Hydraulic Eng.* **125**, pp. 686–695, 1999.

MANUEL PASTOR and PABLO MIRA
Centro de Estudios y Experimentación de Obras Públicas
Alfonso XII, 3
E-28014 Madrid, SPAIN
E-mail: `Manuel.Pastor@cedex.es`
        `Pmira@cedex.es`

MANUEL QUECEDO and JOSÉ A. FERNÁNDEZ-MERODO
ETS de Ingenieros de Caminos, UPM
Ciudad Universitaria s/n
E-28040 Madrid
SPAIN

E-mail: Mquegut@ciccp.es
       Jose.A.Fernandez@cedex.es

LI TONGCHUN and LIU XIAOQING
Hohai University
Building 16-406
Xikang Road, 3
210024 Nanjing
CHINA

E-mail: Tongchun@jlonline.com